THEORETICAL ASPECTS OF HETEROGENEOUS CATALYSIS

Progress in Theoretical Chemistry and Physics

VOLUME 8

The titles published in this series are listed at the end of this volume.

Theoretical Aspects of Heterogeneous Catalysis

edited by

Marco Antonio Chaer Nascimento

Instituto de Química,
Universidade Federal do Rio de Janeiro,
Rio de Janeiro, Brazil

KLUWER ACADEMIC PUBLISHERS
DORDRECHT / BOSTON / LONDON

A C.I.P. Catalogue record for this book is available from the Library of Congress.

ISBN 1-4020-0127-4

Published by Kluwer Academic Publishers,
P.O. Box 17, 3300 AA Dordrecht, The Netherlands.

Sold and distributed in North, Central and South America
by Kluwer Academic Publishers,
101 Philip Drive, Norwell, MA 02061, U.S.A.

In all other countries, sold and distributed
by Kluwer Academic Publishers,
P.O. Box 322, 3300 AH Dordrecht, The Netherlands.

Printed on acid-free paper

Printed in the Netherlands.

Progress in Theoretical Chemistry and Physics

A series reporting advances in theoretical molecular and material sciences, including theoretical, mathematical and computational chemistry, physical chemistry and chemical physics

Aim and Scope

Science progresses by a symbiotic interaction between theory and experiment: theory is used to interpret experimental results and may suggest new experiments; experiment helps to test theoretical predictions and may lead to improved theories. Theoretical Chemistry (including Physical Chemistry and Chemical Physics) provides the conceptual and technical background and apparatus for the rationalisation of phenomena in the chemical sciences. It is, therefore, a wide ranging subject, reflecting the diversity of molecular and related species and processes arising in chemical systems. The book series *Progress in Theoretical Chemistry and Physics* aims to report advances in methods and applications in this extended domain. It will comprise monographs as well as collections of papers on particular themes, which may arise from proceedings of symposia or invited papers on specific topics as well as initiatives from authors or translations.

The basic theories of physics – classical mechanics and electromagnetism, relativity theory, quantum mechanics, statistical mechanics, quantum electrodynamics – support the theoretical apparatus which is used in molecular sciences. Quantum mechanics plays a particular role in theoretical chemistry, providing the basis for the valence theories which allow to interpret the structure of molecules and for the spectroscopic models employed in the determination of structural information from spectral patterns. Indeed, Quantum Chemistry often appears synonymous with Theoretical Chemistry: it will, therefore, constitute a major part of this book series. However, the scope of the series will also include other areas of theoretical chemistry, such as mathematical chemistry (which involves the use of algebra and topology in the analysis of molecular structures and reactions); molecular mechanics, molecular dynamics and chemical thermodynamics, which play an important role in rationalizing the geometric and electronic structures of molecular assemblies and polymers, clusters and crystals; surface, interface, solvent and solid-state effects; excited-state dynamics, reactive collisions, and chemical reactions.

Recent decades have seen the emergence of a novel approach to scientific research, based on the exploitation of fast electronic digital computers. Computation provides a method of investigation which transcends the traditional division between theory and experiment. Computer-assisted simulation and design may afford a solution to complex problems which would otherwise be intractable to theoretical analysis, and may also provide a viable alternative to difficult or costly laboratory experiments. Though stemming from Theoretical Chemistry, Computational Chemistry is a field of research in its own right, which can help to test theoretical predictions and may also suggest improved theories.

The field of theoretical molecular sciences ranges from fundamental physical questions relevant to the molecular concept, through the statics and dynamics of isolated molecules, aggregates and materials, molecular properties and interactions, and the role of molecules in the biological sciences. Therefore, it involves the physical basis for geometric and electronic structure, states of aggregation, physical and chemical transformations, thermodynamic and kinetic properties, as well as unusual properties such as extreme flexibility or strong relativistic or quantum-field effects, extreme conditions such as intense radiation fields or interaction with the continuum, and the specificity of biochemical reactions.

Theoretical chemistry has an applied branch – a part of molecular engineering, which involves the investigation of structure–property relationships aiming at the design, synthesis and application of molecules and materials endowed with specific functions, now in demand in such areas as molecular electronics, drug design or genetic engineering. Relevant properties include conductivity (normal, semi- and supra-), magnetism (ferro- or ferri-), optoelectronic effects (involving nonlinear response), photochromism and photoreactivity, radiation and thermal resistance, molecular recognition and information processing, and biological and pharmaceutical activities, as well as properties favouring self-assembling mechanisms and combination properties needed in multifunctional systems.

Progress in Theoretical Chemistry and Physics is made at different rates in these various research fields. The aim of this book series is to provide timely and in-depth coverage of selected topics and broad-ranging yet detailed analysis of contemporary theories and their applications. The series will be of primary interest to those whose research is directly concerned with the development and application of theoretical approaches in the chemical sciences. It will provide up-to-date reports on theoretical methods for the chemist, thermodynamician or spectroscopist, the atomic, molecular or cluster physicist, and the biochemist or molecular biologist who wish to employ techniques developed in theoretical, mathematical or computational chemistry in their research programmes. It is also intended to provide the graduate student with a readily accessible documentation on various branches of theoretical chemistry, physical chemistry and chemical physics.

TABLE OF CONTENTS

Preface

Heterogeneous catalysis is a subject of outmost importance in the present days. Heterogeneous catalytic processes play an essential role in the manufacture of a wide range of products, from gasoline and plastics to fertilizers, which would otherwise be unobtainable or prohibilively expensive. Apart from manufacturing processes, heterogeneous catalysis is also finding other very important applications in the fields of pollution and environment control.

The immense technological and economic importance of heterogeneous catalysis, and the inherent complexity of the catalytic phenonema, have stimulated theoretical and experimental studies by a broad spectrum of scientists, including chemists, physicists, chemical engineers and material scientists.

Nowadays, due to the great sophistication achieved by many of the experimental techniques used in surface science studies, a great deal of information can be obtained, at the atomic scale, about the mechanism of heterogeneous catalytic processes under well controlled conditions. In spite of that, it is still difficult, and sometimes impossible, to obtain a precise picture of the mechanism of the reaction, at the molecular level, without any theoretical support.

This book aims to illustrate and discuss the subject of heterogeneous catalysis and to show the current capabilities of the theoretical and computational methods for studing the various steps (diffusion, adsoprtion, chemical reaction, etc.) of a heterogeneous catalytic process.

M.A. Chaer Nascimento
July 2001

THEORETICAL STUDY OF REACTIONS CATALYZED BY ACIDIC ZEOLITE.

Xavier Rozanska, Rutger A. van Santen and François Hutschka[#]
Schuit Institute of Catalysis, Technical University of Eindhoven, PO. Box 513, 5600MB Eindhoven, The Netherlands
[#]TotalFinaElf, European Research and Technical Center, B.P. 27, 76700 Harfleur, France

Abstract After a general introduction on zeolites and their properties, we will show how the current quantum chemical methods allow nowadays the simulation of catalytic models that become more and more realistic. Predictability of activity, selectivity and stability based on known structures of catalysts can be considered the main aim of the theoretical approach to catalysis. Here for a particular class of heterogeneous catalysts, the acidic zeolites, and for a particular class of reactions, the aromatics isomerization reactions, we will describe a quantum mechanic study of a reaction pathway investigation. The cluster approach method as well as electronic structure periodic method will be used for this purpose.

1. INTRODUCTION

1.1 Zeolite Crystals

The Greek name zeolite has been given to a class of mineral. Zeolite means "boiling stone" in allusion to the behavior of some silicate minerals on heating. This relates to a well-known property of a zeolite that is to adsorb large quantity of water and gases. This is permitted by the microporous structure of zeolite, which can accept an important amount of guest molecules. Zeolites are silicon oxides crystals that show various and well defined microporous structures (see Figure 1). It is this characteristic which makes them interesting molecular sieves for which application they have been employed for many years.[1]

The zeolite framework is constituted of the assembly of SiO_4 tetrahedra which link together by sharing an oxygen atom. The potential energy surface of the zeolite Si-O-Si angle is rather flat,[2] which explains the large variety of accessible structure that can be accommodated by zeolites.[3]

The zeolites have been first used as catalyst in the 1960s for alkane cracking reactions in petroleum industry.[3b] They replaced favorably previously employed alumina based catalysts because of their better thermal and mechanical stability. Moreover, they showed higher selectivity. The selectivity finds its source in the zeolite micropore structure with different

M.A. Chaer Nascimento (ed.), Theoretical Aspects of Heterogeneous Catalysis, 1–28.

consequences on the course of a reaction[3b,4]

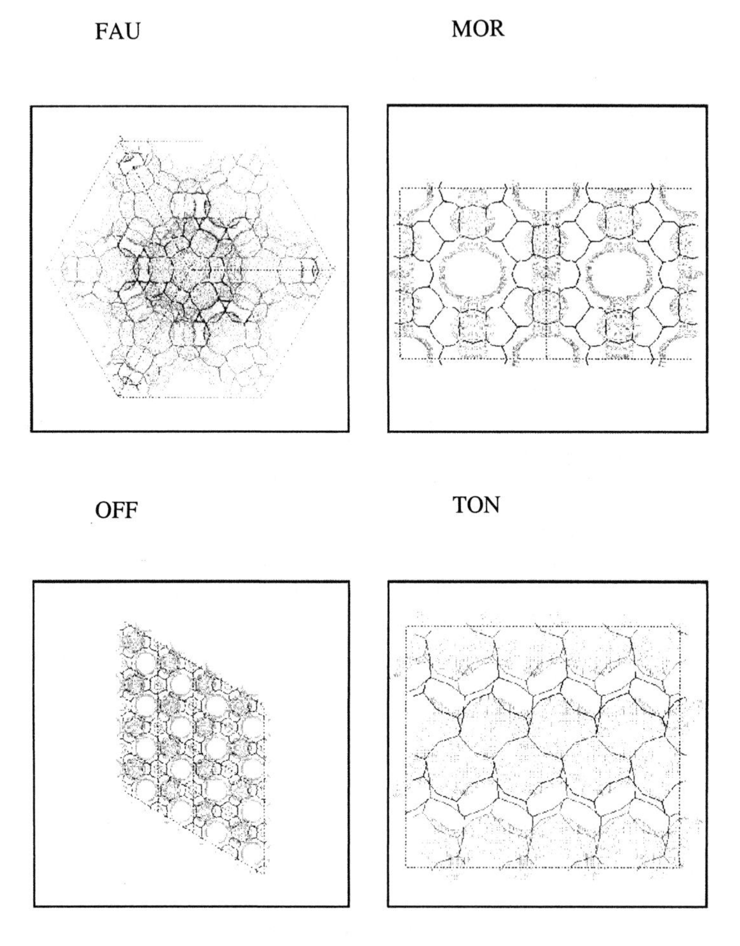

Figure 1. Some examples of zeolite structures (the dots depict the Connolly surfaces).

A first, rather obvious reason is the reactant selectivity. As the zeolite micropore channels have a well defined diameter, reactants bigger than this diameter cannot enter within the micropores to react, in case external catalytic active sites of zeolite have been deactivated.[4a] Smaller reactants than the micropores will be the only ones to get involved in reactions.

Once a reactant molecule has adsorbed within the zeolite mouth, it needs to diffuse toward the active sites in contact of which reactions will occur. This diffusion can be very dependent on the size and shape of the zeolite micropores as well as on the size of the reactants or products. This becomes especially true when the reactants or products have similar size to the micropores diameter.[4a] After reaction, it is the turn of the products to diffuse away from the micropores.

These stereo-selectivity properties do not only apply to the case of zeolite catalysts but also in the case of zeolite molecular sieves.

After reactants have adsorbed within the zeolite mouth and have diffused towards the actives sites, they must react. Selective adsorption depends on the local topology of the active site, its immediate neighborhood, and on size and shape of the reactant. One may say that a similar behavior as the lock and key principle of molecular recognition known from biochemistry applies. In this sense zeolites can be viewed as analog of enzymatic catalysts.[5] This adsorption selectivity of the reactants to the active sites favors preferential reaction pathways.

Elementary reaction steps proceed via transition states. Transition state requires a specific geometry that differs for each reactions step. Reaction steps can be favored or prohibited as a function of the available space around the catalytic active site.[4] This effect is designated as the transition state selectivity.

Whereas selective diffusion can be better investigated using classical dynamic or Monte Carlo simulations,[6] or experimental techniques,[7] quantum chemical calculations are required to analyze molecular reactivity. Quantum chemical dynamic simulations provide with information with a too limited time scale range (of the order of several hundreds of ps)[8] to be of use in diffusion studies which require time scale of the order of ns to s.[6,7] However, they constitute good tools to study the behavior of reactants and products adsorbed in the proximity of the active site, prior to the reaction. Concerning reaction pathways analysis, static quantum chemistry calculations with molecular cluster models, allowing estimates of transition states geometries and properties, have been used for years.[9] The application to solids is more recent.

But before continuing the discussion on the quantum chemical simulations, we will give a deeper description of the current understanding of physical chemistry properties of zeolites. We will be using experimental as well as theoretical studies to support the discussion.

1.2 Zeolite Properties

It has already been mentioned that zeolites are shape selective with respect to molecular adsorption. This property relates to their micropores structure. The zeolite framework shows a limited flexibility, which is essential. For instance, Yashonath et al.[10] have shown in their classical dynamic simulations study of molecular diffusion within zeolite micropore that the zeolite framework flexibility affects significantly diffusion when the molecules have a size comparable with the micropore size. To get an idea of the order of magnitude of this flexibility, one can consider the hybrid semi-empirical DFT periodic study of chabazite zeolite of Ugliengo et al.[11]. They introduced in the unit cell of chabazite Brønsted acidic sites which are known to induce an increase of the volume of around 10 $Å^3$.[2,12] This increase of the volume relates with the difference of volume between a SiO_4 tetraheron and a

$AlO_3(OH)$ tetrahedron. Ugliengo et al. found that a linear relationship between the volume of the unit cell as a function of the number of introduced Brønsted acidic sites is not anymore followed only above a Si/Al ratio of 3.

In this last example, we mentioned that zeolite framework silicon atoms could be substituted. This substitution is naturally found in zeolites. Silicon atoms can be replaced with +III valence atoms, such as Al^{+III} or Ga^{+III}, or +V valence atoms, such as P^{+V}. Then, cations or anions are introduced to neutralize the charge of the framework (see Figure 2). The only known limitation of the silicon substitutions is the Lowenstein rule which states that substitutions cannot be found into two adjacent tetrahedra.[13] The presence of cations or anions within zeolites allows the zeolite crystals to be used in ion exchanged processes.[14]

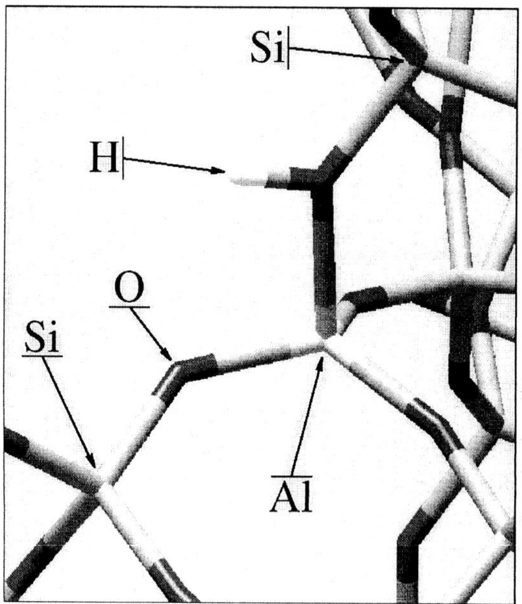

Figure 2. Proton bonded to bridging O-atom in the neighborhood of a silicon atom substitution with an aluminum atom in H-Mordenite.

Zeolite crystals are insulators. However, they are ionic crystals, and the radii of the zeolite oxygen and silicon atoms is close to that of the ones observed for O^{-II} and Si^{+IV} ions from others crystals.[15] Despite this, long range electrostatic contributions play a limited role, and the bonding within the zeolite framework is mainly covalent with the ionic bonding contributing only for around 10%.[16] Zeolite crystals are characterized with small dielectric constant: ε_1 is generally between 2 to 7 for full silicon zeolites.[17] Therefore, the short range electrostatic contributions dominate over the interaction energy of molecules with the zeolite host, where dispersion Van der Waals contribution plays the main role.[18] As the zeolitic oxygen atoms have a large radius, adsorbing molecules experience interactions only with the zeolitic oxygen atoms (see Figure 3).

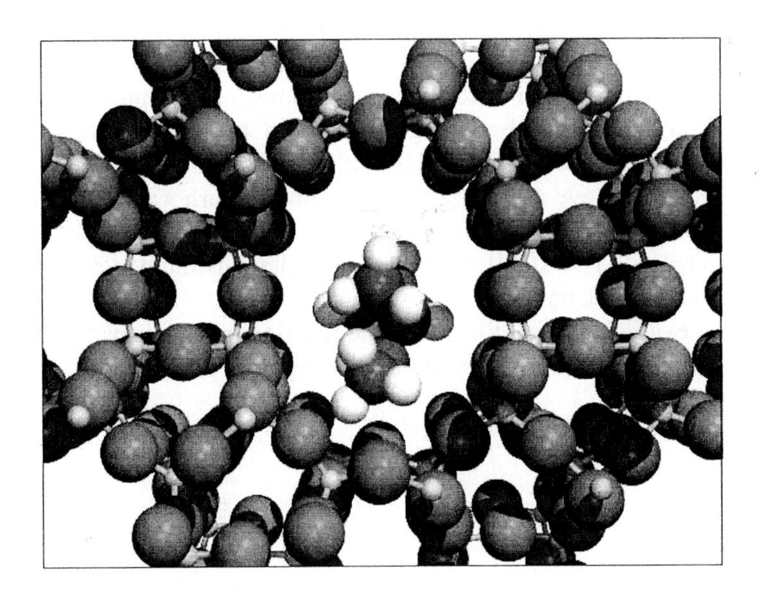

Figure 3. Hexane occlude in a mordenite zeolite micropore. The hydrocarbon molecule can only interact with zeolitic oxygen atoms.

Kiselev et al.[19] defined a classical force field for which zeolite silicon atoms are not explicitly considered. The dynamic simulations they performed using this force field succeeded well in describing molecules diffusion within zeolite micropores.[19,20]

We will now consider more specifically the properties of zeolite catalysts.

1.3 Zeolite Catalysts

Zeolites can be used as inert support for small metallic clusters catalysts or can be used themselves as catalysts. For the later case, silicon substitutions by another cation have to occur. The catalytic activity of zeolites will be dependent upon the nature of the introduced cations (viz. Fe^{3+}, Zn^{2+}, Fe^{2+}, Cu^{2+}, H^+, Li^+, Na^+, etc...) or anions (viz. F^-, OH^- etc...).[21] Lewis basic and/or acid sites and/or acid Brønsted sites are created. We will only describe the zeolites with silicon substituted by Al and a proton as cation to neutralize for the framework charge. In this case, a zeolite is a solid acid catalyst, which acts as more environmentally friendly acid catalyst than the classical ones. Such catalysts are obtained when the M^+ cation of a $M_x[Al_xSi_{1-x}O_2]$ nH_2O zeolite is ion exchanged with NH_4^+. The heating of the NH_4^+ exchanged zeolite induces NH_3 to desorb, and a proton is left behind. The proton binds to an oxygen atom that bridges an aluminum atom and a silicon atom (see Figure 2). As for other zeolitic bonds, the proton - zeolite oxygen atom shows a strong covalent behavior.[22]

Solid acid zeolites, beside inducing reactions known from superacids,[23] are indeed moderate acids.[24] The moderate acidity allows zeolite catalyzed reactions to not suffer from side chains reactions.[4] Enthalpies of reaction compare actually more closely with gas phase reactions than with reactions catalyzed by homogeneous high dipolar acids.[25] However, zeolite micropores have been shown to have a stabilizing effect on the carbocationic transition states with respect to gas phase results, which can reach 10 to 30 % of the activation energies.[26] This stabilization has been demonstrated to be mainly of short range electrostatic nature.[26a]

A famous example, which illustrates the reaction mechanisms that happen with the acidic zeolite catalysts, is the chemisorption reaction of olefins (see Figure 4).

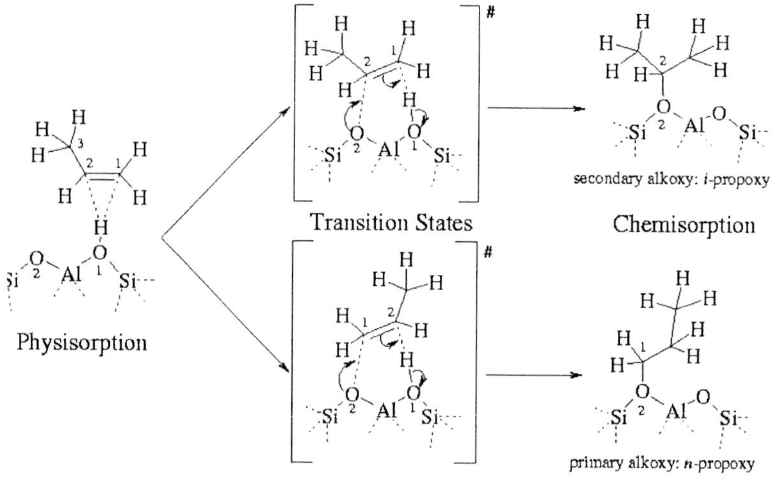

Figure 4. Reaction steps involved in the chemisorption of propylene catalyzed by an acidic zeolite.

This reaction has been the subject of many experimental,[24,27] and theoretical[28] studies. The reaction, which initiates from a propylene physisorbed to the acidic proton, leads to the formation of a more stable chemisorbed propylene, or alkoxy species. Such reaction has been shown experimentally to occur readily at room temperature within an acidic zeolite.[27] Whereas this reaction in principle can produce two different alkoxy species (viz. a primary and a secondary alkoxy species), experiment reports that only the secondary alkoxy species can be formed.[27] This is explained by the fact that the formation of a transient primary carbenium ion is energetically more demanding than the formation of a secondary carbenium ion.[28c] As already mentioned, the zeolite framework stabilizes partially the carbocationic transition state. This stabilization is however not sufficient to allow carbocations to exist as stable intermediates, and carbocations become covalently bonded to a zeolite framework oxygen atom.

Solid acid zeolites are used as catalysts for a large range of reactions on hydrocarbons.[29] They can for instance induce alkylation, transalkylation, isomerization and cracking reactions. Moreover, they are also used to achieve fine organic reactions.[5,30] We will not describe all these mechanisms, and we will only consider the case of aromatics isomerization reactions catalyzed by acidic zeolite. These reactions have been used for years as a support in the discussion on the zeolite catalysts selectivity.[4,29] They allow us to show how modern quantum chemical tools can be used to describe zeolite framework effects on the course of a reaction. These effects include the zeolite framework steric constraints and electrostatic contributions. Before discussing the reaction, we will first give a short description of the available quantum chemistry tools.

1.4 Quantum Chemistry Applied to Zeolite Catalysis

In this part, we will not give an overview of the quantum chemistry theory that has led to the current state of affairs, as excellent introductions can be found elsewhere.[8b,31] We will rather shortly describe which methodology can be followed to describe acid zeolite catalyzed reactions.

Because of the size of the reaction centers to be considered, a breakthrough in quantum-chemistry has been necessary to make computational studies feasible on systems of catalytic interest. The Density Functional Theory (DFT) has provided this breakthrough.[32] Whereas in the Hartree-Fock based methods, mainly used before the introduction of Density Functional Theory, electron-exchange had to be accounted for by computation of integrals that contain products of four occupied orbitals, in Density Functional Theory these integrals are replaced by functionals that only depend on the electron density. This allows a consequent reduction of the computational costs. An exchange-correlation functional can be defined that accounts for exchange as well as correlation-energy. Correlation energy is the error made in Hartree-Fock type theories by the use of the mean-field approximation for electronic motion.

The unresolved problem yet is the determination of a rigorously exact functional. For this reason all of the currently used functionals are approximate. Furthermore, one of the main problem of DFT methods is that Van der Waals dispersion contribution is not explicitly considered in the equations, which leads to some severe errors for the description of the zeolitic guest-host interaction.[33] We will give further details on this point later in the discussion.

1.4.1 The Cluster Approach

We mentioned that bonding in zeolites is dominated by covalent bonding. Therefore, zeolite properties can be described as being mainly locally dependent. For instance, it is known that Brønsted acidic sites induce important distortions in the zeolite framework (see Figure 3). However, these distortions remain local.[2,34] A direct application of this property is the validity of the use of small fragments to describe catalytic active sites (see Figure 5).

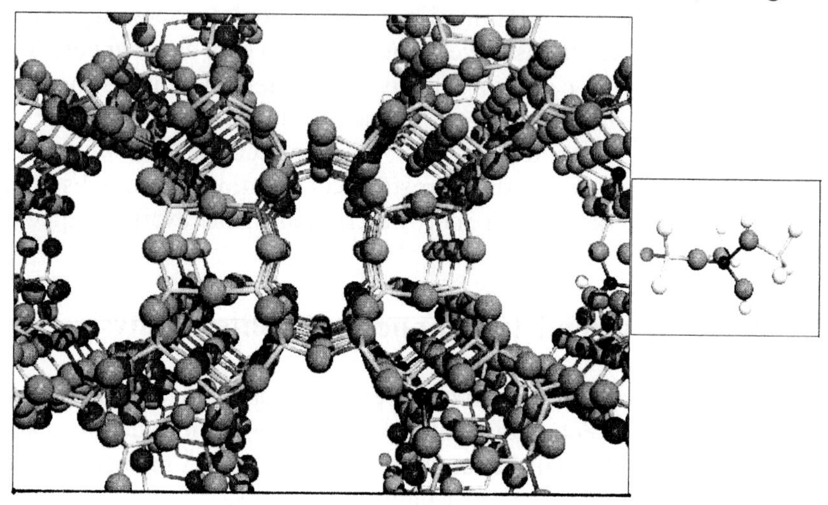

Figure 5. The principle of the cluster approach method. The small cluster model (right) aims to model an acidic zeolite catalyst (left).

This model approximation is known as the cluster approach.[35] The fragment extracted from a zeolite framework is terminated with hydrogen atoms. Such method has been used with success for the estimate of properties such as frequencies of vibration,[36] NMR shielding constants,[37] or reactivity.[2] Another mandatory reason of the use of this method was the too heavy computational cost to simulate larger systems.

In recent years, progress in computer power as well as in theoretical methods allowed studies with increasingly larger clusters.[38] A large cluster model can partly allow for the zeolite framework description. Especially, Zygmunt et al.[26a] provided with large cluster studies which led to a deeper insight into reactions catalyzed by acidic zeolites. These large cluster models can however partly described the effect of the zeolite framework over the properties of interest.[39] If long range electrostatic contributions have a limited impact within zeolite, they still do exist, and should not be simply not considered as they can alter some properties. On the other hand, even small cluster calculations give valuable and useful information, as it will be described in a future section. Alternatively to the cluster approach, others methods have been developed.

1.4.2 Beyond the Cluster Approach

A way to describe the zeolite framework at a low computational cost is to use quantum mechanic - molecular mechanic methods (QM/MM) (see Figure 6).[39] With QM/MM, only the site of interest (viz. reactants and catalytic active site) are treated at a quantum mechanic level, whereas the zeolite framework is described using force field equations of molecular mechanic.[40]

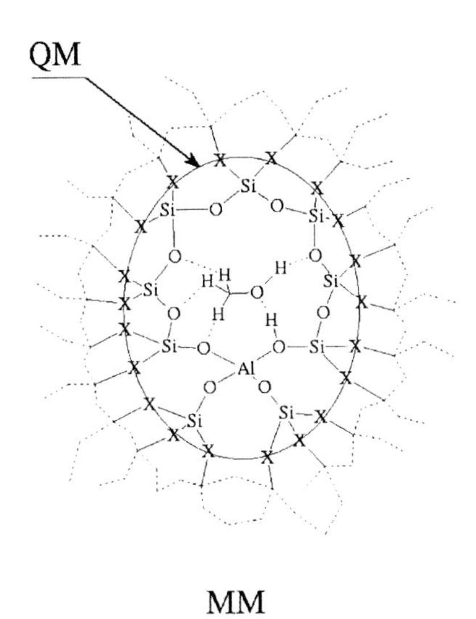

Figure 6. Principle of the QM/MM method applied to a zeolitic case. Here, the adsorption of methanol on a Brønsted site proton within a 8-membered ring zeolite.

The quality of these methods depends on the force fields parameters, the way the QM and MM parts are linked, and how QM and MM parts affect each other. The main advantages of QM/MM are to present a limited increase of required computer power as a function of the size of the system. The MM part can be constituted by up to thousands of atoms.[12,28b,38] A drawback is that it is not easy to define a priori what should be the size of the QM and MM parts. Ramanchandran et al.[41] observed in their periodic study that during transition state electron delocalization from the Brønsted site to others zeolite framework oxygen atoms was an important phenomenon. Then, large QM part is required which makes more costly calculations. Furthermore, another drawback of QM/MM is the complexity of the tuning which can lead to misleading results.[28b]

The periodic electronic structure calculations methods constitute the other available approach.[42] These methods require a relative high computational effort, beside progress in computer power allows calculations on systems of size of interest.[43] The advantage of periodic approach method is that the entire system is described at a quantum chemical level. The disadvantages are the heavy computational cost which limits currently the size of the unit cell to systems below 300 atoms, and the usual artifacts that are associated with periodic boundary conditions.[44] Next, others problems originate from the DFT,[33] and they will be explained in a next section.

We will give further details about the periodic electronic structure calculations method as we will use results obtained with this method to support the discussion.[45] First, we will describe how cluster approach can be used to investigate reactions catalyzed by acidic zeolites. Next, periodic electronic structure calculations will be used to enlighten the effects of the approximations of the cluster approach. These approximations relate mainly with the missing description of the zeolite framework contributions on the molecules involved in the reactions. Finally, by increasing the size of the aromatics involved in the isomerization reactions, we will show how steric constraints affect the course of a reaction.

Before this, we will give details of the methods employed for the calculations realized in this study.

1.4.3 Methods and Models Employed in this Study

The cluster approach calculations have been performed using *Gaussian98*[46] with the B3LYP method.[47] This DFT method appears to be the optimum choice for treatment of zeolite cluster systems and presents results comparable with MP2 method.[48] In order to describe the Brønsted acidic site, a 4 tetrahedra cluster ($Al(OHSiH_3)(OSiH_3)_2(OH)$) has been chosen. We selected the basis set d95. Geometry optimization calculations have been carried out to obtain local minima for reactants, adsorption complexes and products and to determine the saddle point for transition states (TS). Frequency calculations have been computed in order to check that the stationary points exhibit the proper number of imaginary frequencies: none for a minimum and one for a transition state. Zero point energy (ZPE) corrections have been calculated for all optimized structures.

The Vienna Ab Initio Simulation Package (VASP) has been used to perform the periodic structure calculations.[42a,b,43] All atoms have been allowed to relax completely within the periodic unit cell. The large 12-membered ring Mordenite has been used for this study as this zeolite has a relatively small unit cell (i.e. 146 atoms). It has previously been studied by Demuth et al.[50]. For our zeolite model, the Si/Al ratio is 23, and the geometry of the unit cell is described by a = 13.648 Å, b = 13.672 Å, c = 15.105 Å, α = 96.792 °, β = 90.003 ° and γ = 90.022 °. With VASP, the energy is obtained solving the Kohn-Sham equation with the Perdew-Zunger exchange-correlation functional.[51] The results are corrected for non-locality within the generalized gradient approximation (GGA) with the Perdew-Wang 91 functional.[52] This functional is currently the only available in VASP. This program uses planewaves and pseudopotentials, which allow a consequent reduction of computational costs. A cut-off of 300 eV and a Brillouin zone sampling restricted to the Γ-point have been used. A quasi-Newton forces minimization algorithm has been employed: convergence was assumed to be reached when forces were below 0.05 eV/Å. The TS search method in VASP is the nudged elastic band (NEB) method.[53] Several images of the system are defined along the investigated reaction pathway. These images are optimized but only allowed to move perpendicularly to the hyper-tangent defined by the normal vector between the neighboring images. We employed up to 8 images to analyze transition states. When forces of the images atoms were below 0.08 eV/Å, the forces of the maximum energy image were minimized separately.

2. TOLUENE ISOMERIZATION

In this part, we will summarize some of our results on the investigation of the toluene intramolecular isomerization pathways.[45,54] Both cluster approach and periodic approach methods have been employed which allow giving an illustration of the consequence of the simplistic model in the cluster approach. H-Mordenite (H-MOR) zeolite is used for the periodic calculations. The toluene molecule does not have a problem to fit within the large 12-membered ring channels of this zeolite.[18] Furthermore, the intramolecular transition states do not suffer from steric constraints.[4] It is known that intramolecular aromatics isomerization can proceed via two different reaction pathways (see Figure 7).[4,55] The first route proceeds through a methyl shift isomerization, whereas the second route involves a dealkylation or disproportionation reaction which results in the formation of a methoxy species and benzene as intermediate.

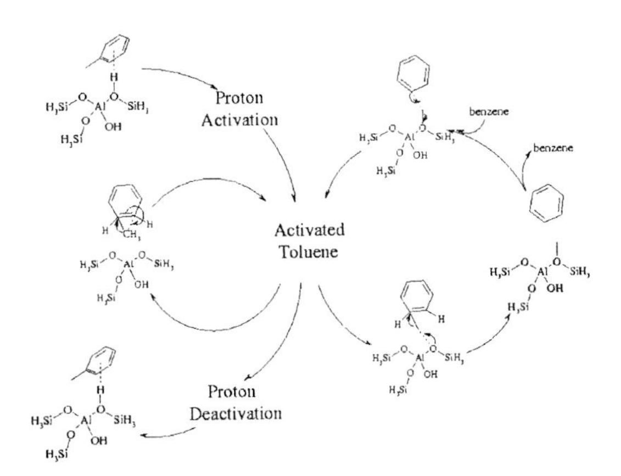

Figure 7. Catalytic cycles and description of the mechanisms of the intramolecular isomerization reactions of toluene catalyzed by an acidic zeolite.[55]

The initial step of this investigation is to analyze the reaction pathways using the low computational cost cluster approach. The small cluster model aims to model a zeolitic Brønsted acidic site, and has been demonstrated to fill successfully this task.[2] On the other hand, a small cluster cannot describe the zeolite framework. By comparison of reaction pathways taking and not taking into account the zeolite framework, we will be able to evaluate this effect on reactivity.

2.1 Cluster Approach

The investigation of the shift isomerization reaction pathway led to the establishment of the reaction energy diagram plotted in Figure 8.[54] The geometries of the intermediates and transition states are also summarized in this figure. Prior to reach the shift isomerization transition state, toluene needs to be activated by proton attack. This activation proceeds via a transition state very similar to the one obtained from propylene chemisorption in acidic zeolite (see Figure 4). It leads to the formation of a phenoxy intermediate which is less stable than the physorbed toluene. This is expected from the chemical properties differences between propylene and toluene.[56]

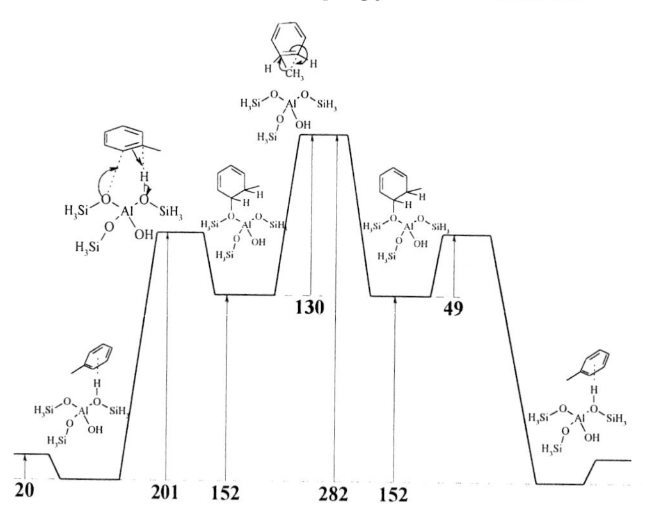

Figure 8. Reaction energy diagram and description of the mechanisms of the shift isomerization reaction of toluene catalyzed by acidic zeolite (all data in kJ/mol).[54]

The phenoxy species is released from the cluster with no activation energy barrier to overcome but a constant increase in energy to a Wheland complex from which shift isomerization transition state takes place. With respect to physisorbed toluene, the activation energy to achieve this transition state is $E_{act} = + 282$ kJ/mol. In the transition state, the shifting methyl group occupies an intermediate position between the aromatic ring carbon atom it was connected to, and the carbon atom it will connect to. The shift methyl group carbon atom is located in a position almost orthogonal with respect to the plane defined by the aromatic ring. Extra weak hydrogen bonds with the cluster oxygen atoms stabilize the transition state with this structure.

For the isomerization of toluene via methoxy and benzene intermediate, the similar proton activation, which results in the formation of a phenoxy intermediate, is achieved (see Figure 9). As previously, the bond between the zeolitic oxygen atom and the aromatic carbon atom stretches out. A "free" Wheland complex is eventually reached which can reorient to favor the position of the toluene methyl group with the demethylation transition state.

The geometry of this transition state can be described as a methenium carbocation sandwiched in between a benzene molecule and a deprotonated Brønsted site. The methenium ion is planar, and its carbon atom is located along the line defined by the zeolitic oxygen atom to which it will be bonded and the aromatic carbon atom to which it was bonded. The activation energy which is required to achieve this reaction is $E_{act} = + 279$ kJ/mol.

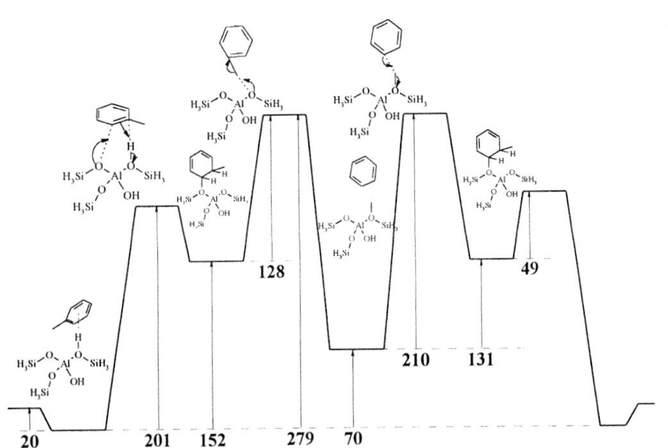

Figure 9. Reaction energy diagram and description of mechanisms of the disproportionation via methyl alkoxy isomerization reaction of toluene catalyzed by an acidic zeolite (all data in kJ/mol).[54]

This transition state results in the formation of a methoxy species and benzene. Benzene can desorb from the methoxy species, or can change its orientation. The energy level of this intermediate is + 70 kJ/mol with respect to physisorbed toluene. The regeneration of toluene from this intermediate follows the reverse transition state.

Interestingly, the two reaction routes require similar activation energy, and further consideration of the entropies of activation cannot decide as well whether a route has preeminence on the other (the entropies of activation are $\Delta_{298K}S_{act}$ = - 13 and - 17 J/mol/K for the shift and via disproportionation isomerizations respectively).

2.2 Periodic Approach

We will now describe shortly our results of the investigation of the toluene intramolecular isomerization catalyzed by H-Mordenite using an electronic periodic structure method.[45] Let us consider first the shift isomerization transition (see Figure 10). As it can be seen in this figure, the geometries of the transition states from cluster approach and from periodic approach are very similar. The large difference between these transition states is that the activation energy from periodic approach is E_{act} = + 179 kJ/mol whereas it is E_{act} = + 282 kJ/mol from the cluster approach. Such a decrease of the activation energy, when a more realistic zeolite model is employed, has already been notified.[28b]

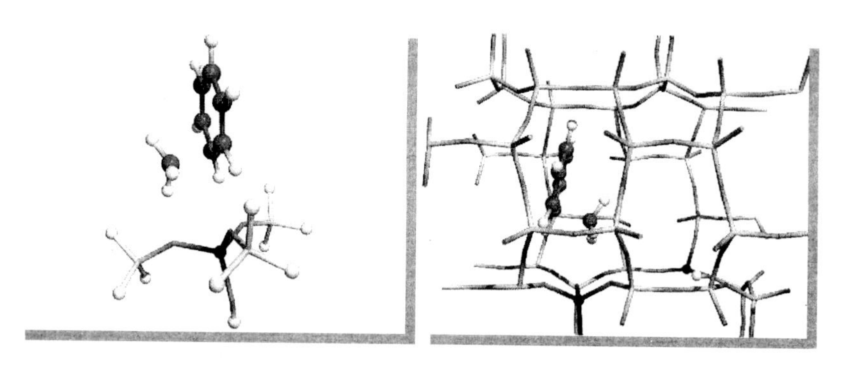

Figure 10. Geometries of the shift isomerization transition states of toluene catalyzed by acidic zeolite as obtained from the cluster approach method (left) and the periodic structure method (right).[45,54]

This important stabilization has other consequences on the reaction pathway of isomerization (see Figure 11). The investigation of the activation of toluene by proton attack reveals a completely different picture than the one obtained with the cluster approach. The protonation step of toluene becomes an inflection point in the reaction pathway, which gives as product a metastable Wheland complex.

The formation of the phenoxy intermediate turns to be unlikely to occur as this intermediate remains at similar energy level as observed from the cluster approach (i.e. + 150 kJ/mol), whereas protonation step and Wheland complex energy levels are around + 110 kJ/mol with respect to physisorbed toluene.

Zeolite framework stabilization effects uniformly all transition states and charged transient intermediates, and does not effect the neutral intermediates. Similar effect of the zeolite framework has been described by Corma et al.[58] for another reaction.

Exactly the same trend is obtained for the second isomerization route. An important consequence of this is that the activation energy barrier for the isomerization via methoxy and benzene transition state is found to be E_{act} = +

182 kJ/mol. This means that a periodic approach method predicts as well as the cluster periodic method that the two alternative isomerization routes are competitive pathways. Qualitatively, cluster and periodic methods give the same result.

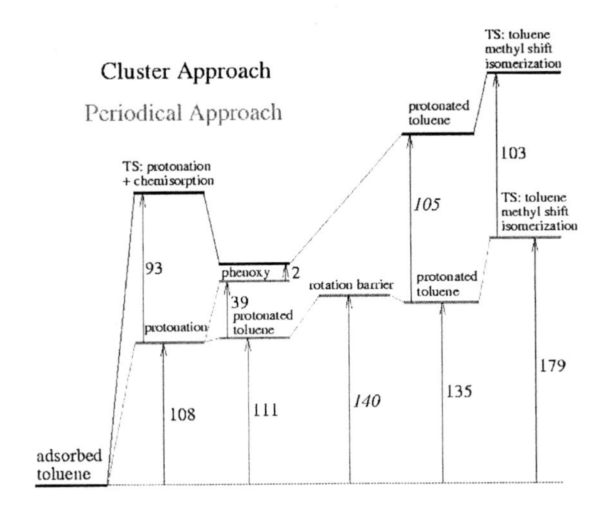

Figure 11. Reaction energy diagrams of the shift isomerization reaction of toluene catalyzed by acid zeolite. The diagrams using black lines refer to the cluster approach results, and the ones using gray lines to the periodic calculations results (in kJ/mol).[45,54,57]

However, the zeolite framework effect on the reaction is not limited only to a stabilization of charged species. We saw already that a transition state from the cluster approach turns to be an inflection point when the zeolite framework contribution is considered. An effect exists also on transition state. In the case of the shift isomerization transition state, it is found an alternative geometry. Before protonated toluene changes its orientation with respect to the deprotonated Brønsted site, the methyl shift reaction step can be achieved (see Figure 12).

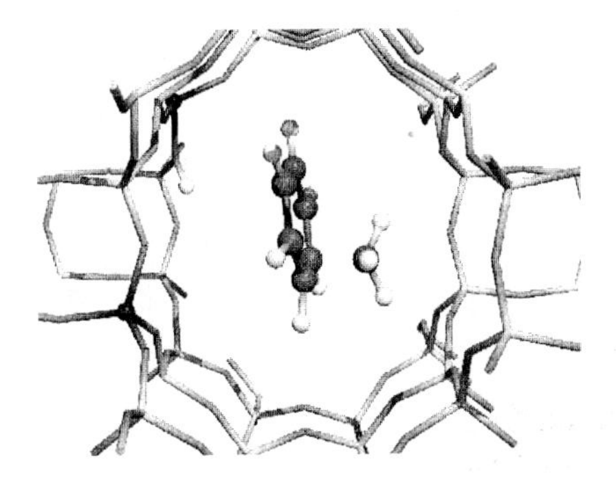

Figure 12. Front view of the alternative geometry of the methyl shift isomerization transition state as obtained from the periodic structure calculations.[45]

For this other transition state, the shifting methyl group is stabilized because of the formation of several weak hydrogen bonds with the zeolitic oxygen atoms located to the opposite side of the channel with respect to the Brønsted site. A striking feature about this methyl shift isomerization transition state is that it has a similar activation energy as the other isomerization transition states with E_{act} = + 179 kJ/mol. Of course, this result is strongly dependent with the local zeolite framework topology, as we will see in the next section.

3. XYLENE ISOMERIZATION

Recently, we investigated the associative alkylation reaction of toluene with methanol catalyzed by an acidic Mordenite (see Figures 13 and 14) by means of periodic ab initio calculations.[43] We observed that for this reaction some transition selectivity occurred, and induced sufficiently large differences in activation energies to explain the small changes in the *para*/*meta*/*ortho* distribution experimentally observed on large pore zeolites.[59] The *para* isomer is the more valuable product as it is an important intermediate for terphthalic acid, an important polymer monomer.[4d] The steric constraints obtained for the transition state structures could be estimated from local intermediates for which the orientations of the toluene molecule were similar as the ones observed for the transition states (see Figure 14).

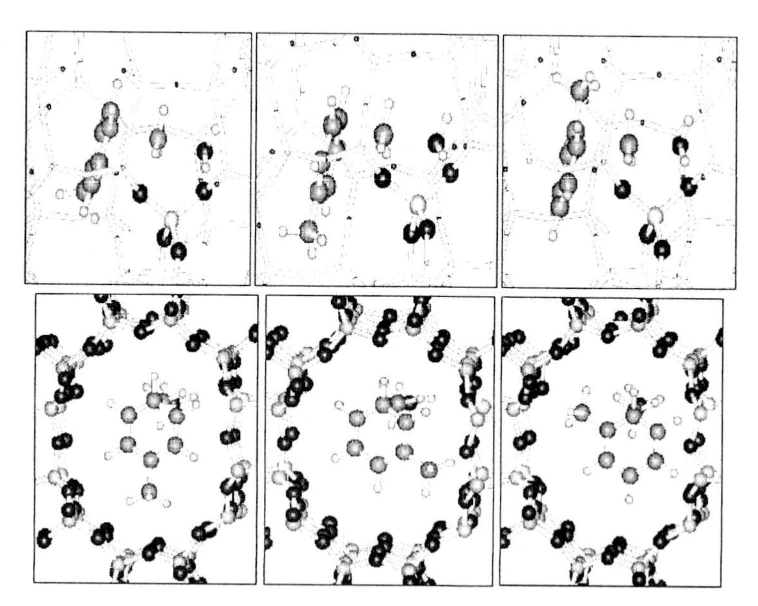

Figure 13. Geometries of the associative mechanism transition states of the alkylation reaction of toluene with methanol catalyzed by H-MOR as obtained from the DFT periodic structure calculations that lead to the formation of the different xylene isomers and water.[43]

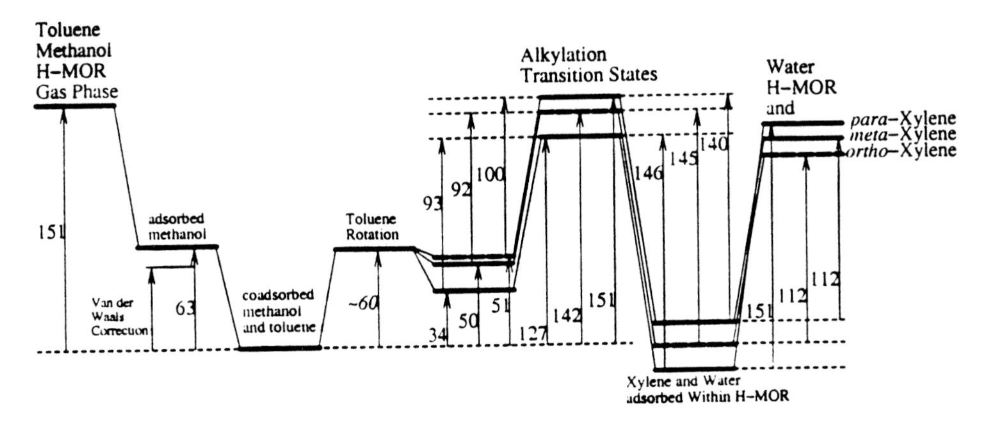

Figure 14. Reaction energy diagrams of the reactions of alkylation of toluene with methanol catalyzed by H-MOR, which give as products *p*-xylene, *m*-xylene, or *o*-xylene, and water (all values in kJ/mol). The values correspond to energies at 0K (American Chemical Society, 2001).[43]

We will use now the same method and Mordenite zeolite model as in the previous part, and investigate the isomerization of xylene isomers.[45] As described in the previous part, this reaction can proceed via two alternative routes, viz. a methyl shift isomerization, and disproportionation reactions. Moreover, we observed than in the case of toluene isomerization, the location of toluene with respect to the Brønsted acidic site for the shift isomerization was of no consequence for the activation energy barrier. We will check these mechanisms for the three xylenes.

It is known from experiment that xylenes do not have different adsorption energies when adsorbed within fully dealuminated mordenite.[18] In our case, we found however that xylenes adsorption to the acidic proton show slight energy differences which are correlated with the local topology of the Brønsted acidic site as well as the geometry of the considered xylene isomer. The computed adsorption energies are - 37 kJ/mol, - 30 kJ/mol, and - 33 kJ/mol for *para*-xylene, *meta*-xylene, and *ortho*-xylene respectively. One notes that these adsorption energies are crude underestimates of the experimental adsorption energies, which have been reported to be around - 130 kJ/mol.[18] This is a direct consequence of the use of DFT method, which are known to not have the ability to describe properly Van der Waals dispersion contribution.[33]

For a study of reactivity within zeolite, it is hopefully a good approximation to not consider Van der Waals dispersion contribution. In a classical dynamic simulation of benzene and toluene within Y zeolite pores, Klein et al.[60] decomposed the guest-host interaction energy. They showed that Van der Waals dispersion contribution was almost constant along diffusion pathways, and that the electrostatic contributions could explain alone the preferred adsorption site locations of aromatics within zeolite. The importance of the electrostatic contribution is enhanced when acidic sites are present. Furthermore, in the case of hydrocarbon molecules reactions within zeolite, it is well understood that transition states are of carbocationic nature.[2]

The main reason of a proper estimate of the adsorption energy is to allow the comparison with experimental. Since the pioneering work of Haag,[61] it is well understood that in zeolites a measured activation energy is actually an apparent activation energy which can be expressed in the case of a first order reaction as:

$$E_{act}^{app} = E_{act} + (1 - \theta)\, E_{ads} \qquad (1)$$

where E_{act} is the activation energy, θ the coverage of the reactant to the active site, and E_{ads} the adsorption energy of the reactant adsorbed to the active site.

To evaluate the Van der Waals interaction energy in the case of our systems, we used the force field parameters provided by Deka et al.,[18] and defined to describe the aromatics adsorption within zeolites. The Van der Waals interaction is then computed between the aromatic guest and the zeolite host,[43,45] and the DFT adsorption energy is corrected with this value as:

$$E_{ads}^{corrected} = E_{ads} + E_{VdW} \qquad (2)$$

In this way, one gets the adsorption energy of the aromatic molecule adsorbed to the acidic proton within the zeolite micropore. It is $E_{ads}^{corrected}$ that is used in (1). Interestingly, the evaluation of E_{VdW} for all our systems gives within few kJ/mol the same value, and we could assume as observed from ref. 60 that the dispersion contribution is a constant contribution for a given adsorbate size. For xylene isomers, this Van der Waals correction is E_{VdW} = - 95 kJ/mol. Then, the adsorption energies for *para*-xylene, *meta*-xylene, and *ortho*-xylene are - 132 kJ/mol, - 125 kJ/mol, and - 128 kJ/mol respectively.

Let us now consider the methyl shift isomerizations that lead from *para*-xylene to *meta*-xylene (see Figure 15).

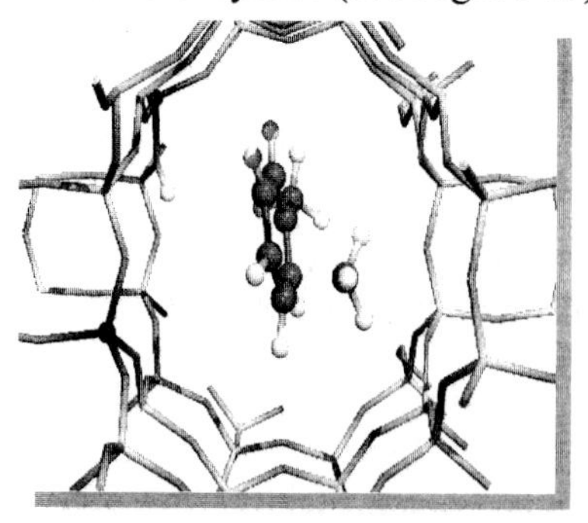
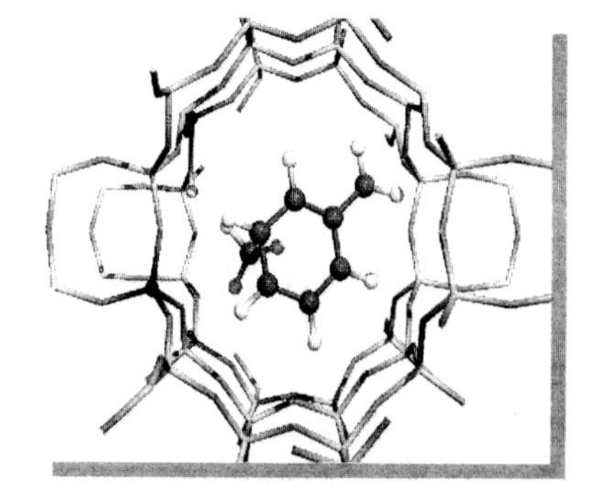

Figure 15. Front views of the transition states of shift isomerization that occur with (right) and without (left) reorientation of the xylene molecule after protonation as obtained from the periodic structure calculations (American Chemical Society, 2001).[45]

As it can be observed in Figure 15, in the case of the methyl shift isomerization transition state for which the orientation of the aromatic ring with respect to the Brønsted site is as the one obtained from the cluster calculations, the non-participating methyl group experiences short interactions with the zeolitic wall. For the transition state structure that is reached without reorientation of the aromatic ring after it became protonated, the non-participating methyl group is oriented in the same direction as the large 12-membered ring channel, and therefore does not suffer from steric constraints with the zeolite wall. This results in two very different activation energies, which prohibit with the achievement of the transition state with an aromatic ring orientation as for the cluster approach transition state. The difference in activation energies is ΔE_{act} = + 63 kJ/mol, and the activation energy for the methyl shift isomerization without reorientation of the aromatic ring after protonation occurred is E_{act} = + 171 kJ/mol.

In the case of the *ortho* to *meta*-xylene methyl shift isomerization transition states, the steric constraints are less important as the non-participating methyl group has more available space because of the ellipsoidal shape of the 12-membered ring channel (see Figure 15). Then, the activation energies are + 168 kJ/mol, and + 184 kJ/mol for the transition state which follows immediately the xylene protonation, and for the transition state which occurs after xylene overcame a rotation energy barrier to change its orientation with respect to the Brønsted site respectively.

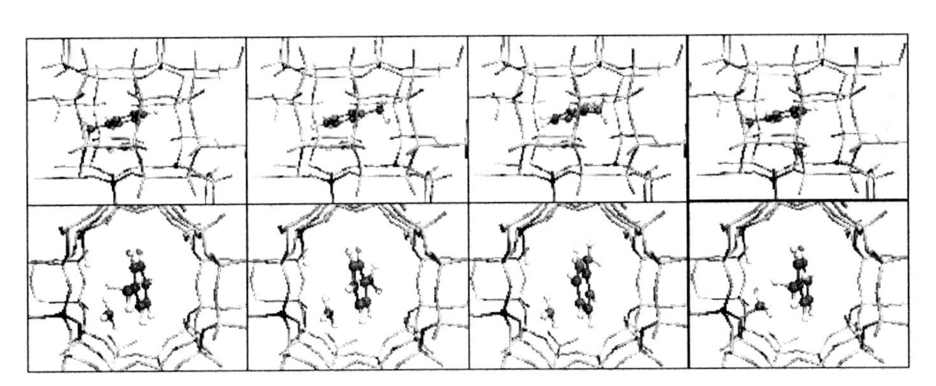

Figure 16. Front and side views of the transition states and intermediate for the isomerization reaction via disproportionation reaction pathway of xylene molecules catalyzed by an acidic Mordenite as obtained from the periodic calculations (American Chemical Society, 2001).[45]

Concerning the disproportionation isomerization reactions, the activation energies follow the predictions of Corma et al.[62] using the HSAB principle of Pearson. In their study, Corma et al. predicted that the activation energies for the alkylation of toluene with methoxy follow the ordering *ortho* < *para*. They estimated these data for zeolitic systems in absence of steric constraints.

It can be seen in Figure 16 that the non-participating methyl group for each of the disproportionation transition states has enough room to avoid steric constraints with the zeolite wall. The activation energies are + 164 kJ/mol, + 174 kJ/mol, and + 187 kJ/mol for an alkylation/dealkylation to the *ortho*, *meta*, and *para* position respectively.

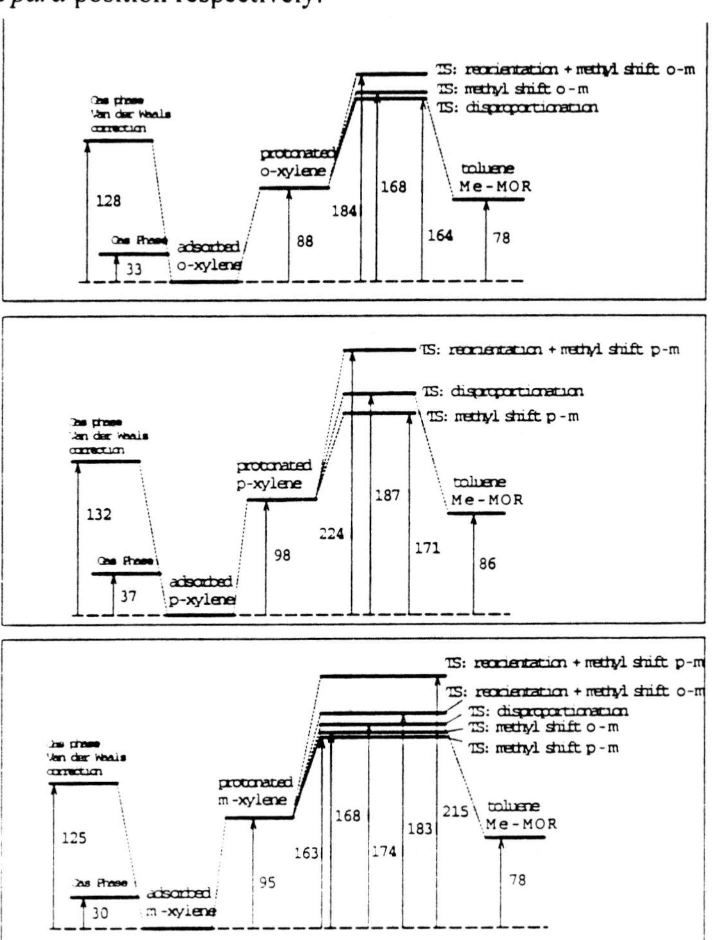

Figure 17. Reaction energy diagrams of the intramolecular isomerization of *ortho*-xylene (top), *para*-xylene (middle) and *meta*-xylene (bottom) catalyzed by H-MOR as obtained from the periodic structure calculations (in kJ/mol) (American Chemical Society, 2001).[45]

These data support the experimental study of Ivanova et al.[63] or the theoretical cluster approach study of Blaszkowski et al.[64]. It is known that the alkylation of toluene with methanol can proceed via two reaction routes. The first one involves an associative protonation of methanol, which induces the methyl jump to toluene (see Figures 13 and 14), whereas the second one is a consecutive reaction pathway, which initiates with the formation of a methoxy species and a water molecule. The second step in the consecutive reaction is an alkylation reaction between toluene and the methoxy species. However, this conclusion has to be toned down as in the present study of the alkylation of toluene with methoxy, the ancillary effect of water is not considered.[5] It has been showed that water has an ancillary and stabilizing effect on transition state. These can decrease the activation energy barrier for carbocationic nature transition state by around 20-30 kJ/mol.[5]

The different isomerization reaction pathways considered in this part are summarized in the reaction energy diagrams in Figure 17.

From Figure 17, one can conclude in total that the differences in activation energies for the isomerization of *ortho*, *meta*, and *para*-xylene are not large enough to induce transition state selectivity. However, transition state selectivity exists and prevents with the isomerization to proceed through some reaction pathways with respect to others. This absence of transition state selectivity when all possible isomerization reaction routes are considered is indeed expected from experimental and dynamic simulation diffusion studies. Deka et al.[18] reported that differences in the diffusion of the three xylene isomers within mordenite are sufficient to explain for the distribution of products as observed from experimental. On the other hand, the computed activation energies provided with this study are in very good match with experimental apparent activation energies after (1) and (2) are applied.[55,65]

In the next part, we will briefly consider a larger aromatic molecule, viz. dimethyldibenzothiophene (DMDBT).[66] This will give us the opportunity to describe the zeolite selectivity that is induced by the adsorption behavior of reactants to the catalytic active site on the products distribution.

4. DMDBT ISOMERIZATION

Hydrodesulfurization (HDS) is an important process in petrochemical refinery as it allows decreasing the sulfur content in diesel fuel.[67] Compounds such as alkylated dibenzothiophene (DBT) raise some problems in HDS. The alkyl groups, especially when located in the 4 and 6 positions of the ring (see Figure 18) make alkylated DBT resistant to classical HDS catalysts because of the methyl groups that prevent the thiophenic sulfur atom to be in contact with the active site of the catalyst.

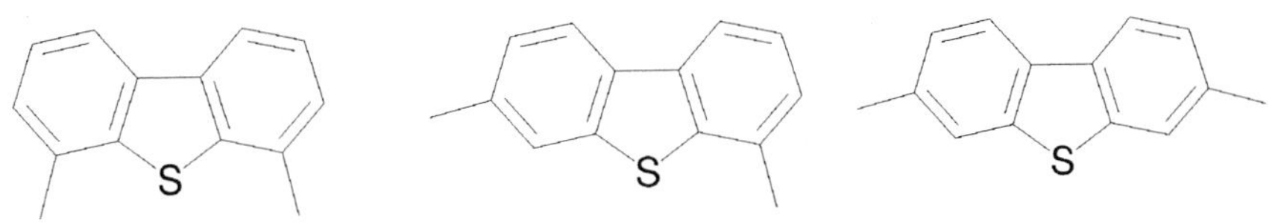

Figure 18. Dimethylated dibenzothiophene derivatives.

However, Michaud et al.[68] and Landau et al.[69] reported that the use of an acid zeolite catalyst and classical HDS catalyst bifunctional catalyst was a very efficient way to desulfurize alkylated DBT. In this case, they observed that acid zeolite catalyst achieves isomerization of the more hindered sulfur atom alkylated DBT (viz. 46DMDBT) to compounds for which the sulfur atom is not anymore hindered by the methyl groups (viz. 37DMDBT).

We will analyze the intramolecular isomerization of 46DMDBT to 37DMDBT catalyzed by H-MOR.[70] Obviously, the large size of DMDBTs prevents with change in the orientation of the molecule within the narrow pore of the zeolite catalyst (see Figure 19). Therefore, only the shift isomerization reactions without reorientation of the DMDBT molecules after they became protonated have been considered.

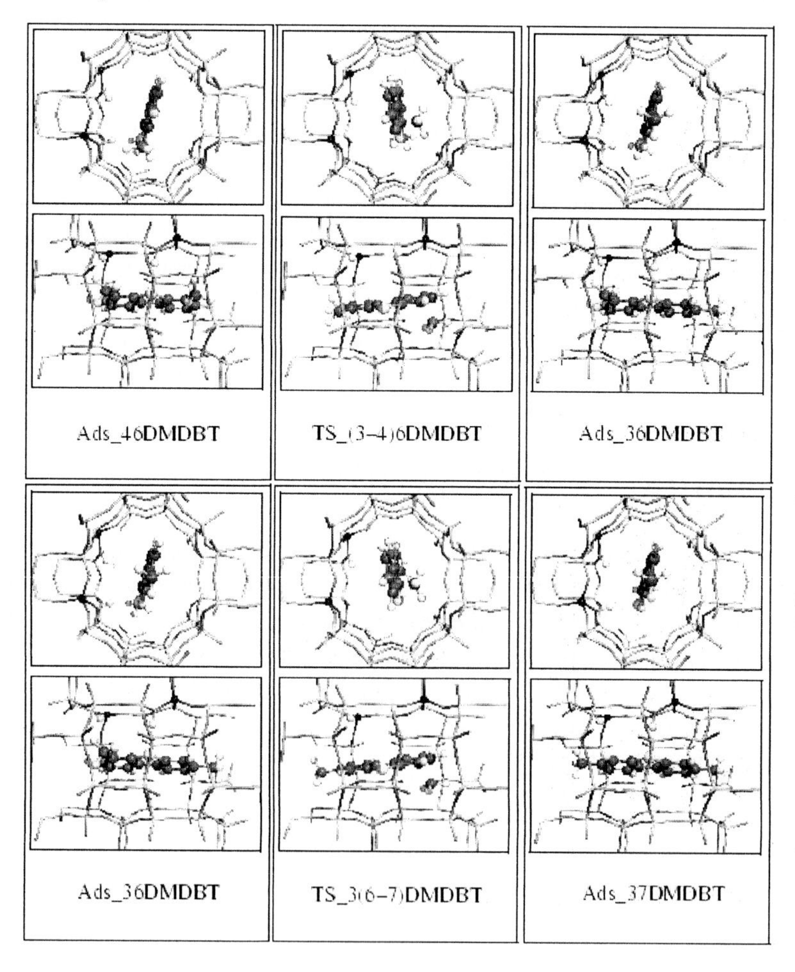

Figure 19. Geometries of 46, 36, and 37DMDBT adsorbed to the acidic site, and of the transition states of the shift isomerization of 46 to 36DMDBT and of 36 to 37DMDBT as obtained from the periodic structure calculations.[70]

The geometries of the shift isomerization transition states of 46 to 36DMDBT and 36 to 37DMDBT and of the physisorbed 46, 36, and 37DMDBT are shown in Figure 19. In the transition states, the shifting methyl groups are located between the carbon atoms it was bonding/will bond. As for the mechanisms obtained for toluene and xylene isomers, here, the zeolitic oxygen atoms that belong to an 8-membered ring side-pocket of MOR stabilize the shifting methyl.

The activation energy of the 46DMDBT to 36DMDBT isomerization with respect to adsorbed 46DMDBT is E_{act} = + 137 kJ/mol. It is E_{act} = + 142 kJ/mol for the 36DMDBT to 37DMDBT reaction with respect to adsorbed 36DMDBT. Interestingly, the activation energies for these two reactions are similar. The same is observed for the isomerizations from 36 to 46DMDBT and 37 to 36DMDBT. In this case, the computed activation energies are E_{act} = + 155 kJ/mol and + 156 kJ/mol respectively.

However, further consideration of the full reaction energy diagram reveals an interesting situation (see Figure 20). The energy levels of the different physisorbed DMDBTs are very different from each other because of more or less sterically hampered DBT methyl groups within the narrow mordenite pore (see Figure 19).

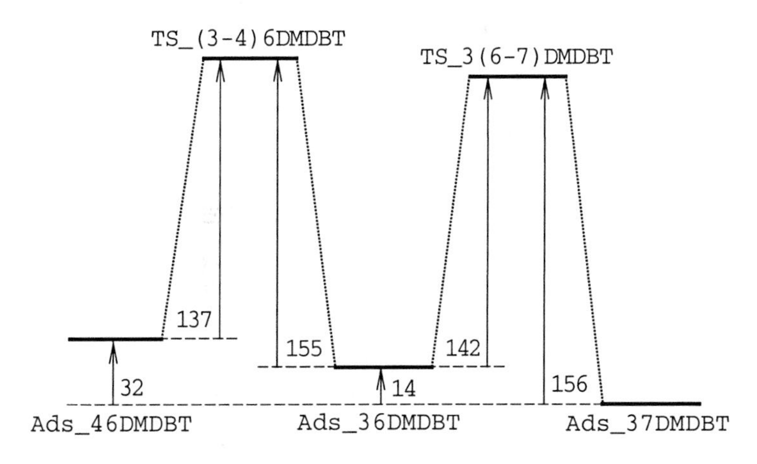

Figure 20. Reaction energy diagram of the intramolecular isomerization reactions of the DMDBT molecules catalyzed by H-MOR as obtained from the periodic structure calculations (all values in kJ/mol).[70]

With respect to adsorbed 46DMDBT energy level, the energy levels of adsorbed 36DMDBT and adsorbed 37DMDBT are - 18 kJ/mol and - 32 kJ/mol respectively. One can estimate using the Polanyi-Brønsted relation,[71] which states that the changes in activation energy are proportional to the changes in reaction energy, that the activation energy for a methyl shift isomerization of DMDBT is E_{act} = + 148 kJ/mol. Then, the selective adsorption behavior of the different DMDBT isomers induces with transition state selectivity. For 36DMDBT, it is more likely that isomerization leads to formation of 37DMDBT than 46DMDBT as the ΔE_{act} is 13 kJ/mol. This is the result that is experimentally observed.[68,69]

5. CONCLUSION

We started this paper with a rather long introduction. This was very important as it allows understanding zeolites minerals and their use as catalysts. Based on experimental information, it becomes possible to define inputs to the calculations. Moreover, we discussed which data can and cannot be accessed by means of quantum chemistry calculations.

The study of reactivity by QM calculations concerns only a small part of a catalytic event, as phenomena such as macro and micro diffusion of the reactants and products outside and inside the zeolite micropores cannot be investigated. On the other hand, QM methods are the only theoretical methods available that provide information on the reactivity.

The shape selectivity of zeolites in diffusion processes is now rather well understood.[72] There remains a gap in the description of the reactive events. With methods such as the small cluster approach, the reactivity is relatively well described. But, such a system cannot model the zeolite framework. Progress in computer power permits for the emergence of increasingly larger cluster studies. The large cluster models provide a useful analysis of the short-range electrostatic effect of the zeolite framework on transition state structures. But it remains questionable whether the large cluster method can describe transition state selectivity properly. In order to keep these molecular clusters in shape, generally fixed geometry models are used. How far can the limited flexibility of the zeolite framework smooth the large energy differences that are due to steric constraints? It is because of this that quantum chemical periodic methods constitute an adequate tool to investigate selective reactivity within zeolites. Of course, QM/MM methods are also useful, especially for a more basic (MM level) analysis of the zeolite selective reactivity. However, considering computational cost criteria, QM/MM methods are often to be preferred over periodic structure QM calculations. Another advantage of QM/MM methods is that Van der Waals dispersion contributions can be modeled.

We have shown here, with the support of selected examples, that periodic calculations have become possible on systems of interest. This allows for a deeper understanding of zeolite catalyst reactivity. The full range of the zeolite selectivity can now be investigated by theoretical methods.

We have shown in the first part what are the consequences of the absence of the zeolite framework on a reaction. This has an important energetic effect. The reaction mechanisms predicted by the cluster approach appear to be very similar to periodic method ones.

Then, we investigated a reaction for which transition state selectivity is important. It appears that several alternative reaction pathways can be followed among which some lead to a minimization of the steric constraints. These steric constraints are strongly dependent on the local topology of the zeolite framework as well as on the geometry of the transition state.

We observed a very good agreement of the computed activation energies with experiment.

Finally, we presented a case for which the large size of reactants and products induces transition state selectivity mainly due to differences in adsorption. The reactivity behavior is shown to follow the Polaniy-Brønsted relation.

ACKNOWLEDGEMENTS

The studies described here have been performed within the European Research Group "*Ab Initio Molecular Dynamics Applied to Catalysis*", supported by the *Centre National de la Recherche Scientifique* (CNRS), the *Institut Français du Pétrole* (IFP) and TotalFinaElf. X. R. thanks TotalFinaElf for the support.

REFERENCES

1. Barrer, R. M. *Zeolite and Clay Minerals as Sorbents and Molecular Sieves*; Ed.; Academic Press: London, New York, San Fransisco, 1978.
2. Van Santen, R. A.; Kramer, G. J. *Chem. Rev.* **1995**, *3*, 637-660.
3. (a) Meier, W. M.; Olson, D. H.; Baerlocher, C. *Zeolites* **1996**, *17*, 1-230. (b) Thomas, J. M.; Thomas, W. J. *Principles and Practice of Heterogeneous Catalysis*; Eds.; VCH: Weinheim, New York, 1997; pp 6-10.
4. (a) Csicsery, S. M. *Zeolites* **1984**, *4*, 202-213. (b) Fraenkel, D.; Levy, M. *J. Catal.* **1989**, *118*, 10-21. (c) Chen, N. Y.; Degnan, T. F. Jr.; Smith, C. M. *Molecular Transport and Reaction in Zeolites, Design and Application of Shape Selective Catalyst*; Eds.; VCH Publishers: New York, 1994; pp 195-289. (d) Tsai, T.-C.; Liu, S.-B.; Wang, I. *Appl. Catal. A* **1999**, *181*, 355-398.
5. Barbosa, L. A. M. M.; Van Santen, R. A. *J. Catal.* **2000**, *191*, 200-217.
6. (a) Frenkel, D. *Proceeding of the Euroconference on Computer Simulation in Condensed Matter Physics and Chemistry, Como 1995*; Binder, K., Ciccotti, A., Eds.; Italian Physical Society: Bolognia, 1995; chapter 7. (b) Schuring, D.; Jansen, A. P. J.; Van Santen, R. A. *J. Phys. Chem. B* **2000**, *104*, 941-948. (c) Smit, B.; Siepmann, J. I. *J. Phys. Chem.* **1994**, *98*, 8442-8452. (d) Smit, B. *Mol. Phys.* **1995**, *85*, 153-172. (e) Smit, B.; Maesen, T. L. M. *Nature* **1995**, *374*, 42-44.
7. Schumacher, R. R.; Anderson, B. G.; Noordhoek, N. J.; De Gauw, F. J. M. M.; De Jong, A. M.; De Voigt, M. J. A.; Van Santen, R. A. *Microporous Mesoporous Mater.* **2000**, *35-36*, 315-326.
8. (a) Jeanvoine, Y.; Ángyán, J. G.; Kresse, G.; Hafner, J. *J. Phys. Chem. B* **1998**, *102*, 7307-7310. (b) Sandré, E.; Pasturel, A. *Mol. Simul.* **1997**, *20*, 63-77.
9. (a) Nicholas, J. B. *Topics Catal.* **1997**, *4*, 157-171. (b) Bernadi, F.; Robb, M. A. *Ab Initio Methods in Quantum Chemistry - I*; Lawley, K. P., Ed.; John Wiley & Sons

Ltd., 1987; pp 155-248.

10. (a) Yashonath, S.; Santikary, P. *J. Phys. Chem.* **1993**, *97*, 3849-3857. (b) Yashonath, S.; Santikary, P. *Mol. Phys.* **1993**, *78*, 1-6. (c) Yashonath, S.; Santikary, P. *J. Chem. Phys.* **1994**, *100*, 4013-4016.
11. Ugliengo, P.; Civalleri, B.; Zicovich-Wilson, C. M.; Dovesi, R. *Chem. Phys. Lett.* **2000**, *318*, 247-255.
12. Brändle, M.; Sauer, J.; Dovesi, R.; Harrison, N. M. *J. Chem. Phys.* **1998**, *109*, 10379-10389.
13. Lowenstein, W. *Am. Mineral.* **1942**, *39*, 92.
14. Barrer, R. M. *Zeolite and Clay Minerals as Sorbents and Molecular Sieves*; Ed.; Academic Press: London, New York, San Fransisco, 1978; pp 18-19.
15. Fricke, R.; Kosslick, H.; Lischke, G.; Richter, M. *Chem. Rev.* **2000**, *100*, 2303-2405.
16. (a) De Man, A. J. M.; Van Santen, R. A. *Zeolites* **1992**, *12*, 269-279. (b) Lee, C.; Parrillo, D. J.; Gorte, R. J.; Farneth, W. E. *J. Am. Chem. Soc.* **1996**, *118*, 3262-3268.
17. (a) Levien, L.; Previtt, C. T.; Weider, D. J. *Am. Mineral.* **1980**, *65*, 925. (b) De Man, A. J. M.; Van Beest, B. W. H.; Leslie, M.; Van Santen, R. A. *J. Phys. Chem.* **1990**, *94*, 2524-2564. (c) Schröder, K.-P.; Sauer, J. *J. Phys. Chem.* **1996**, *100*, 11043-11049.
18. Deka, R. C.; Vetrivel, R.; Miyamoto, A. *Topics Catal.* **1999**, *9*, 225-234.
19. Bezuz, A. A.; Kiselev, A. G.; Loptakin, A. A.; Quang Du, P. *J. Chem. Soc., Faraday Trans. II* **1978**, *74*, 367-379.
20. Bezuz, A. A.; Kocirik, M.; Kiselev, A. V.; Lopatkin, A. A.; Vasilyena, E. A. *Zeolites* **1986**, *6*, 101-106.
21. Coombs, D. S.; Alberti, A.; Armsbruster, T.; Artioli, G.; Colella, C.; Galli, E.; Grice, J. D.; Liebau, F.; Mandarino, J. A.; Minato, H.; Nickel, E. H.; Passaglia, E.; Peacor, D. R.; Quartieri, S.; Rinaldi, R.; Ross, M.; Sheppard, R. A.; Tillmanns, E.; Vezzalini, G. *Am. Mineral.* **1998**, *83*, 917-935.
22. (a) Nicholas, J. B. *Topics Catal.* **1997**, *4*, 157-171. (b) Rigby, A. M.; Kramer, G. J.; Van Santen, R. A. *J. Catal.* **1997**, *170*, 1-10. (c) Van Santen, R. A. *Catal. Today* **1997**, *38*, 377-390.
23. (a) Olah, G. A.; Donovan, D. J. *J. Am. Chem. Soc.* **1977**, *99*, 5026-5039. (b) Olah, G. A.; Prakash, G. K. S.; Sommer, J. *Superacids*; Eds.; Wiley Interscience: New York, 1985. (c) Olah, G. A.; Prakash, G. K. S.; Williams, R. E.; Field, L. D.; Wade, K. *Hypercarbon Chemistry* ; Eds.; Wiley Interscience: New York, 1987.
24. (a) Haw, J. F.; Richardson, B. R.; Oshiro, I. S.; Lazo, N. D.; Speed, J. A. *J. Am. Chem. Soc.* **1989**, *111*, 2052-2058. (b) Haw, J. F.; Nicholas, J. B.; Xu, T.; Beck, L. W.; Ferguson, D. B. *Acc. Chem. Res.* **1996**, *29*, 259-267. (c) Haw, J. F.; Xu, T.; Nicholas, J. B.; Goguen, P. W. *Nature* **1997**, *389*, 832-835.
25. Gorte, R. J. *Catal. Lett.* **1999**, *62*, 1-13.
26. (a) Zygmunt, S. A.; Curtiss, L. A.; Zapol, P.; Iton, L. E. *J. Phys. Chem. B* **2000**, *104*, 1944-1949. (b) Sauer, J.; Sierka, M.; Haase, F. *Transition State Modeling for Catalysis*; Truhlar, D. G., Morokuma, K., Eds.; ACS Symp. Ser. No 721; American Chemical Society: Washington, 1999; pp 358-367.
27. Derouane, E. G.; He, H.; Hamid, S. B. D.-A.; Ivanova, I. I. *Catal. Lett.* **1999**, *58*, 1-18.
28. (a) Kazansky, V. B.; Senchenya, I. N. *J. Catal.* **1989**, *119*, 108-120. (b) Sinclair, P. E.; De Vries, A.; Sherwood, P.; Catlow, C. R. A.; Van Santen, R. A. *J. Chem. Soc., Faraday Trans.* **1998**, *94*, 3401-3408. (c) Rigby, A. M.; Frash, M. V. *J. Mol. Catal. A* **1997**, *126*, 61-72.
29. (a) Hölderich, W. F.; Van Bekkum, H. *Introduction to Zeolite Science and Practice*; Van Bekkum, H., Flaningen, E. M., Jansen, J. C., Eds.; Elsevier: Amsterdam, 1991; pp 631-726, vol. 58. (b) Venuto, P. B. *Microporous Mater.* **1994**, *2*, 297-411.
30. (a) Fernàndez, L.; Martí, V.; García, H. *Phys. Chem. Chem. Phys.* **1999**, *1*, 3689-3695. (b) Laszlo, P. *J. Phys. Org. Chem.* **1998**, *11*, 356-361.
31. Schleyer, P. v. R. *Encyclopedia of Computational Chemistry*; Ed.; John Wiley &

Sons: New York, 1998.

32. Perdew, J. P. *Density Functional Theory, a Bridge Between Chemistry and Physics*; Geerlings, P., De Proft, F., Langenaeker, W., Eds.; VUB University Press: Brussels, 1999; pp 87-109.
33. (a) Kristyan, S.; Pulay, P. *Chem. Phys. Lett.* **1994**, *229*, 175-180. (b) Sauer J.; Ugliengo, P.; Garrone, E.; Saunders, V. R. *Chem. Rev.* **1994**, *94*, 2095-2160. (c) Lein, M.; Dobson, J.F.; Gross, E. K. U. *J. Comp. Chem.* **1999**, *20*, 12-22. (d) Pelmenschikov, A.; Leszczynski, J. *J. Phys. Chem. B* **1999**, *103*, 6886-6890.
34. Kramer, G. J.; De Man, A. J. M.; Van Santen , R. A. *J. Am. Chem. Soc.* **1991**, *113*, 6435-6441.
35. (a) Sauer, J. *Chem. Rev.* 1989, 89, 199-255. (b) Gale, J. D. *Topics Catal.* **1996**, *3*, 169-194. (c) Frash, M. V.; Van Santen, R. A. *Topics Catal.* **1999**, *9*, 191-205.
36. (a) Datka, J.; Broclawik, E.; Gil, B.; Sierka, M. *J. Chem. Soc., Faraday Trans.* **1996**, *92*, 4643-4646. (b) Ugliengo, P.; Ferrari, A. M.; Zecchina, A.; Garrone, E. *J. Phys. Chem.* **1996**, *100*, 3632-3645. (c) Frash, M. V.; Makarova, M. A.; Rigby, A. M. *J. Phys. Chem. B* **1997**, *101*, 2116-2119. (d) Zygmunt, S. A.; Curtiss, L. A.; Iton, L. E.; Erhardt, M. K. *J. Phys. Chem.* **1996**, *100*, 6663-6671.
37. (a) Koller, H.; Overweg, A. R.; Van Santen, R. A.; De Haan, J. W. *J. Phys. Chem. B* **1997**, *101*, 1754-1761. (b) Koller, H.; Meijer, E. L.; Van Santen, R. A. *Solid State Nuc. Magn. Res.* **1997**, *9*, 165-175. (c) Koller, H.; Engelhardt, G.; Van Santen, R. A. *Topics Catal.* **1999**, *9*, 163-180. (d) Bull, L. M.; Bussemer, B.; Anupõld, T.; Reinhold, A.; Samoson, A.; Sauer, J.; Cheetham, A. K.; Dupree, R. *J. Am. Chem. Soc.* **2000**, *122*, 4948-4958.
38. Sauer, J.; Eichler, U.; Meier, U.; Schäfer, A.; Von Arnim, M.; Ahlrichs, R. *Chem. Phys. Lett.* **1999**, *308*, 147-154.
39. (a) Sherwood, P.; De Vries, A. H.; Collins, S. J.; Greatbanks, S. P.; Burton, S. A.; Vincent, M. A.; Hillier, I. H. *Faraday Discuss.* **1997**, *106*, 79-92. (b) Sierka, M.; Sauer, J. *Faraday Discuss.* **1997**, *106*, 41-62.
40. Allen, M. P.; Tildesley, D. J. *Computer Simulation of Liquids*; Eds.; Oxford Science Publications: New York, 1987.
41. Ramachandran, S.; Lenz, T. G.; Skiff, W. M.; Rappé, A. K. *J. Phys. Chem.* **1996**, *100*, 5898-5907.
42. (a) Kresse, G.; Furthmüller, J. *Comput. Matter. Sci.* **1996**, *6*, 15-50. (b) Kresse, G.; Furthmüller, J. *Phys. Rev. B* **1996**, *54*, 11169-11186. (c) Civalleri, B.; Zicovich-Wilson, C. M.; Ugliengo, P.; Saunders, V. R.; Dovesi, R. *Chem. Phys. Lett.* **1998**, *292*, 394-402.
43. Vos, A. M.; Rozanska, X.; Schoonheydt, R. A.; Van Santen, R. A.; Hutschka, F.; Hafner, J. *J. Am. Chem. Soc.* **2001**, *123*, 2799-2809.
44. Makov, G.; Payne, M. C. *Phys. Rev. B* **1995**, *51*, 4014-4022.
45. Rozanska, X.; Van Santen, R. A.; Hutschka, F.; Hafner, J. to appear.
46. Frisch, M. J.; Trucks, G. W.; Schlegel, H. B.; Scuseria, M. A.; Robb, M. A.; Cheeseman, J. R.; Zakrzewski, V. G.; Montgomery, J. A.; Stratmann, R. E.; Burant, J. C.; Dapprich, S.; Millam, J. M.; Daniels, A. D.; Kudin, K. N.; Strain, M. C.; Farkas, O.; Tomasi, J.; Barone, V.; Cossi, M.; Cammi, R.; Mennucci, B.; Pomelli, C.; Adamo, C.; Clifford, S.; Ochterski, J.; Petersson, G. A.; Ayala, P. Y.; Cui, Q.; Morokuma, K.; Malick, D. K.; Rabuck, D. K.; Raghavachari, K.; Foresman, J. B.; Cioslowski, J.; Ortiz, J. V.; Stefanov, B. B.; Liu, G.; Liashenko, A.; Piskorz, P.; Komaromi, I.; Gomperts, R.; Martin, R. L.; Fox, D. J.; Keith, T.; Al-Laham, M. A.; Peng, C. Y.; Nanayakkara, A.; Gonzales, C.; Challacombe, M.; Gill, P. M. W.; Johnson, B. G.; Chen, W.; Wong, M. W.; Andress, J. L.; Head-Gordon, M.; Replogle, E. S.; Pople, J. A. Gaussian 98 (revision A.1) Gaussian, Inc.: Pittsburg PA, 1998.
47. (a) Becke, A. D. *Phys. Rev. A* **1988**, *38*, 3098-3100. (b) Lee, C.; Yang, W.; Parr, R. G. *Phys. Rev. B* **1988**, *37*, 785-789. (c) Becke, A. D. *J. Chem. Phys.* **1993**, *98*, 5648-5652.

48. Zygmunt, S. A.; Mueller, R. M.; Curtiss, L. A.; Iton, L. E. *J. Mol. Struct.* 1998, 430, 9-16.
49. (a) Kresse, G.; Hafner, J. *Phys. Rev. B* **1993**, *48*, 13115-13126. (b) Kresse, G.; Hafner, J. *Phys. Rev. B* **1994**, *49*, 14251-14269. (c) Kresse, G.; Hafner, J. *J. Phys. Condens. Matter.* **1994**, *6*, 8245-8257.
50. Demuth, T.; Hafner, J.; Benco, L.; Toulhoat, H. *J. Phys. Chem. B* **2000**, 104, 4593-4607.
51. Perdew, J. P.; Zunger, A. *Phys. Rev. B* **1981**, *23*, 5048-5079.
52. Perdew, J. P.; Burke, K.; Wang, Y. *Phys. Rev. B* **1996**, *54*, 16533-16539.
53. Mills, G.; Jónsson, H.; Schenter, G. K. *Surf. Sci.* **1995**, *324*, 305-337.
54. Rozanska, X.; Saintigny, X.; Van Santen, R. A.; Hutchska, F. to appear.
55. (a) Kumar, R.; Ratnasamy, P. *J. Catal.* **1989**, *116*, 440-448. (b) Mirth, G.; Cejka, J.; Lercher, J. A. *J. Catal.* **1993**, *139*, 24-33. (c) Cortes, A.; Corma, A. *J. Catal.* **1978**, *51*, 338-344. (d) Young, L. B.; Butter, S. A.; Kaeding, W. W. *J. Catal.* **1982**, *76*, 418-432. (e) Beschmann, K.; Riekert, L. *J. Catal.* **1993**, *141*, 548-565. (f) Pérez-Pariente, J.; Sastre, E.; Fornés V.; Martens, J. A.; Jacobs, P. A.; Corma, A. *Appl. Catal.* **1991**, *69*, 125-137. (g) Morin, S.; Gnep, N. S.; Guisnet, M. *J. Catal.* **1996**, *159*, 296-304. (h) Guisnet, M.; Gnep, N. S.; Morin, S. *Microporous Mesoporous Mater.* **2000**, *35-36*, 47-59.
56. Paukshtis, E. A.; Malysheva, L. V.; Stepanov, V. G. *React. Kinet. Catal. Lett.* 1998, 65, 145-152.
57. Rozanska, X.; Van Santen, R. A.; Hutschka, F.; Hafner, J. *Stud. Surf. Sci. Catal.* to appear.
58. Boronat, M.; Zicovich-Wilson, C. M.; Corma, A.; Viruela, P. *Phys. Chem. Chem. Phys.* **1999**, *1*, 537-543.
59. Corma, A.; Sastre, G.; Viruella, P. M. *J. Mol. Catal. A* **1995**, *100*, 75-85.
60. Klein, H.; Kirschhoch, C.; Fuess, H. *J. Phys. Chem.* **1994**, *98*, 12345-12360.
61. Haag, W. O. *Zeolites and Related Materials, State of the Art 1994*; Weitkamp, H. G., Karge, H., Pfeifer, H., Hölderich, W., Eds.; Elsevier Science: Amsterdam, 1994; pp 1375-1394.
62. Corma, A.; Llopis, F.; Viruela, P.; Zicovich-Wilson, C. *J. Am. Chem. Soc.* **1994**, *116*, 134-142.
63. Ivanova, I. I.; Corma, A. *J. Phys. Chem. B* **1997**, *101*, 547-551.
64. Blaszkowski, S. R.; Van Santen, R. A. *Transition State Modeling for Catalysis*; Truhlar, D. G., Morokuma, K., Eds.; ACS Symp. Ser. No. 721; American Chemical Society: Washington, DC, 1999; Chapter 24
65. Norman, G. H.; Shigemura, D. S.; Hopper, J. R. *Ind. Eng. Chem., Prod. Res. Dev.* **1976**, *15*, 41-45.
66. Rozanska, X.; Van Santen, R. A.; Hutschka, F.; Hafner, J. to appear.
67. Weber, T.; Prins, R.; Van Santen, R. A. *Transition Metal Sulphides, Chemistry and Catalysis*; Eds.; NATO ASI Series 3, 1998; Vol. 60.
68. Michaud, P.; Lemberton, J. L.; Pérot, G. *Appl. Catal. A* **1998**, *169*, 343-353.
69. Landau, M. V. *Catal. Today* **1997**, *36*, 393-429.
70. Rozanska, X.; Van Santen, R. A.; Hutschka, F. to appear.
71. Van Santen, R. A. *J. Mol. Catal. A* **1997**, *115*, 405-419.
72. Bates, S.; Van Santen, R. A. *Adv. Catal.* **1998**, *42*, 1-114.

Quantum Chemical Modeling the Location of Extraframework Metal Cations in Zeolites

G. N. VAYSSILOV
Faculty of Chemistry, University of Sofia, 1126 Sofia, Bulgaria; e–mail: gnv@chem.uni–sofia.bg

1. INTRODUCTION

Due to the presence of three-valent aluminum in tetrahedral framework positions instead of tetra-valent silicon, the zeolite framework of the alumosilicate zeolites is negatively charged and its charge is compensated by different cations. Since these cations are outside the lattice, they are easy exchangeable, which allows a wide variation of zeolite properties. In H-forms of zeolites, the charge-compensating cations are protons which are attached to framework oxygen atoms, thus forming bridging hydroxyl groups (Al-OH-Si). In alkali forms of zeolites as well other zeolites, which exchanged with metal cations, it is expected to find the metal ions located near Al positions of the lattice because of the local negative charge there.

Cation exchanged zeolites are successfully applied as catalysts or selective sorbents in separation technologies.[1-3] For both catalytic and sorption processes a concerted action of polarizing cations and basic oxygen atoms is important. In addition, transition metal cation embedded in zeolites exhibit peculiar redox properties because of the lower coordination in zeolite cavities compared to other supports.[4-7] Therefore, it is important to establish the strength and properties of active centers and their positions in the zeolite structure. Various experimental methods and simulation techniques have been applied to study the positions of cations in the zeolite framework and the interaction of the cations with guest molecules.[1-4,8] Here, some of the most recent theoretical studies of cation exchanged zeolites are summarized.

Since the acid-basic properties as well as the local arrangement at individual zeolite rings depend strongly on the aluminum content of the ring, it is necessary to establish the actual distribution of rings with different numbers of aluminum atoms in a given zeolite sample.[9] This is a problem for zeolites with high Al content, e.g. faujasites, while zeolites with high Si content contain at most one Al center per ring. However, there is still no certified method to measure this distribution. For zeolites with more complicated framework structure as MOR, MFI, FER, arises the problem

M.A. Chaer Nascimento (ed.), Theoretical Aspects of Heterogeneous Catalysis, 29–38.

with various possibilities of Al substitution in the framework positions and, respectively, various possible locations of the extraframework cations.

2. ALKALI AND ALKALINE-EARTH CATIONS

2.1 Sodium cation

The optimized sodium cation positions in a six-ring of FAU zeolite structure containing two Al atoms in para-position (denoted as Na-Al-2p)[10] is shown in Figure 1. This position of the cation is representative for SII cation position in Y and X zeolites,[11-13] as well as the position of Na^+ in Na-EMT zeolite. As expected, for this and the other zeolite model structures, Na^+ prefers positions near to oxygen centers bonded to Al atoms rather than those of Si-O-Si bridges. Also, the cation is far from oxygen centers which are connected to compensating cations, an additional proton in this case.

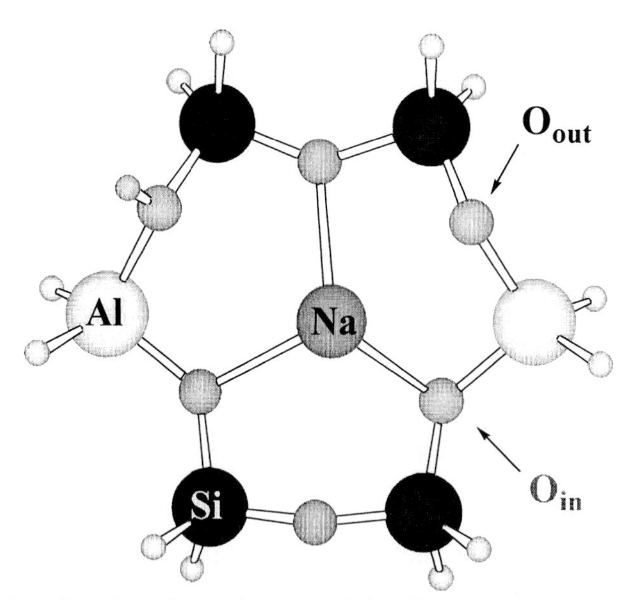

Figure 1. Location of a sodium cation at a model zeolite six-ring. Zeolite oxygen centers directed toward inside and outside the cluster are also shown.

The mean Na-O distances of the rings with two or three aluminum atoms are close to the value obtained for the simpler model $Al(H)(OSiH_3)_3$.[6] When Na^+ is present in the ring, the oxygen centers move slightly inward the ring and it intereacts stronger with the oxygen centers oriented toward the inside the zeolite ring, denoted in Fig. 1 as O_{in} centers.

The calculated BE of a sodium ion at six-rings with one, two or three aluminum atoms are 493-519 kJ/mol. The binding energies of the cation at six-rings are calculated larger than those at the four-rings. For instance, the Na^+ BE at the four-ring with two aluminum atoms is by 40 kJ/mol lower than at the six-rings with two Al centers. This is in agreement with the experimental observation that in faujasites sodium cations prefer locations at six-rings (SII and SI' sites) over SIII sites, corresponding to four-ring.[11-13] The binding energy of Na^+ at the SIII position is probably underestimated to some extent in our model calculations because we did not account for the cation interaction with neighboring four-rings.

It is interesting to compare the relative energies of the two isomeric clusters with two aluminum atoms, Na-Al-2p and Na-Al-2m. Both for the initial negatively charged cluster and the clusters with Na^+, the ring with aluminum atoms in *para* position is more stable by 21-23 kJ/mol. This agrees with the Dempsey's rule[14] applied to six-rings of faujasites.

The sodium cation in the cluster with one Al atom, Na-Al-1, is located almost in the plane of the ring, near the aluminum atom, at 234 and 235 pm from the oxygen atoms of the Al-O-Si bridges. After coordination of probe molecules as carbon monoxide[10] or methanol,[15] the sodium ion shifts higher, along the axis perpendicular to the ring, 32 pm for CO and 42-47 pm for methanol. This could indicate that the potential energy surface for Na^+ motion perpendicularly to the ring is flat since even such a weak interaction as that with CO can shift the sodium ion position. After CO adsorption the sodium cation is above the plane of T-atoms and the BE is reduced only by 1 kJ/mol compared to the optimized position. The energy gain from carbon monoxide adsorption (13 kJ/mol) easily compensates this loss of binding energy. Similar shift of the Na^+ cation is observed also after adsorption of probe molecules at a cation located at the other model rings.

The geometry optimizations of the two clusters containing two aluminum atoms were performed for rings whose negative charge excess is compensated by a proton. For both structures, Na-Al-2m and Na-Al-2p, two possible positions of the sodium ion were found, one at each side of the ring. The positions are denoted by reference to the oxygen atoms directed inward the ring (O_{in}), *syn* or *anti*. The *anti* cation position is located almost in the plane of the oxygen atoms O_{out} which are directed toward the outside of the ring; it lies 59 pm above the plane of T-atoms for Na-Al-2p (Fig. 2b) and 80 pm above the reference plane for Na-Al-2m. Despite the fact that Na^+ is on the side of the oxygen atoms O_{out}, it is closer to the oxygen atoms O_{in} because they are directed inward the ring. The position of the O_{in} centers with respect to the ring is important not only because they are near the center of the ring but also because their non-bonding lone-pair orbitals are directed inward the ring.

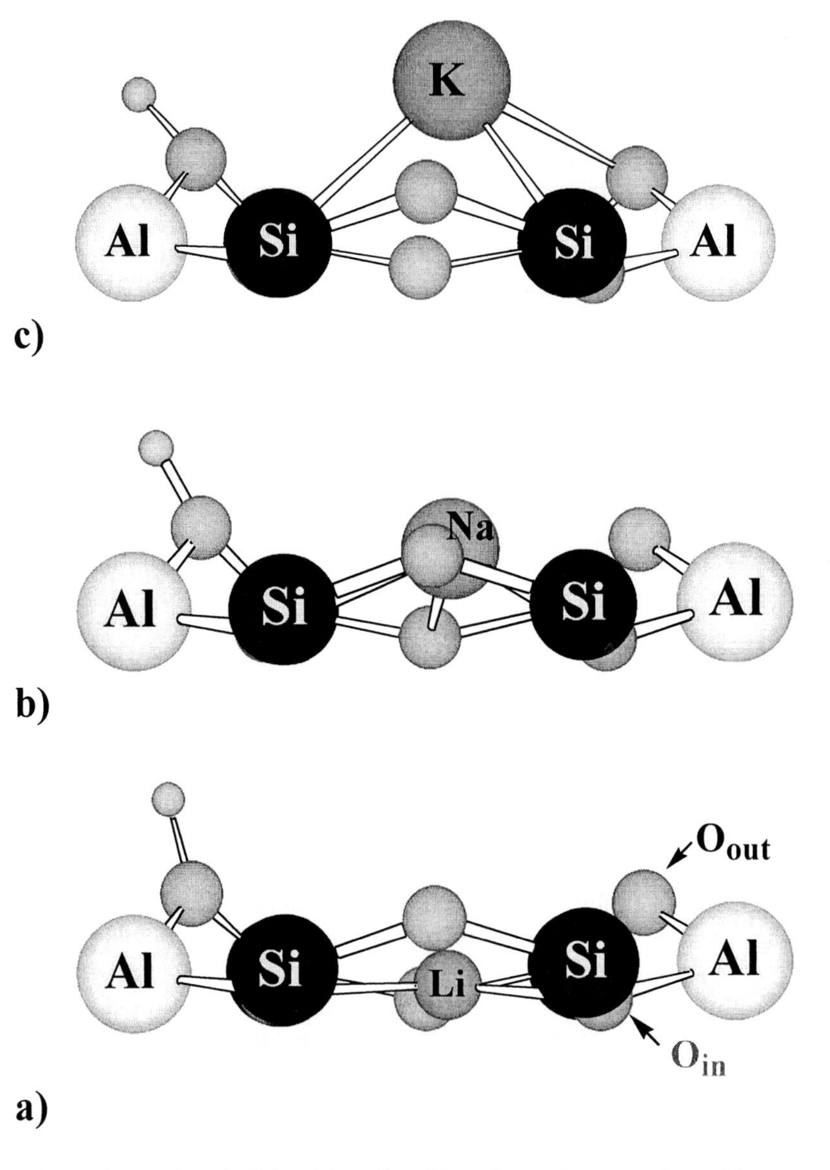

Figure 2. Location of a litium (a), sodium (b), and potassium (c) cations with respect to the plane of the T-atoms of the model six-ring.

The lone-pair density of the O_{out} centers is mainly directed toward the outside of the six-rings studied. Therefore, the centers O_{out} can take part in coordination of cations outside the ring or they can play a role of basic centers (on the walls of the supercage) in catalytic reactions or adsorption. Indeed, the sites with the highest proton affinity in the clusters studied correspond to oxygen centers, directed outward the ring.[16]

The sodium position *syn* is on the side of the centers O_{in}, at 112 pm below the plane of T-atoms for the cluster Na-Al-2p or 129 pm for Na-Al-2m. This is 60-70 pm beyond the plane of the oxygen atoms at the same side of the ring. In the *syn* configuration, Na^+ is by more than 40 pm farther from the O_{out} atoms at the other side of the ring, compared to the *anti* position. The distances Na-O_{in} to the closer oxygen centers are by 1-5 pm shorter. The two sodium positions have almost the same energy for both clusters.

The positions of a sodium cation at the cluster containing three aluminum atoms are similar to that with two Al centers. The first one (*anti*) is 33 pm above the ring, the second one (*syn*) is 119 pm below the plane of T-atoms. In this ring all oxygen atoms participate in Al-O-Si bridges, but due to the charge compensation two of them have each a proton attached. In the *anti* position, Na^+ is closer to the plane of T-atoms than in clusters with two aluminum centers, and the cation interacts not only with two of the O_{in} atoms (at distances of 225-227 pm), but also with one of the O_{out} oxygen centers (at 234 pm) which, in this cluster model, is also oriented inward the ring. In the *syn* position, Na^+ is almost at the same distance to the O_{in} atoms, but much farther from the others (the mean Na-O_{out} distance increases by 40 pm).

Similarly to the six-ring, the four-ring with one aluminum atom has only two oxygen atoms connected to Al. The four-ring is smaller, Na^+ is located farther from the ring, and its interaction with both oxygen atoms of the Al-O-Si bridges is weaker. For this reason the binding energy of Na^+ is the lowest of all clusters investigated. In the four-ring with two Al centers, the sodium cation is at 233-235 pm from two of the oxygen atoms. The position is shifted in the direction opposite to the charge-compensating cation.

In order to gain information on the potential energy barrier for the transfer of a sodium cation between the two local minima observed for three of the six-rings, several intermediate sodium positions at the cluster Na-Al-2p were partially optimized. For each structure the Na^+ position was fixed relative to the ring and the positions of the oxygen centers and the charge compensation proton were optimized. We found a clear an energy barrier for crossing the ring of only about 10 kJ/mol. The barrier occurs when Na^+ is in the plane of the oxygen atoms directed inside the ring (at about -50 pm) because of the short Na-O distances. During the transfer of the cation across the ring, the O_{in} centers shift slightly outward to keep optimal distances. A similar energy check for the sodium motion perpendicular to the ring with one aluminum center did not reveal other stationary points except the minimum described earlier.

2.2 Other cations

The position of the metal cation with respect to the zeolite ring depends strongly on the ionic radius of the cation, as is evident from x–ray diffraction (XRD) data.[17] The other alkali and alkaline-earth cations[18] studied show trends similar to Na^+ cation. They interact mainly with (i) the oxygen centers of Si–O–Al bridges and (ii) the oxygen centers directed inwards the model ring (O_{in}); also, the cation was found far from the bridging OH group.

The positions of the smaller cations, lithium at the Al–2p ring and magnesium at the Al–3 ring, are near the plane of T–atoms of the six–ring, shifted somewhat toward the plane of the oxygen centers O_{in}. The shortest Li–O distance is 190 pm (Fig. 2a) and Mg–O distance is 203 pm. The Mg–O distances to the farther oxygen centers O_{in} are almost the same, 220–223 pm. The position of Mg^{2+} near the center of the six–ring agrees with the XRD data for dehydrated MgNaX zeolite (for Mg at site SII),[17] while the presence of water molecules results in a shift of the cation.[17] Geometry optimization with different starting positions of Li^+ or Mg^{2+} at the six–rings Al–2p and Al–3, respectively, lead to the same optimized structures, described above, i.e. there are no *syn* and *anti* location as observed for sodium cation.

The position of Ca^{2+} at the six–ring Al–3 is very similar to that of a sodium cation at the same six–ring; this has been expected from the similarity of the ionic radii of the two cations, 97 and 99 pm for Na^+ and Ca^{2+}, respectively.[19] The calcium cation is at 30 pm above the plane of T–atoms of the ring with a shortest Ca–O distance of 229 pm.

Two local minima were found for K^+ at the six–ring K-Al–2p, similar to the *syn* and *anti* positions of a sodium cation, but farther away from the plane of T–atoms. The cation is located 203 pm from the plane of the ring when K^+ is at the side of the O_{in} oxygen atoms, i.e. in *syn* position, and 171 pm from the plane, when in anti position – at the opposite side of the ring (Fig. 2c). At variance with Na^+, both positions of K^+ were also observed at the six–ring K-Al–1, because the cation is too big to move to the center of the ring as does the sodium cation. The shortest K–O distances of the *syn* position at the six–ring K-Al–2p are 260 and 278 pm, while for the *anti* position these values are slightly larger, 270 and 279 pm.

The Lewis acidity of the sodium cation in the two positions at the rings with two or three Al centers is different judged both by the Mulliken charge and the frequency shift of an adsorbed probe molecule carbon monoxide. The cation is more acidic in *syn* position, q(Na) = 0.61 e, and

$\Delta\nu(CO) = 38\ cm^{-1}$, while for the *anti* position the charge is by 0.07–0.10 e smaller and so is the CO frequency shift (by 6–10 cm^{-1}). At variance with these results, the Lewis acidity of the potassium cation at both positions is essentially the same: the charge varies by 0.02 e and the CO frequency shift by 2 cm^{-1}. Since Rb^+ and Cs^+ cations are farther than K^+ from the zeolite ring due to their larger ionic radii,[19] one should expect even smaller differences between the properties of these cations in the two possible positions. This comparison suggests that the existence of two cationic positions with respect to an individual six–ring with distinguishable properties is a peculiarity of Na^+ among the alkali cations (as mentioned above, only one position was observed for Li^+).

3. TRANSITION METAL CATIONS

3.1 Copper

Copper exchanged zeolites and especially, Cu(I)-MFI zeolite, are very promising catalysts for DeNOx processes in environmental protection and by this reason there is a great experimental[20-22] and theoretical[23-29] interest to these materials. Sauer and coworkers[23-26] studies in details the siting, coordination, and spectral properties of Cu^+ ions in MFI and FER zeolite using combined quantum mechanics / interatomic potential function technique. Various possible locations of the cation have been examined first my molecular modeling and after that by the hybrid technique. The Cu^+ ion prefers to occupy the open space in the channel intersection and it is coordinated to two oxygen atoms of the AlO_4 tetrahedron when Al is at the edge of the main and sinusoidal channels. The largest binding energy of the cation in MFI zeolite was found for Cu^+ located inside a distorted six-membered ring (Z6 structure, Z denotes zig-zag channels) on the wall of the sinusoidal channel, where it can coordinate to three or four oxygen atoms of the zeolite framework. The cation-oxygen distances in this case are 212 – 227 pm.[23] The structure I2 (I denotes channel intersection) where the cation is coordinated to only two oxygen centers of one Al atom is considered to be the most reactive position of the cation. Especially interesting is the behaviour of the excited cations in singlet and triplet states[25] in different binding sites of Cu^+. In the triplet state, the coordination of the Cu^+ ion to the zeolite framework is significantly different from that in the singlet state. The Cu^+ ion moves away from the wall of zeolite channels, and it is coordinated to oxygen atoms of only one AlO_4 tetrahedron. This behaviour of the excited states explains the appearance of two bands in the photoluminescence spectra of Cu(I)-MFI zeolites. The peak with smaller emission energy are found for Al located at the intersection of two channels, while the higher energy peak corresponds to the Cu^+ sites inside one of the channels. Similar results were observed also for Cu(I)-FER zeolite,[26] however, here the most stable Cu^+ site are located inside the zeolite channels because of smaller pore diameter of these materials.

Pierloot et al.[28] considered coordination of Cu^{2+} ion at cluster models representing six-ring sites with different Al contents. As expected, the optimized structures indicated a strong preference of the Cu^{2+} ion for coordination to oxygen centers connected to Al rather than Si. They have a planar four-fold oxygen coordination in the six-rings. Depending on the number and relative positions of the aluminum atoms in the ring, two distinct coordination modes were distinguished with respect to the positions of the aluminium atoms in the ring. The optimized structures were used in clarification the peculiarities of the UV and ESR spectra of Cu^{2+}-A, Cu^{2+}-Y and Cu^{2+}-ZK4 zeolites.

3.2 Zinc

The Zn-exchanged zeolites was studied theoretically by van Santen and coworkers using different zeolite models.[30-32] The Zn^{2+} cations were located at four-, five-and six-membered rings, as well as on two four-rings with a common oxygen bridge. The binding energy of the cation is higher to the six-ring and lower to the four-ring. The Zn^{2+} cation is the most exposed to probe molecules when situated at the four-ring of zeolites. A structure of the type $[Zn\text{-}O\text{-}Zn]^{2+}$ was found to be one of the stable forms of the Zn cation on zeolites with a low Si/Al ratio. This site is more thermodynamically stable than Zn^{2+} at the five-ring, which suggests that $[Zn\text{-}O\text{-}Zn]^{2+}$ species likely exist in MFI zeolites. Owing to its size, this oxide cluster may only be stable in special positions inside the pores.

3.3 Rhodium

Coordination of Rh^+ in FAU zeolites was modeled by Goellner et al.[33] Three types of cationic sites typical for faujasites were considered - a four-ring, a six-ring, or a three-hollow position next to the Al center, denoted as T5 and modeled as a four-membered ring with an additional $OSiH_3$ group attached at the Al atom. The calculations show that at four-rings and at six-rings, a Rh^+ ion is bonded to the two oxygen centers of the ring connected to the Al center, as a consequence of their high basicity;[16] the calculated Rh–O distances are 220–225 pm. At the four-ring, Rh^+ is at a distance of 220 pm from the plane of T-atoms and interacts with two of the oxygen centers that are on the same side of the ring. At the six-ring the cation is exactly in the plane of the T-atoms since this is the only position that allows interaction with the oxygen centers of both Al–O–Si bridges in the ring (because the oxygen centers are located at different sides of the ring). At the cluster T5, Rh^+ is located in a three-hollow position, close to the oxygen center of the Al–O–Si bridge, at a Rh–O distance of 222 pm, and the two closest oxygen centers of the four-ring are at distances from the Rh^+ ion of 241 and 245 pm.

The binding energy of a Rh^+ ion at each of the model clusters is about 600 kJ/mol (with respect to free Rh^+ and a negatively charged zeolite cluster).[33] The highest BE value was found for the six-ring, 613 kJ/mol. The value is greater than those representing the other clusters because of the additional interaction of the cation with the oxygen centers of Si–O–Si bridges. The lowest BE was found for the T5 cluster, 577 kJ/mol, probably as a consequence of the rather short Rh–Al distance, 264 pm.

3.4 Other cations

Rice et al.[34] investigated the coordination of various divalent metal cations (Co^{2+}, Cu^{2+}, Fe^{2+}, Ni^{2+}, Pd^{2+}, Pt^{2+}, Ru^{2+}, Rh^{2+}, and Zn^{2+}) to MFI zeolite using DFT. They considered coordination of the cations at both isolated charge-exchange sites with a single Al atom and pairs of charge-exchange sites with two Al atoms in the model clusters, although the later situation is unlikely in the low-aluminum MFI zeolites. The results indicate that Cu^{2+}, Co^{2+}, Fe^{2+}, and Ni^{2+} are coordinated preferentially to five-membered rings containing two Al atoms, which are located on the walls of the sinusoidal channels, whereas Pd^{2+}, Pt^{2+}, Ru^{2+}, Rh^{2+}, and Zn^{2+} are coordinated preferentially to distorted six-rings located on the walls of the sinusoidal channels.

ACKNOWLEDGEMENT.

The author gratefully acknowledges the computer donation of the Alexander von Humboldt Foundation.

REFERENCES

1 Barthomeuf, D. *Catal. Rev.* **1996**, *38*, 521.
2 Zecchina, A.; Otero Arean, C. *Chem. Soc. Rev.* **1996**, *25*, 187.
3 Knözinger, H.; Huber, S. *J. Chem. Soc., Faraday Trans.* **1998**, *94*, 2047.
4 Sauer, J.; Ugliengo, P.; Garrone, E.; Saunders, V. R. *Chem. Rev.* **1994**, *94*, 2095.
5 Hadjiivanov, K.; Knözinger, H. *J. Phys. Chem. B* **2001**, *105*, 4531.

6 Ferrari, A. M.; Neyman, K. M.; Rösch, N. *J. Phys. Chem. B* **1997**, *101*, 9292.
7 Vayssilov, G. N.; Hu, A; Birkenheuer, U.; Rösch, N. *J. Molec. Catal. A* **2000**, *162*, 135.
8 Hadjiivanov, K.; Knözinger, H. *Chem. Phys. Lett.* **1999**, *303*, 513.
9 Klinowski, J.; Ramdas, S.; Thomas, J. M. *J. Chem. Soc., Faraday Trans. 2* **1982**, *78*, 1025.
10 Vayssilov, G. N.; Staufer, M.; Belling, T.; Neyman, K. M.; Knözinger, H.; Rösch, N. *J. Phys. Chem. B* **1999**, *103*, 7920.
11 Eulenberger, G. R.; Shoemaker, D. P.; Keil, J. G. *J. Phys. Chem.* **1967**, *71*, 1812.
12 Olson, D. H. *Zeolites*, **1995**, *15*, 439.
13 Fitch, A. N.; Jobic, H.; Renouprez, A. *J. Phys. Chem.* **1986**, *90*, 1311.
14 Dempsey, E.; Kühl, G. H.; Olson, D. H. *J. Phys. Chem.* **1969**, *73*, 387.
15 Vayssilov, G. N.; Lercher, J. A.; Rösch, N. *J. Phys. Chem. B* **2000**, *104*, 8614.
16 Vayssilov, G. N.; Rösch, N. *J. Catal.* **1999**, *186*, 423.
17 Anderson, A. A.; Shepelev, Y. F.; Smolin, Y. I. *Zeolites* **1990,** *10*, 32.
18 Vayssilov, G. N.; Rösch, N. *J. Phys. Chem. B* **2001**, *105*, 4277.
19 *Handbook of Chemistry and Physics*, 69th edition, Weast, R. C., Ed., CRC Press, Boca Raton, **1989**, p. F-152.
20 Spoto, G.; Bordiga, S.; Scarano, D.; Zecchina, A. *Catal. Lett.* **1992**, *13*, 39.
21 Wichterlova, B.; Dedecek, J.; Vondrova, A. *J. Phys. Chem.* **1995**, *99*, 1065.
22 Hadjiivanov, K.; Knözinger, H. *Phys. Chem. Chem. Phys.* **2001**, *3*, 1132.
23 Nachtigallova, D.; Nachtigall, P.; Sierka, M.; Sauer, J. *Phys. Chem. Chem. Phys.* **1999**, *1*, 2019.
24 Nachtigallova, D.; Nachtigall, P.; Sauer, J. *Phys. Chem. Chem. Phys.* **2001**, *3*, 1552.
25 Nachtigall, P.; Nachtigallova, D.; Sauer, J. *J. Phys. Chem. B* **2000**, *104*, 1738.
26 Nachtigall, P.; Davidova, M.; Nachtigallova, D. *J. Phys. Chem. B* **2001**, *105*, 3510.
27 Berthomieu, D.; Krishnamurty, S.; Coq, B.; Delahay, G.; Goursot, A. *J. Phys. Chem. B* **2001**, *105*, 1149.
28 Pierloot, K.; Delabie, A.; Groothaert, M. H.; Schoonheydt, R. A. *Phys. Chem. Chem. Phys.* **2001**, *3*, 2174.
29 Treesukol, P.; Limtrakul, J.; Truong, T.N. *J. Phys. Chem. B* **2001**, *105*, 2421.
30 Barbosa, L. A. M. M.; van Santen, R. A., *Catal. Lett.* **1999**, *63*, 97.
31 Barbosa, L. A. M. M.; Zhidomirov, G. M.; van Santen, R. A. *Phys. Chem. Chem. Phys.* **2001**, *3*, 1132.
32 Shubin, A. A.; Zhidomirov, G. M.; Yakovlev, A. L.; van Santen, R. A. *J. Phys. Chem. B* **2001**, *105*, 4928.
33 Goellner, J. F.; Gates, B. C.; Vayssilov, G. N.; Rösch, N. *J. Am. Chem. Soc.* **2000**, *122*, 8056.
34 Rice, M.J.; Chakraborty, A.K.; Bell, A.T. *J. Phys. Chem. B* **2000**, *104*, 9987.

Chemical Reactions of Alkanes Catalysed by Zeolites.

E. A. FURTADO and M. A. CHAER NASCIMENTO.
Insttituto de Química.
Departamento de Físico-Química.
Universidade Federal do Rio de Janeiro.
Cidade Universitária, CT Bloco A sala 412.
Rio de Janeiro.R.J. 21949-900. Brazil.

Abstract: The chemical activation of light alkanes by acidic zeolites was studied by a combined Classical Mechanics/Quantum Mechanics approach. The diffusion and adsorption steps were investigated by Molecular Mechanics, Molecular Dynamics and Monte Carlo simulations. The chemical reactions step was studied at the DFT (B3LYP) level with 6-311G** basis sets and 3T and 5T clusters to represent the acid site of the zeolite.

1. Introduction

Zeolites are inorganic microporous materials, mainly aluminosilicates, which exhibit crystalline structures containing pores and channels large enough to allow the diffusion of organic molecules. As a result of such unusual crystalline structure, zeolites have been widely used as molecular sieves [1,2] and also for ion exchange processes [2], as they usually present loosely bound compensation cations in their structures which can be easily exchanged with other cations from appropriate solutions. However, it is in the field of the heterogeneous catalysis that the most important applications of these materials can be found [3]. Certain positions in the internal walls of the zeolite micropores behave as active sites where catalyzed conversions may take place [4,5]. Moreover, since these micropores are very uniform and in the same size range of most

M.A. Chaer Nascimento (ed.), Theoretical Aspects of Heterogeneous Catalysis, 39–75.

of the small molecules (4 - 12 Å), zeolites can also present shape selectivity [3].

In their protonated forms, zeolites are widely employed in the oil and petrochemical industries, in processes such as the conversion of alcohols to gasoline, catalytic cracking, isomerization and alkylations of hydrocarbons [3]. These chemical reactions most probably involve proton transfer from the acidic site of the zeolite to the organic substrate. In the case of hydrocarbons, this transfer gives rise to carbenium (I) or carbonium (II) ions as intermediates or transition states.

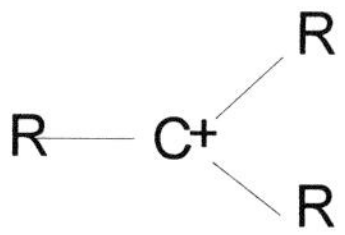

Carbenium (I) 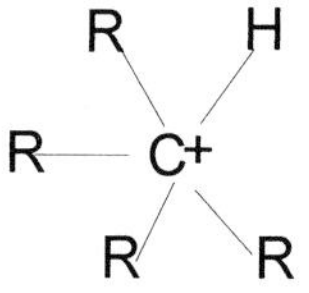 Carbonium (II)

In spite of the enormous technological importance of these reactions, very little is known about their mechanisms. The understanding of the mechanism of these reactions, at the molecular level, would be extremely important not only to establish better experimental conditions for increasing the yield in some desired product, but also to develop new selective catalysts of industrial interest.

Catalytic heterogeneous processes can be very complex as they necessarily involve at least the following five steps:

(a) diffusion of the reagent(s) towards the catalytic active site;
(b) adsorption of the reagent(s);
(c) chemical reaction at the active site;
(d) desorption of the product(s);
(e) diffusion of the product(s) out of the active site region.

Thus, for a complete understanding of the catalytic process one should take into account all these steps, inasmuch as in principle any of them could be the rate determining of the reaction.

Irrespective of which of the above steps is the rate determining one, the catalytic process takes place mainly inside the zeolitic cavity and therefore it would be important to characterize the diffusion of the molecules in order to understand, among other things, the phenomenon of shape selectivity. As the molecules diffuse, they constantly interact with the walls of the zeolite and for understanding the chemical step it would be important to determine the nature of the most favorable adsorption sites. In fact, the diffusion trajectories can also provide useful information about the trapping sites of the zeolitic cavity, where most probably the substrate molecules would be adsorbed. Once these sites are determined, a more quantitative study can be made to calculate the adsorption energies. Finally, with the active sites well characterized, one could try to model them in order to study the chemical reaction step.

In principle, the diffusion steps (a) and (e) could be studied through molecular dynamics simulations as long as reliable forces fields are available to describe the zeolite structure and its interaction with the substrates. Also, if the adsorption takes place without charge transfer between the reagents/products and the zeolite, steps (b) and (d) could also be investigated either by molecular dynamics or Monte Carlo simulations. Step (c) however can only be followed by quantum mechanical techniques because the available force fields cannot yet describe the breaking and formation of chemical bonds.

Among the chemical reactions of interest catalyzed by zeolites, those involving alkanes are specially important from the technological point of view. Thus, some alkane molecules were selected and a systematic study was conducted, on the various steps of the process (diffusion, adsorption and chemical reaction), in order to develop adequate methodologies to investigate such catalytic reactions. Linear alkanes, from methane to n-butane, as well as isobutane and neopentane, chosen as prototypes for branched alkanes, were considered in the diffusion and adsorption studies. Since the chemical step requires the use of the more time demanding quantum-mechanical techniques, only methane, ethane, propane and isobutane were considered.

The diffusion step was investigated by molecular dynamics (MD) simulations. For the adsorption step, molecular mechanics (MM), molecular dynamics (MD) and Monte Carlo (MC) techniques were used, under the same simulation conditions, in order to compare their performance. The influence of the force-field, the loading of alkane molecules and the relaxation of the zeolite framework on the adsorption energies were also investigated. For the chemical reaction step, the acidic site of the zeolite was represented by model–clusters containing three (3T) and five (5T) tetrahedral centers. The calculations were performed at the DFT (B3LYP) level with 6-31G** and 6-311G** basis sets.

2. The Diffusion Step

Molecular dynamics has been used by several authors to study the diffusion of light hydrocarbons (C_1-C_6) in different zeolitic structures, but except for references [6-8], all the other calculations assumed that the molecules could be represented either by point particles or rigid entities. In their study of the diffusion of methane and ethane in ZSM-5, Catlow *et al.* [6] used a moderate size cluster (576 atoms) to represent the zeolite structure and long trajectories (up to 120 ps). Nicholas *et al.* [7] used a much lager cluster (7776 atoms) but much shorter trajectories (40ps) for propane. Dumont and Bougeard [8] used a moderate size cluster and short trajectories (40ps). In references [6] and [7] elaborate force fields were used to represent the zeolitic structure and its interaction with the organic molecules, while in Reference [8] a much simpler force field was used.

2.1 Methodology

In all simulations the ZSM-5 zeolitic structure was represented by a model-cluster containing four crystallographic cells (1152 atoms) superimposed along the c axis, with cell parameters a= 20.076 Å, b= 19.926 Å and c= 13.401 Å [9]. This structure contains two interconnecting channel systems. The straight channels run parallel to the b (010) crystallographic axes and present elliptical cross-sections with minor and major axes of 5.4 Å and 5.6 Å, respectively. The sinusoidal (zigzag) channels run along the a (100) axis and present a more circular but still elliptical cross-section (5.1 Å , 5.2 Å) slightly smaller than the one of the straight channels. The periodic interconnection between these two channel systems allows molecules inside the cavity to diffuse also along the c (001) direction.

Similarly to the previous diffusion studies, the cluster-model used in this work is one of a purely siliceous zeolite, silicalite, isostructural with ZSM-5. This simplifies the calculation because the simulated lattice is neutral and no counter-ions need to be considered.

The absence of aluminum atoms in the cluster-model is a reasonable assumption inasmuch as ZSM-5 can only tolerate low levels of aluminum substituent. Besides, the comparison with experimental ZSM-5 data is still possible because the diffusion constants seem to be very insensitive to the Si /Al ratio [10].

The zeolite framework was described by a specific force field developed by van Santen *et al.* [11] while the hydrocarbon molecules and their interaction among themselves and with the zeolite lattice were described by the generic force field Dreiding II [12]. All the internal coordinates of the alkane molecules were allowed to fully relax. The nonbonded interactions (electrostatic and van der Waals) were computed for all atoms within a cutoff-radius of 12Å. Periodic boundary conditions were imposed along the three axes of the zeolite model to simulate an infinite crystal.

The molecular dynamics runs were performed within the microcanonical (NVE) and canonical ensembles, at 300 K. For the NVE simulations, energy exchange between the framework and the organic molecules occurs only via intermolecular collisions. The equations of motion were integrated using the Verlet (summed Verlet) algorithm [13], with an integration time step of 1 fs, which is small enough to ensure energy conservation. Loads of two to eight molecules per unit cell were considered and the initial velocities were assigned according to a Maxwell-Boltzmann distribution corresponding to 300K. The system was then thermalized during 20-30 ps. Then longer (up to 120 ps) simulations were performed and the trajectory information was saved every 15 steps for subsequent analysis.

From the coordinate data stored in the trajectory files, the mean-square displacements for all the movable atoms were calculated, and the diffusion coefficient, D_{msd}, computed using Einstein's relation

$$D_{msd} = \frac{1}{6Nt}\left(\sum_i [r_i(t+t_0) - r_i(t_0)]^2\right) ,$$

where N is the number of moving particles and t the simulation time. Strictly speaking, the above equation is valid for Brownian motion in a tridimentional homogeneous medium, while the motion of the molecules diffusing in the zeolitic channels is constrained by the framework geometry, which is not isotropic. However, most experimental measurements of the diffusion coefficient are conducted in polycrystalline or powdered samples, and over a length scale at which the microscopic structure of the zeolite is not resolved. Therefore, the resulting measurements represent averages over all orientations and a comparison with the results obtained from the simulations is reasonable.

2.2 Results and Discussion

The results presented in table 1 clearly show that our calculated diffusion coefficients are in fair agreement with the available experimental data. No appreciable changes are observed if the simulations are performed within the (NVE) or the canonical ensemble. However, the results from different simulations mentioned above differ considerably and a comparison can be very instructive.

Table 1. Diffusion coefficients (in 10^{-5} cm^2/s) compared to other calculations and experimental results, at 300K.

Molecule	Our results	Ref. [8]	Ref. [7]	Ref. [14]	Other Theoretical Results	Experimental Results
Methane	6.02	16	5.0	6.2	3.6 ± 1.9^a; 5.41^b	6.5 ± 1.0^c
Ethane	2.71	5.9	-	4.7	-	3.8^d, 3.0^e
Propane	1.53	1.9	2.5	4.1	-	2.8^d, 1.3^e
n-Butane	0.92	-	-	-	2.9^f; 2.4 ± 0.79^g; 2.7^h	0.1^i; 0.04^j
i-Butane	0.06	-	-	-	-	-

a) Ref.[6]; b) Ref.[15]; c) Ref.[16]; d) Ref.[17]; e) Ref.[18]; f) Ref.[21]; g) Ref.[22]; h) Ref.[23]; i) Ref.[19]; j) Ref.[20].

For methane our result and the ones by Catlow *et al.* [6] and Nicholas *et al.* [7] are all within the experimental error in spite of the differences in the simulations. On the other hand, Dumont and Bougeard [8] obtained a much larger value for the diffusion coefficient of methane. These authors [8], comparing their results with the ones of Nicholas *et al.* [7], attributed the large difference to the rigidity of the methane molecules in the latter calculation. However, our result and the one by Catlow *et al.* [6], obtained with flexible methane molecules, do not differ appreciably from the one by Nicholas *et al.* [7]. Therefore, the much larger value obtained by Dumont and Bougeard cannot be attributed to the flexibility of the methane molecules. Catlow *et al.* [6] also considered the zeolitic framework flexibility. Although the flexibility of the lattice could influence the diffusion process in several ways, the close agreement between our result and the one by Catlow *et al.* [6] seems to indicate that, at least for methane, the consideration of the lattice flexibility is not an important factor.

For the ethane molecule, the result of Dumont and Bougeard [8] differs from ours by a factor of 2.81. The two simulations differ in many respects. We used a larger cluster to model the zeolite, longer trajectories and a specific force field to represent the zeolite. Although we believe that the main reason for such discrepancy resides on the difference in quality of the force fields employed, a combination of factors cannot be ruled out.

For propane, there is a much better agreement among all the calculations, which considered flexible molecules. The results of Novak *et al.*[14], obtained will both the zeolite lattice and the alkane molecules rigid, show the largest discrepancy.

On the other hand, for n-butane the range of experimental values is much wider (from 1×10^{-6} cm^2/s, obtained with frequency response techniques at 332K [19], to 3.7×10^{-8} cm^2/s from membrane permeation measurements at 334K [20]) and the comparison with the theoretical values is not straightforward because there are no measurements at 300K. Hernandez and Catlow [21] reported a value of 2.9×10^{-5} cm^2/s for a loading of 4mpuc at 287.2K, a value close to the one by June *et al.* ($2.4 \pm 0.79 \times 10^{-5}$ cm^2/s) at the same loading [22]. Molecular dynamics was used in both simulations but with different force fields. Curiously, a value very close (2.7×10^{-5} cm^2/s) to the one of ref.[21] was found by Fried and Weaver [23] using the burchart-Dreiding force field but with just 0.5mpuc, a much lower loading. Our results compare well with the ones by June *et al.* [22] at the same loading.

Finally our result for isobutane, when compared to the other molecules, is consistent with the fact that its shape should make the diffusion process more difficult. In any case, it follows the expected trend if one considers the values of diffusion coefficient obtained for the other alkanes.

The results obtained for the set of molecules studied follow the expected trend, based on the differences of mass and shape of the molecules. Moreover, the results, when compared to the available experimental data, indicate that the simulations can indeed provide a realistic representation of the microscopic process of the diffusion of light hydrocarbons in the pores of zeolites. However, the fact that simulations conducted in very different conditions gave similar results also indicate that a more systematic study is required to establish the influence of the different factors (cluster size, force field, trajectory length, flexibility of the zeolite lattice) in the final results.

3. The Adsorption Step

In order to understand the chemical step in the zeolitic catalysis process, it is important to determine the nature of the most favorable adsorption sites. A variety of computational techniques has been used to approach this problem. As we have previously seen, the molecular dynamics approach provides detailed dynamical information on the diffusion process. The generated trajectories can be used to compute diffusion coefficients as well as to determine the most favorable adsorption sites for the molecules and their respective adsorption energies. Monte Carlo simulations have also been used to locate sorption sites and to compute adsorption energies as well as sorption equilibrium. The molecular mechanics method, more commonly used to compute the structure and stability of zeolites, has also been employed to locate the lowest energy site for sorbed hydrocarbons [24]. Although more information can be obtained from MD and MC calculations, these techniques usually require large amounts of computer time what makes them less suitable than MM for investigating larger molecules and the effect of the zeolitic framework relaxation on the adsorption energies. However, the MM approach may become unpractical if many different sites, closer in energy, are available for the molecule. In this case it would be important to consider that at a given temperature the molecules would be distributed over a range of different sites.

Although for light alkanes the reported adsorption energies, computed by the three different methodologies (MM, MC and MD), do not differ appreciably, the fact that quite different conditions and force fields were used in the simulations precludes any comparative evaluation of the merits of these methodologies.

Smit and den Ouden [25] used MC to study the adsorption of methane on ZSM-5 with both the zeolite and the methane kept rigid and interacting through a simple Lennard-Jones (LJ) plus a coulombic potential. June et al., [26] examined methane on silicalite using MD. The zeolite and the methane molecule were also considered as rigid bodies interacting through a LJ potential, but coulombic terms were neglected. In another article [27] those authors used MC to study the adsorption of methane, n-butane and three hexane isomers in silicalite. The zeolite was again kept rigid but, except for methane, the hydrocarbon molecules were partially relaxed by allowing rotations about the carbon-carbon bonds. An intrinsic torsional potential was added to the previous LJ potential in order to account for the hydrocarbon relaxation. In another paper those authors [28] reinvestigated the adsorption of n-butane and n-hexane in silicalite, using MD. The zeolite was once more kept rigid but bond angle deformations and rotations around non-terminal C-C bonds were allowed for the hydrocarbon molecules. A more systematic study was performed by Titiloye *et al.*, [24] who used MM to investigate the adsorption of linear hydrocarbons, C1-C8, in silicalite, H-ZSM5 and faujasite. A much more sophisticated force field was employed and both the zeolite and the hydrocarbon molecules were allowed to fully relax during minimization. Nicholas *et al.*, [29], using MD, investigated the adsorption of methane and propane in silicalite. The zeolite and the methane molecule were kept rigid but in the case of propane the internal coordinates were relaxed. The effect of different force fields on the adsorption energy of methane was also examined and from this study the MM2 force field [30] was selected for the simulations with propane. More recently Smit and Siepmann [31] used the so-called configurational-bias Monte Carlo method to simulate the adsorption of linear (C1-C12) alkanes in silicalite. The zeolite was once more kept rigid but bond bending and torsions were allowed for the alkane molecules while keeping the bond lengths fixed at 1.54Å.

In addition, those studies also differ in some other respects such as the size of the cluster representing the zeolite, the cut-off radius for non-bonded interactions and the number of hydrocarbon molecules used in the simulations. Therefore, considering such a variety of different conditions used in the previous calculations, it is quite difficult to make any judgements about the relative accuracy of those three methods, in computing adsorption energies. Moreover, the distribution of alkane molecules over the zeolite channels seems to be quite dependent on the method and conditions of simulation. Thus, Tiltiloye *et al.*, [24] concluded that the hydrocarbons prefer the straight channels while June et al., [26-28] found that normal butane and hexane have equal probability of being in both types of channels. The results of Nicholas *et al.*, [29] indicate that propane adsorbs preferentially at the sinusoidal channels. Finally, Smit and Siepmann [31] concluded that short linear alkanes ($n \leq 7$, where n is the number of carbon atoms in the chain) have nearly equal probability of being in both the sinusoidal and straight channels while for larger alkanes ($n=8$) the straight channels are favored.

3.1 Methodology

Molecular mechanics (MM), molecular dynamics (MD), and Monte-Carlo (MC) methods were employed to simulate the adsorption of methane, ethane, propane and isobutane on silicalite and HZSM-5. The silicalite was simulated using the same cluster-model adopted in the diffusion calculations. The H-ZMS-5 structure was constructed according to the procedure suggested by Vetrivel et al. [32], which consists in replacing one Si^{4+} atom at the channel intersection by Al^{3+} and protonating the oxygen atom bridging the T_2 and T_8 sites in order to preserve the lattice neutrality.

In a first series of simulations, the force field description of the system was the same as the one used for diffusion. In order to study the effect of the force field on the adsorption energies, a second set of simulations was performed using Dreiding II to represent both the zeolite and the hydrocarbon molecules.

For silicalite all the simulations (MC, MD and MM) were carried out with the zeolite structure held rigid inasmuch as, according to the results of Tiltiloye *et al.*, [24], for $n \leq 4$ the effect of the zeolite relaxation is negligible (< 0.2 kcal/mol). In addition, full relaxation was allowed to all hydrocarbon molecules. However, for ZSM-5 the effect of the framework relaxation on the adsorption energy is already noticeable. Therefore, because the computer time requirements for including the framework relaxation are quite large, fewer molecules were considered in the simulations with ZSM-5.

The Monte Carlo simulations were performed in the NVT ensemble and, in order to minimize the possibility of sampling regions of the zeolitic structure not accessible to the molecules, the initial distribution of the hydrocarbon molecules were chosen to be identical to the corresponding one of the thermalized configuration, at 300 K. The Metropolis [33] algorithm was then used to generate up to 8000 configurations. Three different steps, with equal probability, were considered: random translation of the center of mass, random rotation of the whole molecule [34] and a perturbation on any of the internal coordinates of the molecules.

Two sets of MM calculations were performed: the first considering only one molecule of hydrocarbon, for comparison with the results of Tiltiloye *et al.*, [24], and the second with the same loading used in the MD and MC calculations. For the first set of calculations, the hydrocarbon molecule was placed in the center of the straight channel ten-membered ring. For the second set of simulations we used the initial configurations as given by the respective thermalized distributions at 300 K.

3.2 Results and Discussion

Table 2 shows the results of adsorption energies in silicalite, for the molecules studied, obtained with different methodologies, compared to the available experimental results and other theoretical data. Table 3 shows the equivalent results for the adsorption process in ZSM-5.

In table 2, the last two values in the *present results* column (D-MM), for each molecule, were obtained using the Dreiding II force field to represent the zeolite. The first of the two values, for instance 11.7 in case of ethane, were obtained considering a loading of just one molecule of alkane, in order to compare with the results of Tiltiloye *et al.*[24]. From table 2 it can be seen that the MD and MC methods furnish comparable results of adsorption energy although the MD values are slightly better when compared with the experiments.

Table 2. Adsorption energies (kcal/mol) in silicalite from molecular mechanics (MM), molecular dynamics (MD) and Monte Carlo (MC) calculations (300K).

Molecule	Present results	Exp. results	Other results
Methane	5.5(MM); 5.2(MC); 4.9(MD); 7.6(D-MM)[a]; 7.3(D-MM)[b]	5.0[c]	5.4(MM)[f]; 4.2(MD)[g]; 4.8(MC)[h]; 5.8(MD)[i]
Ethane	8.1(MM); 7.4(MC); 7.1(MD); 11.7(D-MM)[a]; 10.9(D-MM)[b]	6.93[c]	8.7(MM)[f]
Propane	11.7(MM); 10.8(MC); 10.1(MD); 14.9(D-MM)[a]; 14.7(D-MM)[b]	9.7[d]	12.1(MM)[f] 10.3(MD)[j]
n-Butane	14.2(MM); 13.3(MC); 12.7(MD); 17.8(D-MM)[a]; 17.1(D-MM)[b]	12.7[c]	15.1(MM)[f]; 16.2(MC); 11.5(MD)
i-Butane	12.3(MM); 12.0(MC); 11.2(MD); 12.8(D-MM)	11.7[e]	–
Neopentane	10.6(MM); 9.5(MC); 9.5(MD); 10.7(D-MM)	–	–

a) MM results with Dreiding II force field representation of the zeolite ad load of one alkane molecule.
b) Same as above with full load.
c)Ref.[35], d)Ref.[36], e)Ref.[37], f)Ref.[24], g)Ref.[26], h)Ref.[27], i)Ref.[28], j)Ref.[29].

For the linear alkanes, except for methane, the MM results are ~ 1-2 kcal/mol higher than either the MC or MD results for a given molecule. This is consistent with the fact that at the MM level only the lowest energy sites are being considered while for T > 0 the molecules will occupy a range of different sites and consequently the average adsorption energy will be reduced.

The importance of using a specific force field to represent the zeolite cluster can be inferred from the comparison of the adsorption energies obtained at the MM and D-MM levels. The D-MM values are ~2-3 kcal/mol higher than the ones obtained when the van Santen *et al.*[38,39] force field is used to describe the zeolite.

For the branched alkanes the differences in the adsorption energies computed with the three methods are much smaller, the theoretical results differing at most by 0.6 kcal/mol (MM) from the experimental one. Also interesting to note is the fact that for the branched alkanes the MM and D-MM results are practically the same. This observation and the fact that the three methods gave very similar results of adsorption energies can be also related to the way the branched alkane molecules distribute themselves on the zeolite channels.

The effect of the loading on the adsorption energy in silicalite is shown in figure 1, for the linear C_1-C_4 alkanes and also for isobutane. The adsorption energy for the linear alkanes raises slightly with increasing the loading. For isobutane it remains almost constant

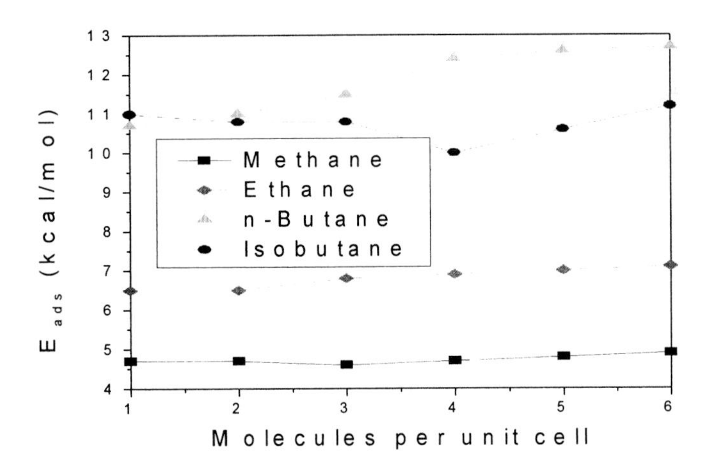

Figure 1. Effect of loading on the adsorption energy in silicalite,at 300 K.

up to a loading of three molecules and drops at higher loading. This difference in behavior can be understood once we learn how these different molecules distribute themselves on the silicalite pores, as will be discussed below.

The results in table 3 indicate that the alkanes adsorb more strongly on ZSM-5 than in silicalite, in agreement with the experimental data and the MM results of Tiltiloye *et al.*[24]. However, the small differences in the adsorption energies, in silicalite and ZSM-5, for all the studied alkanes, is a clear indication that the adsorption process is dominated by dispersion forces. For methane and ethane, the difference in adsorption energy computed at the MM and MD is much smaller than the one for silicalite. When compared to the available experimental data, the results obtained indicate that, at least for methane and ethane molecules, the MM and MD methodologies are equally accurate in predicting adsorption energies.

Table 3. Adsorption energies (kcal/mol) in ZSM-5 from (MM) and (MD) calculations (300K).

Molecule	Present results	Exp. results	Other results
Methane	6.3(MM), 5.9 (MD)	6.1 – 6.7[a]	5.7 (MM)[b], 5.3 (MC)[c]
Ethane	10.6 (MM); 9.2 (MD)	8.9[d]	9.9 (MM)[b]
Propane	12.3 (MD)	-	-
n-Butane	14.2 (MD)	15.5	18.5 (MD)
i-Butane	13.5 (MD)	-	-

a) Ref. [68], b) Ref. [24], c) Ref. [25], d) Ref. [35].

Figure 2 shows the effect of the ZSM-5 framework relaxation on the adsorption energy of the linear alkanes. Contrary to what is observed for silicalite, the effect of the zeolite relaxation is noticeable even for the smaller alkanes. That implies the use of specific force fields to represent the ZSM-5 structure in order to avoid its colapse during the simulations.

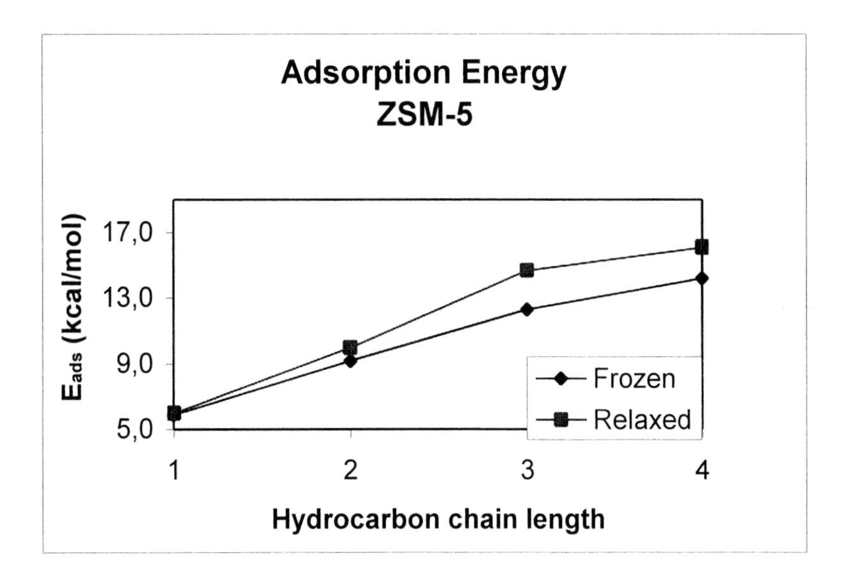

Figure 2. Effect of the zeolite framework relaxation on the adsorption energy.

Another important aspect of the problem, which can also be addressed using computer simulations, has to do with the distribution of the alkane molecules over the zeolite channels. If one takes into consideration the fact that a zeolite such as ZSM-5, for instance, has 48 different acidic sites, with distinct acidic strengths, the catalytic activity of the zeolite towards the different alkanes will be certainly related to the way the substrate molecules are distributed within the zeolite network. As mentioned in the last section, the previous simulations [24,26-29,31] predicted quite distinct distributions, but considering the variety of different simulation conditions employed, no clear conclusion could be reached. On the contrary, we have used exactly the same conditions (force fields, cluster size, loading, initial distribution of molecules, etc.) with the three methodologies, except in the case of the MM calculations with a single alkane molecule.

Figure 3 illustrates two typical distributions in silicalite, along the *a* axis, for linear and branched alkanes respectively. There is no reason to believe that these distributions in ZSM-5 will differ significantly from those in silicalite. For the sake of clarity not all the molecules are shown. Regardless of the method used, the distributions do not differ appreciably from the ones shown in the figure. From Figure 3 one sees that the linear alkanes, represented by propane molecules, show equal probability of being found in both types of channels in agreement with the calculations of Smit and Siepmann [31]. However, the branched alkane, represented by isobutane, prefers the channel intersections. If only one guest molecule is considered the linear alkanes do indeed prefer to stay at the straight channels, as observed by Tiltiloye *et al.*[24]. However as we increase the loading they tend to distribute almost equally between the two types of channels. The branched alkane, on the other hand, even at the minimum loading clearly shows a tendency for occupying the more spacious channel intersections. Although their critical diameters are smaller than both channels cross-sections, steric effects most certainly make it more difficult for the molecules to reenter either the straight or the sinusoidal channels once they are positioned at an intersection. Thus, the fact that they tend to stay most of the time at the channel intersections precludes them of visiting many different zeolitic sites and therefore, irrespective of which methodology is used, the average interaction energy between the alkane and the zeolite will be very similar. This is probably the reason why the adsorption energies computed with the three methods are similar for the branched alkanes. Also, because the branched alkanes tend to concentrate at the intersections, the alkane-alkane interactions will dominate over the alkane-zeolite ones. In this case one should expect that a generic force field which describes well the dominant interactions, such as Dreiding II, should be as effective as the specific force field in predicting adsorption energies, as shown by the results in table 2.

Since the alkanes considered are practically non-polar molecules, the interaction between any of them and the zeolite structure will be dominated by dispersions forces, which are known to decrease rapidly with distance. For the linear alkanes the first molecule does prefer to stay at the straight channels. However, as one increases the loading they tend to equally. occupy the sinusoidal channels where they will more strongly interact with the zeolite framework thus increasing the adsorption energy. Of course, for the larger (n > 4) linear alkanes, the possibility of penetrating the sinusoidal channels will be smaller but once that happens, steric effect should start contributing to a reduction of the adsorption energy. In fact, this reduction should also occur even for the smaller linear alkanes at higher loadings, up to a point where they start to stick together preferentially to interacting with the zeolite framework. In the case of isobutane as the channel intersections becomes congested, the molecules will either try to penetrate the sinusoidal channels or move to the straight channels. In both cases one should expect a reduction in the adsorption energy either because of steric effects (in the sinusoidal channels) or a less effective interaction with the zeolite framework (in the straight channels).

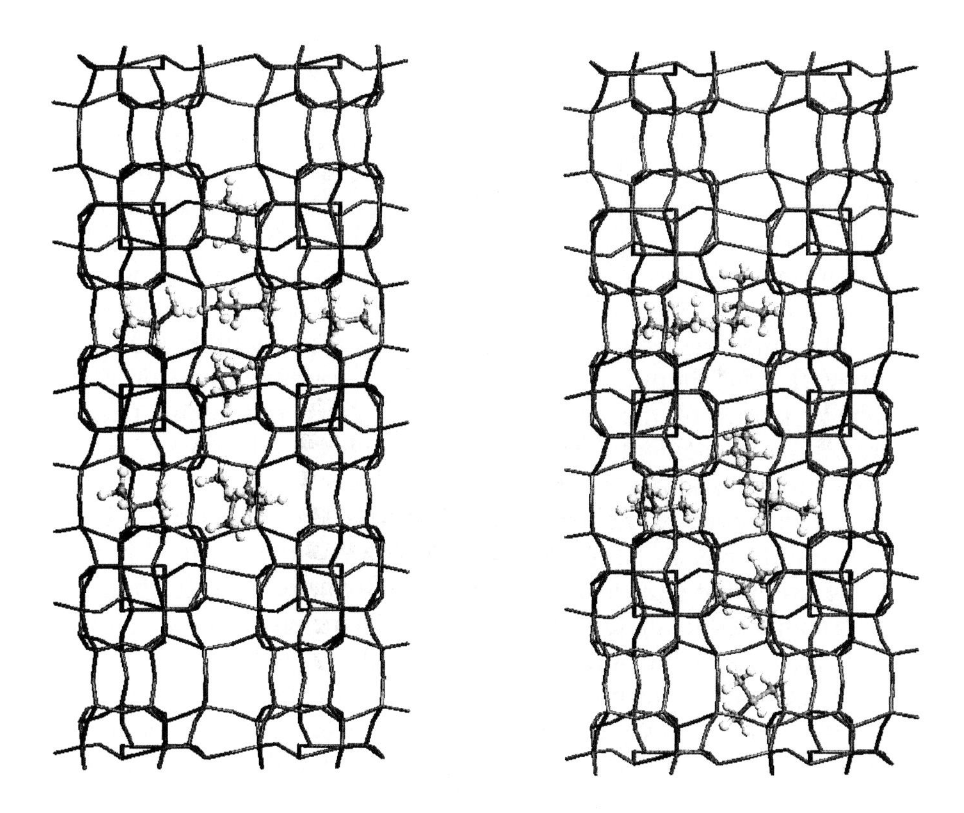

Figure 3. Distribution of propane (left) and isobutane (right) molecules in silicalite, at 300K, along the a axis, as obtained from MD calculations.

Finally, the examination of the geometry of the adsorbed molecules shows that the packing of the alkane molecules inside the zeolite cavity does not cause drastic geometry changes. Thus bond distances show increases of at most 0.01Å while some HCC and CCC angles are slightly reduced (0.6° and 2.0°, respectively).

4. The Chemical Reaction Step

Once the most favorable adsorption sites have been determined we shall examine the possible chemical reactions occurring at these sites. The ultimate goal is to determine the mechanism of all possible reactions of a given substrate. This involves determining the nature of the transition states (TS) for each type of reaction, the respective activation energies and possibly the reaction paths from each TS towards the products. Eventually one should also consider calculating the rate constants of the reactions.

Contrary to the previous steps of the catalytic process, we cannot use force-field-based techniques because the available force fields are unable to describe the breaking and formation of chemical bonds. Thus, the chemical reaction step must be investigated by quantum mechanical techniques. Right away this imposes some limitations on the size of the cluster to be used in the calculations. In principle, since the catalytic sites are well localized within the zeolite framework, one should expect the chemical reactions to occur at very localized points of the zeolitic structure. Thus, one could think of representing the acid sites by much smaller clusters than the ones used in the diffusion and adsorption studies.

The choice of the size of the cluster representing the acid site of the zeolite has to be made taking into account both the chemical and the computational aspects of the problem: the cluster should be large enough as to include the most important characteristics of the chemical process but simultaneously small enough to allow high-quality quantum mechanical calculations to be performed. The smallest cluster which minimally satisfies these criteria is the 3T cluster shown below. Of course, this represents a drastic reduction in size relative to the clusters previously used. Next in size one could have the 5T and the 20T clusters, the latter one representing a full ring of the zeolite. These clusters are amenable to high-quality calculations using the presently available workstations. All the results to be presented and discussed were obtained using the clusters below.

Before proceeding, it would be important to discuss possible limitations imposed by the use of such clusters and how they could influence the final results. As previously mentioned in the discussion of the adsorption step, zeolites present many different types of acid sites. For instance, ZSM-5 has 48 different sites with distinct acid

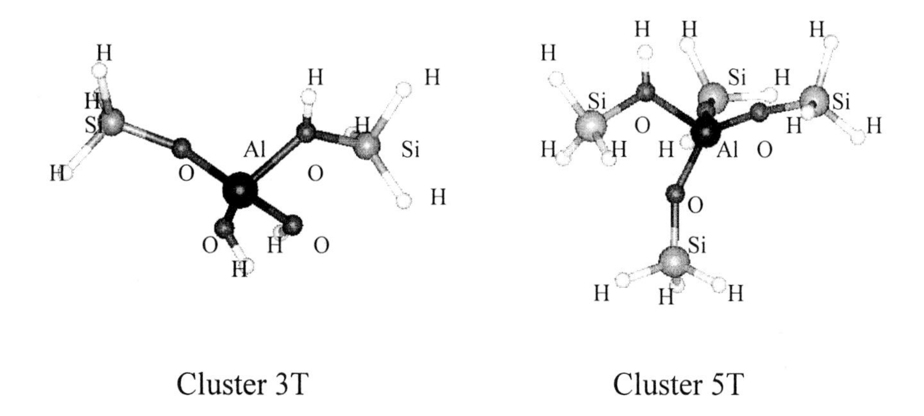

Cluster 3T Cluster 5T

strengths, while faujasite has only 4. However, with the 3T and 5T cluster-models, only one type of acid site can be modeled. Therefore, the calculations performed with such clusters will not distinguish either the different acid sites of a given zeolite or the type of zeolite being used. Another problem has to do with steric hindrance. The micropores of the zeolites have finite dimensions and even for reactions occurring at the larger cavities, the substrate should experience some steric hindrance as it approaches any acid site. Moreover, some sites are more hindered than others and should be more difficult to access. However, the structure of the 3T and 5T cluster-models is such that they behave as acidic sites of a zeolite with very large pores. Finally, as some of the TSs may present ionic character, they should be affected by the electrostatic potential of the zeolite framework. This effect will be totally neglected in calculations with 3T and 5T clusters and only partially included in calculations with the 20T cluster.

Let us now consider how the limitations imposed by the size of these clusters may influence the results for the chemical reaction step. In spite of the fact that only one type of acid site can be modeled by the 3T and 5T clusters, it must be remembered that the difference in acidic strength of the different sites is much smaller than the activation energy of most reactions. Thus, in the absence of other effects, the fact that the clusters used cannot distinguish different types of sites should not be a major problem as far as the computation of activation energies (E_a) is concerned. However, it may drastically affect the results for the rate constants since they vary exponentially with E_a. Steric effects, of course, may play a major role, depending on the structure of the substrate. This effect should be more pronounced for the more hindered sites and bulkier substrates. As shown in the previous section, steric effects control the distribution of alkane molecules over the zeolite channels. However, these two effects are not independent of each other because a given substrate will adsorb preferentially on the less hindered sites. It is the combination of these two effects - type of site and steric hindrance – which will define the sites where the reactions will more easily take place.

For the linear alkanes studied (methane, ethane and propane), the fact that only one type of acid site can be represented with the 3T and 5T clusters should not be a major problem. As shown by the MD studies, because of their sizes, steric effects are of minor importance and these molecules have equal probability of visiting all the distinct sites of the zeolite. In another words, for these molecules, as far as steric effects are concerned, the acid sites are all alike. Thus, the interaction between any of these molecules and the zeolite will depend mainly on the sites' acidic strengths, which do not differ very much from each other. Therefore, for these molecules it is a reasonable approximation to treat all the acid sites alike. However, for isobutane steric effects are more important and the molecule should be more sensitive to the type of the acid site. It will be easier for the isobutane molecule to approach the acid sites represented by 3T and 5T clusters than the one at the channels intersection, in the real zeolite, where it preferentially adsorbs. Therefore, for isobutane and other branched alkanes (and most probably for the large n-alkanes), the chemical reactions at the 3T and 5T clusters may take place artificially easier than in the real zeolite.

As previously mentioned, in calculations with 3T and 5T clusters, the effect of the electrostatic potential of the zeolite cavity is totally neglected. Thus, for chemical reactions presenting ionic TS's, neglecting this effect should increase the activation barrier, other things being equal. On the other hand, for neutral or almost-neutral TSs, this effect should be much less pronounced.

After these considerations, let us briefly examine the H-D exchange reaction and in more detail the dehydrogenation and cracking reactions of light alkanes.

4.1 The Exchange Reaction

This is the simplest reaction which may take place between an alkane substrate and the acid site of the zeolite. Not really important from the chemical point of view, this reaction nevertheless is simple enough to serve as a case–study for testing different methodologies and for establishing the level of calculation which must be used to obtain the TSs and reliable values of activation barriers. Besides, for the methane exchange, experimental results of activation energy and reaction rates are available [38] which can be helpful in selecting the appropriate methodology and level of calculation to be used in the chemical step of the catalytic process.

The isotope exchange reaction was investigated both experimentaly [38] and theoretically [38-40], by Kramer and coworkers. The calculations were performed at the Hartree-Fock (HF) and single-double configuration interaction (CI-SD) levels, using a 3T cluster. This reaction was also investigated by Evleth *et al.* [41] at the HF and Møller-Plesset second-order perturbation theory (MP2) levels but using a much simpler 1T cluster. DFT calculations using the Beck-Perdew (BP) functional have also been performed [42]. Except for ref. [41], where the 1T cluster was used, all the other calculations indicate a TS which resembles a carbonium ion, with the carbon atom symmetrically coordinated to five hydrogen atoms. This picture has been confirmed by more recent calculations employing embedded clusters [43].

In order to compare the performance of the different methods that can be used, test calculations were performed using the same 3T cluster and basis set (6-31G**). The results presented in table 4 indicate that they are all equivalent within the range of experimental values. The TS geometries are also very similar, except for some differences in the Si-O-Si angles obtained with DFT ($< 3^o$ - 5°) and non-DFT methods. Based on these results the B3LYP method was selected for the studies on the chemical reaction step.

Following these test-calculations, a systematic study was conducted for the exchange reaction of light alkanes [44]. The relevant results are shown in table 5 and figure 4. The enthalpy of activation does not show significant variations either with the carbon chain length or the type of hydrogen atom being exchanged. The slightly higher value of $\Delta H^{\neq}$ for the exchange at the tertiary center of isobutane can be attributed to some steric hindrance due to the methyl substituents. Also from table 5 one sees that except for the attack at the methine position of isobutane, an increase in the electronic density on the attacked carbon atom in the TS was observed, relative

Table 4. Activation barriers - E_a (kcal/mol) and reaction coordinate - RC (cm^{-1}) for the methane exchange reaction.

Method	**RC**	$\mathbf{E_a}$
HF/MP2	- 1871.6	31.11
GVB + CI	- 1517.9	27.83
DFT (BP)	- 1609.7	29.92
DFT (B3LYP)	- 1703.1	30.03
Exp. [11]	———	29.2 – 31.1

to the ground state. This implies that the TS should be quite insensitive to the electronic effect of substituents, reflecting an almost constant activation energy for all C-H bonds. From figure 4 one sees that the TS for all the reactions resembles a pentacoordinated carbonium ion, in agreement with the previous calculations for methane. No appreciable changes, either in the TS geometry or enthalpy of activation, are observed by increasing the size of the cluster (5T) or the basis set (6-311G**).

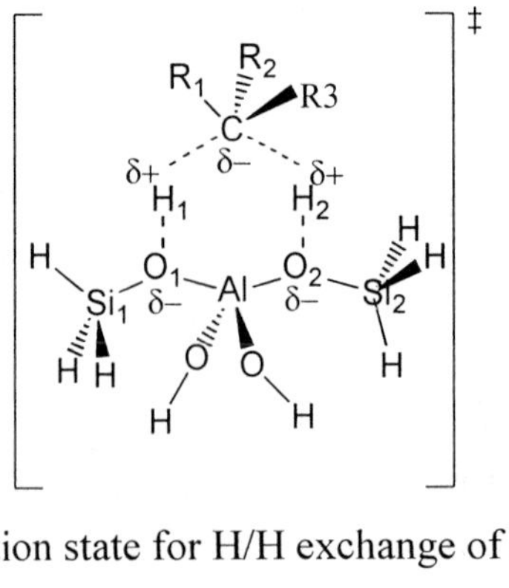

Figure 4. Transition state for H/H exchange of alkanes with zeolites.

Table 5. Calculated enthalpy of activation ($\Delta H^{\neq}$) in kcal/mol, corrected for zero-point energy (ZPE) and 298.15 K, reaction coordinates, and electrostatic potential derived (CHELpG) charges for the H/H exchange with the 3T cluster.

Reaction[a]	**$(\Delta H^{\neq})^b$**	**$(\Delta H^{\neq})^c$**	**RC (cm^{-1})^c**	**(δq)^{c,d}**
$C\mathit{H}_4$ + HT3	31.1	32.3	- 1703	- 0.187
$C_2\mathit{H}_6$ + HT3	31.4	32.3	- 1590	- 0.058
$CH_3CH_2C\mathit{H}_3$ + HT3	30.6	32.2	- 1590	- 0.068
$CH_3C\mathit{H}_2CH_3$ + HT3	30.6	33.3	- 1484	- 0.042
$(C\mathit{H}_3)_3CH$ + HT3	29.8	32.3	- 1567	- 0.043
$(CH_3)_3C\mathit{H}$ + HT3	31.5	36.2	- 1412	+ 0.047

a) The italic hydrogen refers to the one being exchanged.
b) MP2/6-31G**//HF/6-31G**; c) B3LYP/6-31G**//B3LYP/6-31G**.
d) δq = (absolute charge on the carbon atom in the TS) – (absolute charge on the carbon atom in the reactant).

4.2 The Dehydrogenation Reaction

The dehydrogenation reactions of methane [42,45] ethane [46] and isobutane [47] have been previously investigated. However, these calculations differ on the type of cluster used to represent the acid site of the zeolite as well as in the choice of the DFT functional and the basis set. Since we are interested in analysing this reaction as a function of the size and type of chain (linear or branched), it is imperative to treat all the substrates at the same level of calculation. The validation of DFT methods to investigate this reaction was established by comparing the activation energies and the TS geometries for the methane dehydrogenation reaction obtained at the GVB + CI and B3LYP levels of calculation [45].

Once established the appropriate level of calculation, a systematic study was conducted for the dehydrogenation reaction of ethane, propane and isobutane, using the B3LYP functional along the 6-31G** and 6-311G** basis sets and both the 3T and 5T clusters to represent the acid site of the zeolite. The geometry optimization of the reagents, products, intermediates and the search for saddle points were performed with the Jaguar 3.5 program [48]. For each optimized geometry, all the frequencies were verified to be real. The transition states (TS's) were characterized by the existence of a single imaginary frequency. When needed, intrinsic reaction coordinate (IRC) calculations were performed with the Gaussian 98 program [49]. No constraints were imposed during all the calculations.

A typical TS obtained with the 5T cluster and 6-311G** basis set is shown in figure 5. The relevant parameters for analysing the structure and nature of the TS's are presented in table 6. From figure 5 and table 6 it is clear that, for all the reactions, the atoms H1 and H2 are moving away from the alkane molecule and the cluster respectively, while simultaneously approaching each other to form an H_2 molecule. For the methane TS the H1-H2 distance is still much larger than in the ground state of the H_2 molecule (0.74Å) but it drops considerably to $R_{H1H2} \cong 0.785$Å as the carbon chain length increases. The geometry and the nature of the TS's do not exhibit any appreciable changes as the basis set and/or the size of the cluster is increased. However, for the isobutane reaction the activation energy drops considerably when the T5 cluster is used along with the 6-311G** basis set, as shown in table 7.

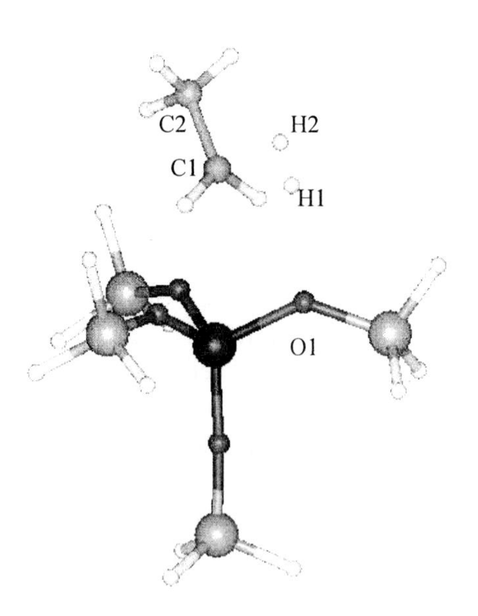

Figure 5. Transition state for the dehydrogenation reaction.

Table 6. Relevant parameters for the transition states of the dehydrogenation reactions of ethane, propane and isobutane with clusters T3 and T5.

		Distances [Å]				
Alkane	**Cluster/Basis**	**H1-H2**	**O1-H2**	**C1-H1**	**C1-O1**	**C1-C2**
Ethane	T3/DZP	0.785	2.008	1.773	2.518	1.467
	T5/DZP	0.783	2.137	1.749	2.512	1.467
	T5/TZP	0.788	1.960	1.777	2.481	1.468
Propane	T3/DZP	0.785	2.023	1.759	2.529	1.466
	T5/DZP	0.781	2.169	1.759	2.544	1.465
	T5/TZP	0.783	2.004	1.799	2.511	1.465
Isobutane	T3/DZP	0.790	1.717	1.904	3.449	1.469
	T5/DZP	0.790	1.707	1.875	3.611	1.482
	T5/TZP	0.788	1.729	1.887	3.692	1.465

Analyzing the vibrational modes which correspond to the imaginary frequencies for the reactions with the linear alkanes, it can be observed that as the hydrogen molecule is being formed the carbocation moves in the direction of the zeolite. Thus, from the reaction coordinate it can be inferred that the products of these reactions are H_2 and the respective alkoxides. For the isobutane reaction, on the other hand, such motion of the *t*-butyl cation is not observed, impling that the alkoxide is not being formed. This observation strongly suggests that the dehydrogenation reactions of linear and branched alkanes follow different mechanisms. The IRC calculations for the isobutane reaction confirm such conclusions and show that the *t*-butyl ion, instead of forming an alkoxide, decomposes into isobutene and a proton which restores the acid site on the zeolite.

It is also interesting to note that the distance between the substrate and the zeolite framework in the TS was much larger for the isobutane reaction, where the attack was on a tertiary center, than for the linear alkanes, where the attack was on primary centers. A similar situation was found in the case of the exchange reactions [44]. Although the nature of the TS was the same for exchanging at primary, secondary and tertiary centers, in the last case the distance from the substrate (isobutane) to the zeolite framework is larger than for the other alkanes. This can be attributed to the steric hindrance of the methyl groups as the isobutane approaches the acid site. Thus, it is possible that in the dehydrogenation reaction of isobutane, the elimination of isobutene is favored with respect to the alkoxide formation because of steric effects.

Table 7. Comparison of the activation energies for dehydrogenation calculated in the present work with available experimental results and other theoretical calculations.

	Activation Energy [kcal/mol]		
	Ethane	**Propane**	**Isobutane**
T3 / 6-31G**	72.80	71.18	58.53
T5 / 6-31G**	72.74	69.74	52.69
T5 / 6-311G**	74.51	73.04	53.35
Experiments	-	22.7[d]	57[b],38.7[c] (ap),23.9[d] (ap)
Other Theoretical Studies	72.84[a]	-	66.8[e],74.7[e]

a) Ref. [46], b) Ref. [50], c) Ref. [51], d) Ref. [52], e) Ref. [47].

The scarcity of experimental data for the reactions with the linear alkanes makes the comparison with our results difficult. However, our mechanism is consistent with similar theoretical studies [45-47,53] on light linear alkanes, which predict the formation of a carbocation-like transition state and alkoxides as products. In the isobutane case, the mechanism differs from the one for the linear alkanes, since the formation of alkoxides is not predicted, the transition state decomposing directly into isobutene and H_2. Therefore the theoretical calculations strongly suggest that once the carbocations are formed, they either eliminate a proton to produce an olefin or get adsorbed in the zeolite framework giving rise to the alkoxides. From the experimental point of view there are also strong evidence that carbocations are not "free" in the cavity of zeolites. This evidence comes from comparative studies of carbocations in liquid superacid solutions and in zeolites, using ^{13}C NMR measurements. In liquid superacids, where carbocations are believed to occur free in solution, a ^{13}C peak in the region of 200 ppm (relative to TMS) has been taken as a signature for free carbocations [54,55]. *In situ* measurements in zeolites do not show the presence of peaks in that region. On the other hand, peaks in the region 80-100 ppm are observed, which are attributed to the alkylalkoxides [56,57]. Curiously enough, no such peak is observed for the *tert*-butyl case. In order to provide extra support to the theoretical prediction, ^{13}C NMR shifts were calculated for the different alkoxides, using the GIAO method [58]. The result of such calculations are shown in table 8 and compared to the available experimental data. From table 8 one can see that the observed ^{13}C peaks in the 80-100 ppm can certainly be attributed to the alkoxides, even for the *t*-butyl case. Thus, if this last peak is not observed, one may conclude that the *t*-butyl alkoxide is not being formed in the reaction, exactly as predicted by our mechanism. The reaction profiles for the dehydrogenation of linear and branched alkanes are shown in figures 6a and 6b.

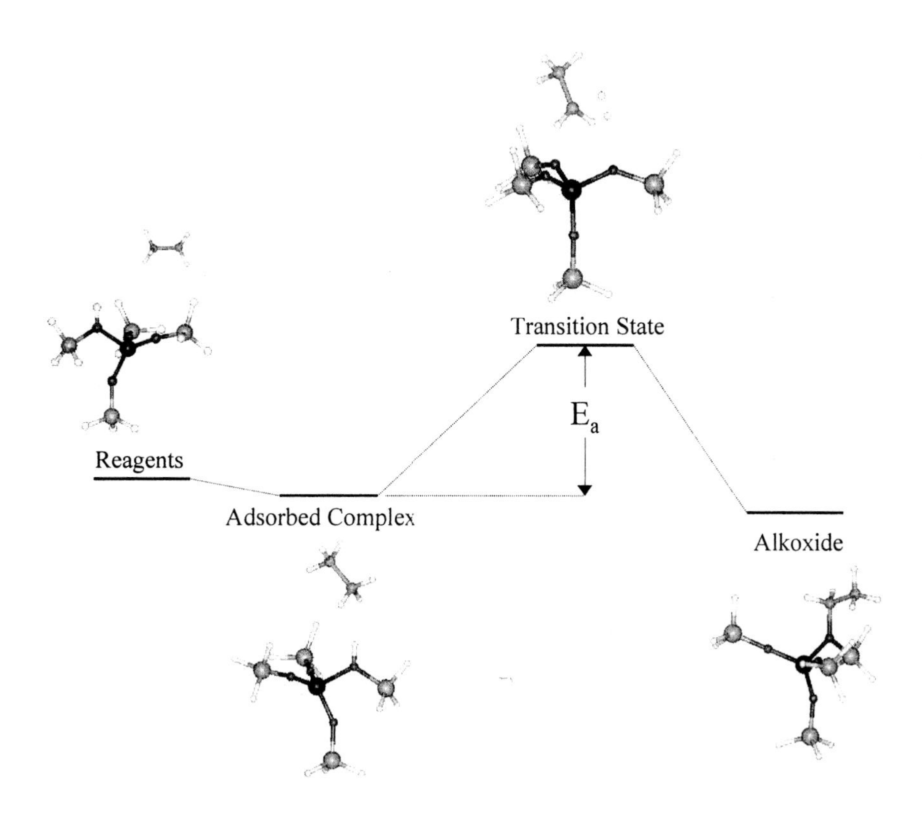

Figure 6a. Reaction profile for the dehydrogenation of ethane.

From the qualitative point of view, the mechanism suggested by the present calculations are in agreement with the experimental studies. However, in spite of fact that the predicted activation energy for isobutane is in close agreement with the available experimental data, we do not claim that the correct energetics for these reactions can be obtained from calculations using 3T and/or 5T clusters. For instance, the good value obtained for the activation energy of the isobutane reaction may result from a fortuitous cancellation of errors due to bad description of the adsorption complex (~10-15 kcal/mol) and the lack of stabilization of the TS by long-range interactions, which could be of the order of 14 kcal/mol according to estimates based on the cracking reaction of ethane [59]. Thus, larger clusters will be needed for quantitative purposes, but as indicated by preliminary calculations using 20T clusters, the mechanism of the reaction for such small alkanes can be well understood at the present level of calculations.

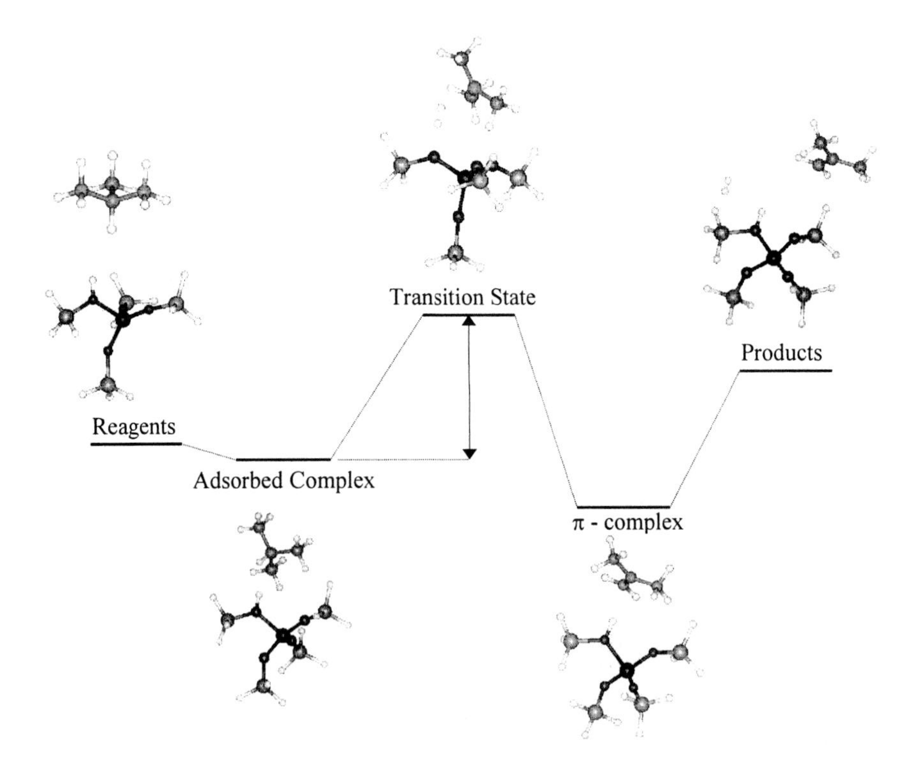

Figure 6b. Reaction profile for the dehydrogenation of isobutane.

Table 8. Calculated ^{13}C NMR chemical shifts for alkyl groups adsorbed on ZSM-5 zeolites compared to experiments.

R	^{13}C Chemical Shift (ppm)				
	T3 cluster[a]		T5 cluster[b]		
	$\delta_{iso,abs}$	$\delta_{isso,rel}$ [c]	$\delta_{iso,abs}$	$\delta_{iso,rel}$ [d]	Exp.
~CH_3	120.55	63.41	122.34	61.70	58 50.1 49
~A–B	A:106.07 B:166.47	77.89 17.49	A:108.65 B:166.52	75.39 17.53	A:68 B:17
~A–B–C	--	--	A: 103.23 B: 155.92 C: 173.83	80.82 28.13 10.22	--
~A(B)(B)	A:89.25 B:159.91	94.71 24.05	A:93.32 B:160.03	90.73 24.02	A:91/88[e]
~A(B)(C)(C)	A:84.40 B:148.81 C:153.01	99.56 35.15 30.95	A:85.12 B:152.76 C:155.33	98.93 31.29 28.72	A:77[a] B: 29[a]

a) Ref. [60]; GIAO/B3LYP/6-311+G(d,p)
b) GIAO/B3LYP/6-311++G(d,p)// B3LYP/6-311G(d,p)
c) $\delta_{iso,TMS}$ [GIAO/B3LYP/6-311+G(d,p)] = 183.96
d) $\delta_{iso,TMS}$ [GIAO/B3LYP/6-311++G(d,p)] = 184.05
e) Ref. [61]

4.3 The Cracking Reaction

Extensive experimental studies of the catalytic cracking of parafins have been performed [62] and their results indicated that the reaction might follow either the protolytic or the chain mechanism. The protolytic cracking is found to be dominant under special conditions such as high temperatures, low conversion and absence of unsaturated impurities in the feed. Several recent theoretical studies were devoted to the protolytic cracking [46,47,59,63-66] but, similarly to the dehydrogenation studies, they differ in many respects. In order to gain understanding about the mechanism of protolytic cracking and of its dependence on the parafins size and shape, we have examined this reaction for ethane, propane and isobutane [67], using the same methodology and level of calculation employed for the dehydrogenation calculations.

A typical TS obtained with the 5T cluster and the 6-311G** basis set is shown in figure 7. The relevant parameters for analysing the structure and the nature of the TS's are presented in table 9. The activation energies for the different clusters and basis sets used are shown in table 10.

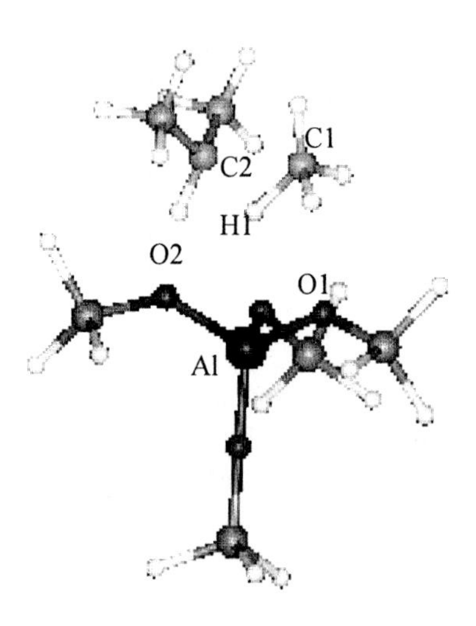

Figure 7. Transition state for the cracking reaction.

The results on tables 9 and 10 show that increasing the size of the cluster and/or the basis set does not change appreciably either the structure of the TS's or the activation energies obtained at a lower level of calculation (3T cluster and DZP basis set). Irrespective of the substrate and level of calculation employed, the results show that the protolytic cracking involves the attack of the zeolitic proton to a carbon atom of the alkane molecule with the simultaneous rupture of a C-C bond, instead of the attack of the proton directly to the C-C bond.

Table 9. Relevant geometric parameters for the cracking transition states of ethane, propane and i-butane.

Alkane	Cluster /Basis	Distances [Å] C1-C2	O1-H1	C2-H1	C1-H1	C1-O1	C2-O1	C2-O2
Ethane	T3/DZP	2.170	2.057	1.351	1.191	3.655	3.052	2.572
	T5/DZP	2.060	2.089	1.276	1.212	3.119	3.118	2.927
	T5/TZP	1.979	1.899	1.242	1.233	2.998	3.609	2.998
Propane	T3/DZP	1.909	2.196	1.282	1.230	2.935	3.066	3.224
	T5/DZP	1.908	2.188	1.280	1.234	2.896	3.085	3.214
	T5/TZP	1.927	2.242	1.284	1.231	2.927	3.152	3.161
i-butane	T5/DZP	1.923	1.761	1.380	1.244	2.725	2.918	3.426

Even for the case of ethane, whose cracking products do not depend on which C atom is attacked, the TS geometry does not exhibit a symmetrical arrangement of the attacking proton relative to the two carbon atoms. Also, from table 9 and figure 7, the results show that for propane and isobutane the attack is preferentially on a primary C atom, in order to form the more stable carbocations, as expected. For the propane molecule, a late transition state was also found, corresponding to the formation of ethane and a methylcarbocation. This process involves a much larger activation energy (76.6 kcal/mol) than the methane channel (56.12 kcal/mol).

Table 10. Calculated activation energies for the cracking reactions compared to the available experimental results and other theoretical calculations.

	Activation Energy		
Our Results	**Ethane**	**Propane**	**Isobutane**
T3 / 6-31G**	68.36	58.14	-
T5 / 6-31G**	67.14	56.75	53.45
T5 / 6-311G**	66.15	56.12	-
Experimental Studies	-	32.53[a]	-

[a] Ref. [52].

As for the dehydrogenation reaction, IRC calculations [67] indicate that the protolytic cracking of linear and branched alkanes follows different mechanisms. For ethane and propane the products of the reaction are methane and the proper alkoxide. For isobutane, as one follows the reaction path towards the products, the t-butyl cation decomposes into propene and a proton which restores the acid site of the zeolite.

5. Conclusions

The diffusion, adsorption and chemical steps for the dehydrogenation and cracking reactions of light alkanes catalyzed by zeolites were studied using a combined classical mechanics (MM, MD and MC) / quantum mechanics approach.

The results for the diffusion coefficients obtained from MD calculations follow the expected trend, based on the differences of mass and shape of the molecules. When compared to the available experimental data, the MD results indicate that the simulations can indeed provide a realistic representation of the microscopic process of diffusion of light alkanes in the pores of zeolites as long as reliable force fields to represented the zeolite and sufficiently long trajectories are used.

For the molecules investigated, the MD and MC methods furnish similar adsorption energies although the MD results are slightly better when compared to the experiments. Since the MC and MD simulations have been performed in different ensembles but with the same force field, the difference in the results of the two simulations may be partly due to finite size effects. However, temperature fluctuations during the MD simulations in the NVE ensemble may also contribute to this difference. For the linear alkanes the MM results are qualitatively correct but only when a specific force field is used to describe the zeolite. However, for the branched alkane the MM results are comparable to the MD and MC ones even when a generic force field, such as Dreiding II, is used to represent the zeolite.

The adsorption energy increases with the carbon chain length for the linear alkanes. For the branched ones, the adsorption energies are smaller than the corresponding ones for the linear molecules with the same number of carbon atoms. The results for ethane, propane and i-butane indicate that the alkanes adsorb more strongly in ZSM-5 than in silicalite, as experimentaly observed. On the other hand, the small differences in the adsorption energies in silicalite and ZSM-5 indicate that the process is dominated by dispersion forces. For the linear alkanes the adsorption energy raises slightly with increasing the loading. For isobutane, on the other hand, it remains almost constant up to a loading of three molecules per unit cell and drops at higher loadings.

The linear alkanes show equal probability of being found in both the straight and the sinusoidal channels while the branched alkane prefers the channel intersections. The packing of the alkane molecules inside the zeolitic cavity does not cause appreciable geometric distortions.

The dehydrogenation reaction proceeds through the simultaneous elimination of the zeolitic proton and a hydride ion from the alkane molecule, giving rise to a transition state which resembles a carbenium ion plus an almost neutral H_2 molecule to be formed. For the linear alkanes, the TS decomposes into an H_2 molecule and the carbenium ion correspondent alkoxide. However, for the isobutane molecule the reaction follows a different path, the TS producing isobutene and H_2. Most certainly the olefin elimination is favored to the alkoxide formation due to steric effects as the t-butyl cation approaches the zeolite framework. The same mechanism is expected to be operative for other branched alkanes.

The protolytic cracking involves the attack of the zeolitic proton to a carbon atom of the alkane molecule and the simultaneous rupture of one its adjacent C-C bond. The carbon atom being attacked and the C-C bond being broken will be preferentially those which produce the most stable carbenium ion. As for the dehydrogenation reaction, the protolytic cracking of linear and branched alkanes also follow different mechanisms, the latter ones producing olefins instead of alkoxides.

Acknowledgements

The authors acknowledge CNPq, FUJB, FAPERJ and PRONEX for financial support.

References

[1] (a) G.T. Kerr., *Sci. Am*,. **261,** 100 (1989),. (b) E.W. Vaughan, *D. Chem. Eng. Prog.*, **84** , 25 (1988).

[2] F.R Ribeiro, A.E Rodrigues,. L.D Rollman, Zeolites: Science and Technology, Martins Nijhoff, (eds.) Amsterdam, (1984).
[3] J.A Cusumano, Perspectives in Catalysis, J.M. Thomas e K. Zamarev IUPAC, Chemistry for the 21st Century, (eds.) Blackwell Scientific Publication, (1992).
[4] (a) D.H.Olson, E. Dempsey, *J.Catal.*, **13**, 221 (1969); (b) W.J. Mortier, J.J. Pluth, J.V. Smith, *J. Catal*, **45**, 367 (1968); (c) Z. Jirák, V. Vratislav, V. Bosacek, *Phys.Chem.Solids* ,**41**, 1089 (1980); (d) M. Czjzek, H. Jobic, A.N. Fitch, T. Vogt, *J.Phys.Chem.*, **96,**1535 (1992).
[5] (a) J. Dubsky, S. Beran, V. Bosacek, *J.Molec.Catal.*, **6**, 321 (1979); (b) P.J. O'Malley, J. e. Dwyer, *J.Phys.Chem.*, **92**, 3005 (1992); (c) K.P. Schröeder, J. Sauer; M. Leslie, C.R.A. Catlow, J.M. Thomas, *Chem. Phys. Lett.*, **188** (3,4), 320 (1992); (d) G.J. Kramer, R.A.van Santen, *J.Am.Chem.Soc.*, **115**, 2887 (1993).
[6] C. R. A. Catlow, C. M. Freeman, B. Vessal, S. M. Tomlinson, M. Leslie, *J.Chem.Soc.Far.Trans.*, **87**, 1947 (1991).
[7] J. B. Nicholas, F. R. Trouw, J. E. Mertz, L. E. Iton, A. J. Hopfinger, *J.Phys.Chem.*, **97**, 4149 (1993).
[8] D. Dumont, D. Bougeard, *Zeolites*, **15**, 650 (1995).
[9] D. H. Olson, G. P. Kokotailo, S. L. Lawton and M. V. Meir, *J. Phys. Chem.*, **85**, 2238 (1981).
[10] W. Heink, J. Karger, H. Pfeiffer, K. P. Datima and A. K. Novak, *J. Chem. Soc. Faraday Trans.*, **88**, 3505 (1992).
[11] B. W. H. van Beest, G. J. Kramer, R. A. van Santen, *Phys.Rev.Lett.*, **64**, 1955 (1990).
[12] S. L. Mayo, B. D. Olafson, W. A. Goddard III, *J.Phys.Chem.*, **94**, 8897 (1990).
[13] W. C. Swope, H. C. Andersen, P. H. Berens, K. Wilson, *J.Phys.Chem.*, **98**, 12938 (1994).
[14] A. K. Novak, C. J. J. den Ouden, S. D. Pickett, B. Smit, A. K. Cheetan, M. F. M. Post, J. N. Thomas, *J.Phys.Chem.*, **95**, 848 (1991).
[15] P. Demontis, E. Fois, G. B. Suffriti and S. Quartieri, *J.Phys.Chem.*, **94**, 4329 (1990).
[16] J. Caro, S. Hocevar, S. Kaerger and L. Riekut, *Zeolites* **6**, 213 (1986).
[17] J. Caro, M. Bulow, W. Schrimer, W. Kaerger, W. Heink, H. Pfeiffer, S. P. Zdanov, *J.Chem.Soc.Far.Trans.*, **81**, 2541 (1985).
[18] H. Jobic, M. Bée, G. J. Kearley, *Zeolites*, **12**, 146 (1992).

[19] J. N. van-den-Begin, C. V. L. Rees Caro, M. Bülow, *Zeolites*, **9** 312 (1989); D. Shen, L. V. C. Rees, J. Caro, M. Bülow, B. Zibrowius, H. Jobic, *J. Chem. Soc., Faraday Trans.* **86** 3943 (1990).
[20] D. T. Hayburst, A. R. Paravar, *Zeolites* **8** 27 (1988).
[21] E. Hernandez, C. R. A. Catlow, *Proc. Royal Soc. London*, **A448** 143 (1995).
[22] R. L. June, A. T. Bell, D. N. Theodorou, *J. Phys. Chem.*, **96** 1051 (1992).
[23] J. R. Fried, S. Weaver, *Compt. Mat. Sci.*, **11** 277 (1998).
[24] J. O. Titiloye, S. C. Parker, F. S. Stone, C. R. A. Catlow, *J.Phys.Chem.*, **95**, 4038 (1991).
[25] B. Smit, C. J. J. den Ouden, *J.Phys.Chem.*, **92**, 7169 (1988).
[26] R. L. June, A. T. Bell, D. N. Theodorou, *J.Phys.Chem.*, **94**, 8232 (1990).
[27] R. L. June, A. T. Bell, D. N. Theodorou, *J.Phys.Chem.*, **94**, 1508 (1990).
[28] R. L. June, A. T. Bell, D. N. Theodorou, *J.Phys.Chem.*, **96**, 1051 (1992).
[29] J. B. Nicholas, F. R. Trouw, J. E. Mertz, L. E. Iton, A. J. Hopfinger, *J.Phys.Chem.*, **97**, 4149 (1993).
[30] N. L. Allinger, *J. Am. Chem. Soc.*, **99**, 8127 (1977).
[31] B. Smit, J. I. Siepmann, *Science*, **264**, 1118 (1994).
[32] R. Vetrivel, C. R. A. Catlow, E. A. Colbourn, *J.Chem.Soc.Far.Trans. II*, **85**, 497 (1989).
[33] N. Metropolis, A. W. Rosenbluth, M. N. Rosenbluth, A. H. Teller, E. Teller, *J.Chem.Phys.*, **21**, 1081 (1953).
[34] M. P. Allen, D. J. Tidesley, Computer Simulation of Liquids, Clarendon Press. Oxford, 1987.
[35] H. Stach, U. Lohse, H. Thamm, W. Schrimer, *Zeolites*, **6**, 74 (1986).
[36] R. E. Richards, L. U. C. Rees, *Langmuir*, **3**, 335 (1987).
[37] J. R. Huffon, D. M. Reuthven, R. P. Danner, *Microporous Mat.*, **5**, 39 (1995).
[38] G.J. Kramer, R.A.van Santen, C. A. Emeis, A. K. Novak, *Nature*, **363**, 529 (1993).
[39] G. J. Kramer, R.A.van Santen, *J.Am.Chem.Soc.*, **115**, 2887 (1993).
[40] G. J. Kramer, R. A. van Santen, *J.Am.Chem.Soc.*, **117**, 1766 (1995).
[41] E. M. Evleth, E. Kassab, L. R. Sierra, *J.Chem.Phys.*, **98**, 1421 (1994).

[42] S. R. Blaszkowski, M. A. C. Nascimento, R. A. van Santen, *J.Phys.Chem.*, **98**, 12938 (1994).
[43] J. M. Volmer, T. N. Truong, J.Phys.Chem B., **104**, 6308 (2000).
[44] P. M. Esteves, C. J. A. Mota, M. A. C. Nascimento, *J.Phys.Chem. B.*, **103**, 10417 (1999).
[45] J. O. M. A. Lins, M. A. C. Nascimento, *Journal of Molecular Structure (Theochem)*, **371**, 237 (1996).
[46] S. R. Blaszkowski, M. A. C. Nascimento, R. A. van Santen, *J.Phys.Chem.*, **100**, 3463 (1996).
[47] V. B. Kazansky, M. V. Frash, R. A. van Santen, *Appl. Catal. A.*, **146**, 225 (1996).
[48] Jaguar 3.5, Schrödinger, Inc.; Portland, OR. (1998).
[49] Gausian 94 (Revision E.2), M. J. Frisch, G. W. Trucks, H. B. Schlegel, P.M.W.Gill, B. G. Johnson, M. A. Robb, J. R. Cheeseman, T. A. Keith, G. A. Peterson, J. A. Montgomery, K. Ragavachari, M. A. Al-Laham, V. G. Zakrewski, J. V. Ortiz, J. B. Foresman, J. Cioslowski, B. B. Stefanov, A. Nanayakkara, M. Challacombe, C. Y. Peng, P. Y. Ayala, W. Chen, M. Wong, J. L. Andres, E. S. Replogle, R. Gomperts, R. L. Martin, D. J. Fox, J. S. Binkley, D. J. Defrees, J. Baker, J. P. Stewart, M. Head-Gordon, C. Gonzales, J. A. Pople, Gaussian, Inc.; Pittsburgh PA. (1995).
[50] C. Stefanadis, B. C. Gates, W. O. Haag, *J.Mol.Catal.*, **67**, 363 (1991).
[51] A. Corma, P. J. Miguel, A. V. Orchiles, *J. Catal.*, **145**, 171 (1994).
[52] T.F. Narbeshuber, A. Brait, K. Sesham, J.A. Lercher, *J. Catal.*, **172**, 127 (1997).
[53] P. Viruela-Martin, C. M. Zicovich-Wilson, A. Corma, *J.Phys. Chem.*, **97**, 13719 (1993).
[54] G.A. Olah, Y. Halpern, J. Shen, Y. K. Huo, *J. Am. Chem. Soc.*, **93**, 1251 (1971).
[55] G.A. Olah, Angew. Chem. Int. Ed. Engl., **12**, 173 (1973).
[56] J.F. Haw, B.R. Richardson, I.S. Oshiko, N.B. Lazo, J.A. Speed, *J. Am. Chem. Soc.*, **111**, 2052 (1989).
[57] M. Zardkoohi, J.F. Haw, J.N. Lunsford, *J. Am. Chem. Soc.*, **109**, 5278 (1987).
[58] K Wolinski, J.F. Hilton, P. Pulay, *J. Am. Chem. Soc.*, **112**, 8251 (1990).
[59] S.A. Zygmund, L.A. Curtiss, P. Zapol, L.E. Ito, *J. Phys. Chem. B*, **104**, 1944 (2000).

[60] M. T. Aronson, R. J. Gorte, W. E. Farneth, D. J. White, *J.Am.Chem.Soc.*, **111**, 840 (1989).
[61] P.M. Esteves, Ph.D. Thesis, Universidade Federal do Rio de Janeiro (1999).
[62] T.F. Narbeshuber, H.Vinek, J.A. Lercher, *J. Catal.*, **157**, 388 (1995), and references there in.
[63] A.M. Rigby, G. J. Kramer, R. A. van Santen, *J. Catal.*, **170**, 1 (1997).
[64] V. B. Kazansky, I.N. Senchenya, M.V. Frash, R.A. van Santen, *Catal. Lett.*, **27**, 345 (1994).
[65] S.J.Collins, P.J.O'Malley, *J. Catal.*, **153**, 94 (1995).
[66] S.J.Collins, P.J.O'Malley, *Chem. Phys. Lett.*, **246**, 555 (1995).
[67] I. Milas, M.A.C. Nascimento (to be published).
[68] H. Papp, W. Hinsen, N. T. Do, M. Bearns, *Thermochim. Acta.*, **82**, 137 (1984).

AB INITIO SIMULATIONS OF ZEOLITE REACTIVITY

János G. Ángyán, Drew Parsons
Laboratoire de Chimie théorique, UMR CNRS 7565
Institut Nancéien de Chimie Moléculaire, Université Henri Poincaré Nancy I,
B.P. 239, F-54506 Vandoeuvre-lès-Nancy, FRANCE

Yannick Jeanvoine
Laboratoire Analyse et Environnement, UMR CNRS 8587
Université d'Evry-Val d'Essonne, Boulevard F. Mitterrand
F-91025 Evry FRANCE

Abstract The ab initio pseudopotential plane wave DFT simulation of the structure and properties of zeolite active sites and elementary catalytic reactions are discussed through the example of the protonation of water and the first step in the protolytic cracking mechanism of saturated hydrocarbons.

1. Introduction

In general, the purpose of theoretical modeling of physicochemical phenomena is a better understanding of the underlying physical laws. The theoretical study of complex systems is particularly difficult, since the corresponding model should be tractable by our tools, nevertheless it should reflect all the essential features of the real system in experimental conditions. Reproduction of experimental data constitutes a necessary step in validating our theoretical tools and models, however it cannot be the final goal of modeling. Usually, theory can be helpful to interpret experimental observations, by allowing us to study those aspects of the phenomenon which are experimentally inaccessible. Furthermore, it can replace experimental measurements by providing us data which for some reason cannot be determined empirically. It is precisely this feature which may be exploited for designing new materials, without the necessity of

M.A. Chaer Nascimento (ed.), Theoretical Aspects of Heterogeneous Catalysis, 77–108.

costly syntheses and experimental testing. The petrochemical industry is one of the technological branches that has recognized the importance of this approach and has begun to call for a contribution of molecular modeling in the design of new and more efficient catalysts and catalytic processes.

Theoretical modeling of the structure and reactivity of zeolitic materials, with special emphasis on the mechanism of catalytic reactions, has been the subject of several exhaustive review articles in the past decade.[1,2,3] Theoretical approaches that have been used to describe such systems range from empirical molecular mechanics calculations to various ab initio methods as well as different variants of the mixed quantum/classical (QM/MM) algorithms. In the present contribution we focus our attention mainly on those studies which were accomplished by ab initio pseudopotential plane wave density functional methods that are able to treat three-dimensional periodic models of the zeolite catalysts. Where appropriate, we attempt a critical comparison of with other theoretical approaches.

Several questions are more or less closely related to the understanding of catalytic reactions in zeolites, each needing a different modeling technique and strategy. First, the structure of the framework itself, the arrangement of the voids and channels of the structure, play a decisive role in the diffusion of the reactants and products inside the catalysts. These phenomena should be described by using sufficiently large models, extending over several unit cells of the zeolite structure. Since the diffusion is mainly governed by steric and non-bonded van der Waals forces, simple empirical potentials are suitable to pursue such studies. However, one should be cautious with over-simplified representations: e.g. the hypothesis of a rigid zeolite framework certainly misses the role of breathing apertures in the cavities in facilitating the diffusion process. While classical molecular dynamics techniques[4] can be suitable for the study of small molecules in the zeolite, specific methods, appropriate to simulate rare events[5,4] are needed to describe the activated process of diffusion of bulky molecules such as branched alkanes.

Physisorption of the substrate molecules on the cavity walls can also be described successfully by intermolecular force fields. Energetics and siting of relatively large molecules in the confined space of zeolite cavities and channels can be simulated by special simulation technics, for instance, the configurational-bias Monte Carlo approach, which is specially well-suited for this purpose.[6,7]

The catalytic activity of the zeolitic framework is strongly dependent on the Si:Al ratio, i.e. the concentration of the potential catalytic sites. This structural feature,[8] as well as the spectroscopic and energetic properties of the Brønsted acid sites,[9] has also been investigated by empirical force field techniques. However, in contrast to the adsorption and diffusion phenomena, the stability of the acid sites, and their acid strength is a result of a subtle balance of covalent and ionic bonding interactions, with an active involvement of shared electron pairs. It is therefore mandatory that the stability and acidity of the various Brønsted sites be revisited also by electronic structure methods. A good compromise is the use of QM/MM style techniques.[10] Whereas the substrate molecules are simply physisorbed on the chemically inactive regions of the internal zeolite surfaces, they can enter in a much stronger interaction with the active site, involving a rearrangement of the electronic structure. Therefore this chemisorption should also be treated by an explicit consideration of the electrons, i.e. by some quantum chemical approach, using either a full electronic structure description of the zeolite framework and the reacting molecule or by invoking QM/MM techniques. The use of quantum chemical methods is also mandatory for modeling the catalytic reaction itself, with various transition structures and intermediates.

Among the numerous theoretical approaches applied to the zeolite reactivity problem, we focus our attention mainly on calculations, which use recently developed ab initio molecular dynamics techniques. After a very brief overview of the main features of this methodology, we discuss some applications taken from the modeling of zeolite chemistry: the characterization of the catalytic sites, the protonation of a water molecule and the mechanism of the protolytic reactions of alkanes.

2. Methods

There are two extreme views in modeling zeolitic catalysts. One is based on the observation that the catalytic activity is intimately related to the local properties of the zeolite's active sites and therefore requires a relatively small molecular model, including just a few atoms of the zeolite framework, in direct contact with the substrate molecule, i.e. a *molecular cluster* is sufficient to describe the essential features of reactivity. The other, opposing view emphasizes that zeolites are (micro)crystalline solids, corresponding to *periodic lattices*. While molecular clusters are best described by quantum chemical methods, based on the LCAO approximation, which develops the electronic wave function on a set of localized (usually Gaussian) basis functions, the methods developed out of solid state physics using plane wave basis sets, are much better adapted for the periodic lattice models.

The obvious difficulty of finding an optimal theoretical approach for the zeolites is precisely due to its intermediary character: the active site, which is local par excellence, is embedded in a periodic lattice, and it is strongly influenced by the whole periodic lattice. Moreover, in a strict sense, one cannot speak about a true periodic system, since the Al-substitutions and the active sites are usually distributed in a disordered, random manner. Therefore, in some sense, we can say that none of the standard approaches, neither the local quantum chemical, nor the periodic solid state physics approach, seem to be fully adapted for the zeolitic systems.

Ab initio density functional methods using pseudopotentials and plane wave basis set are naturally well-adapted to treat periodic lattice models of the zeolite catalyst. Although, in the light of the above discussion, it does not correspond entirely to the real situation of a zeolite framework with substitutional disorder, such a model still offers the most of the guarantee that none of the physico-chemically important interactions are missed by our calculations.

Ab initio molecular dynamics appeared in the context of Car-Parrinello molecular dynamics (CPMD), first proposed in 1985.[11] These are density functional electronic structure calculations, using a plane wave basis set to expand the density of valence electrons and core pseudopotentials to represent inner-shell electrons. The ingenious idea of Car and Parrinello consisted in coupling the classical dynamics on the nuclei with the resolution of the electronic structure problem. The free parameters of the wave functions are considered as dynamical variables associated with a fictitious mass and they participate in the dynamical evolution of the system together with the nuclei. The combined equations of motions are solved by a finite difference algorithm, with a time-step which should be small enough for the fast electronic degrees of freedom, typically one tenth of the time-step appropriate for the nuclei. The Car-Parrinello algorithm ensures that the electronic system remains always in the close vicinity its ground state, i.e. the Born-Oppenheimer surface,[12] even for relatively long (several picosecond) simulations.

This advantage of the CPMD procedure is that it avoids the iterative solution of the self-consistent field equations for each new nuclear configuration, which would be too costly to perform for the several thousands of times in the course of a molecular dynamics trajectory. An alternative way to circumvent the electronic structure bottleneck in ab initio molecular dynamics was opened by the application of more efficient conjugate gradient minimization techniques for the electronic system.[13] In this case the time-step of the molecular dynamics can be chosen according to the nature of nuclear motions and the electronic structure is converged to the Born-Oppenheimer surface at each new nuclear configuration along the trajectory. This technique will be referred to simply as ab initio molecular dynamics (AIMD).

The quality of the electronic wave function is a crucial point in order to apply these methodologies to chemical problems. Early calculations were done in the local density approximation (LDA) using norm-conserving pseudopotentials. Before the appearance of gradient corrected functionals (GGA, or generalized gradient approximation) bond energies and total energy differences could not be determined with chemical accuracy. Furthermore, the plane wave description of first-row elements such as oxygen, nitrogen, carbon, having rapidly varying valence electron densities would have required plane wave basis sets with so high cutoff energy that would have been prohibitive for routine CPMD or AIMD calculations. This problem was solved at the one hand by the norm non-conserving, ultrasoft pseudopotentials (US-PS) of Vanderbilt[14] and on the other hand by the projector augmented plane wave (PAW) technique of Blöchl.[15] The equivalence of these two approaches was demonstrated by Kresse and Joubert[16] and this unified formalism was implemented in the VASP (Vienna Ab initio Simulation Package) code.[17]

3. Properties of the Brønsted acid sites

Zeolitic catalysts are constituted of $[TO_4]$ tetrahedrons which are assembled by sharing oxygens to form one of several topologically distinct framework structures.[18] In the case of alumino-silicates, T=Si with some Al substitutions, while in silico-alumino-phosphates (SAPO), T sites are occupied by alternating Al and P atoms with some Si substitutions. The substitutions destroy the electro-neutrality of the framework and necessitate the presence of compensating cations. While mono- or bivalent metallic cations usually occupy crystallographically well-characterized sites in the cavities, protons are chemically bonded to framework oxygens.

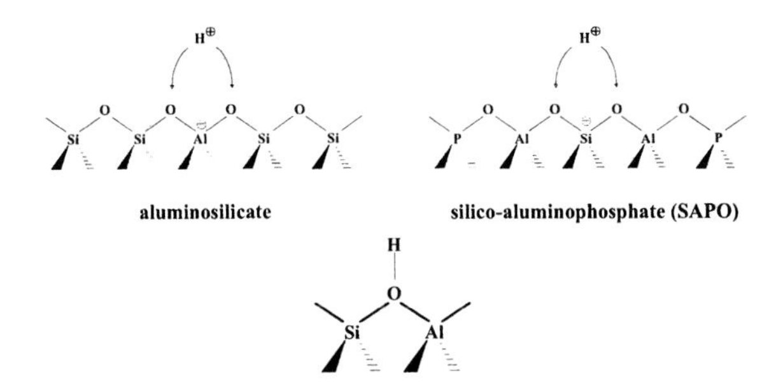

Figure 1. Common minimal model of the Brønsted acid site in different zeolitic structures.

The proton is always bonded to one of the oxygens belonging to the substituted T-site: Al in alumino-silicates or Si in silico-alumino-phosphates. Since the two vicinal tetrahedral sites are practically never substituted, the immediate neighborhood of the protonated framework oxygen is identical in both alumino-silicates and SAPOs, as indicated in the above scheme. This bridging OH group or *Brønsted acid site* is at the heart of the catalytic activity of zeolites and SAPOs.

There are relatively few experimental techniques which allow one to obtain direct information about the localization and structure of the acid sites. The adsorbate-free acid site can be studied by MAS-NMR, IR spectroscopy and X-ray or neutron diffraction. Even if neutron diffraction makes it possible to localize the protonation sites in deuterium-exchanged zeolite samples, the substitutional disorder makes it impossible to measure local geometrical parameters in the neighborhood of an acid site. Theoretical modeling is certainly the only approach which is able to provide information about the fine details of the properties of the acid sites.

3.1 Stability of the Brønsted sites

Up till today, relatively few systems have been studied using the most direct technique, the neutron diffraction of deuterated zeolite samples: faujasite,[19, 20] HSAPO-37,[21], HSAPO-34,[22] chabazite (HSSZ-13),[23] ferrierite[24] and mordenite.[25] Indirect methods to assess the Brønsted site locations involve NMR spectroscopy[26] and IR spectroscopy.[27]

Several theoretical works dealt with the relative stability, geometrical and vibrational properties of the acid sites in the the HSAPO-34 and HSSZ-13, belonging to the CHA zeotype. In these systems the proton can be attached to four crystallographically different oxygen sites. The numbering system that we use in this discussion, following of Ito et al.[28], is illustrated on Fig. 2. Neutron diffraction studies on HSSZ-13[23] indicate localized protons on the O1 and O2 framework oxygens. The most stable protonated structures found in the empirical force field calculation of Sastre and Lewis had the protons localized on O3 and O1 sites.[29] Brändle et al.[30] compared empirical shell model potential results, with periodic Hartree-Fock (CRYSTAL[31]) total energies as well as with embedded cluster (QMPot[32]) calculations. Both CRYSTAL and QMPot predict, in agreement with experiment, that O1 and O2 are the most stable sites. The same order of stabilities was found in US-PS PW calculations using GGA,[33, 34] as shown in Table 1.

site	Shell[29]	Shell[30]	QMPot[30]	CRYSTAL[30]	AIMD[33]	AIMD[34]	exp.[23]
O1	0.46	0.41	0.0	0.0	0.0	0.0	*
O2	6.45	3.75	3.92	3.08	0.92	1.25	*
O3	0.00	0.00	2.08	3.01	2.07	1.43	
O4	4.60	4.49	3.92	3.08	1.61	2.10	

Table 1. Relative energies of HSSZ-13 (chabazite) protonated at the O1 – O4 oxygen sites, as predicted by different theoretical approaches.

In the case of HSAPO-34 , the empirical force field results[35] are in good agreement with experimentally established protonation sites at O2 and O4,[22] while in ab initio calculations[33, 34] a slightly different, O1 < O2 < O4, order of stability was found, similar to that in the HSSZ-13.

In the faujasite framework there are also four crystallographically distinct oxygens. The experimental proton occupation in H-faujasite is 8:2:4 for the sites O1:O2:O3.[20] This order of stability has been quite well reproduced both by shell model calculations and embedded cluster methods.[36]

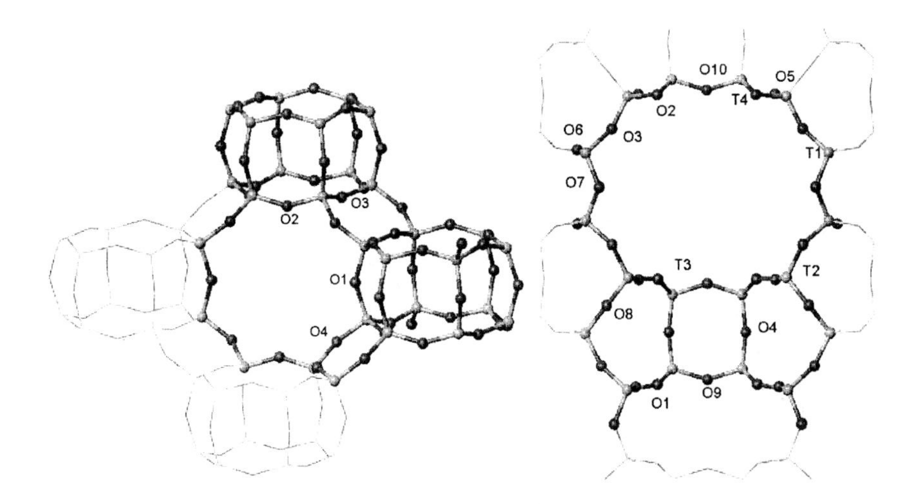

Figure 2. Numbering of oxygen and T atoms in the CHA and MOR frameworks.

The situation is somewhat more complicated in the mordenite structure. There are four crystallographically different T and ten O atoms. O1, O2, O3 and O4 are bonded to two crystallographically different tetrahedral atoms (T1-O1-T3, T2-O2-T4, T1-O3-T2, T3-O4-T4). Depending on the Al-substitution, this can lead to 8 different protonation sites. The six remaining oxygens have symmetrical environments, therefore we have a total of 14 possibilities for protonation. Sites O2, O3, O7 and O10 head toward the center of the large 12-ring, O1 and O9 point toward the 8-ring, O6 is in a side-pocket. O5, which is at the boundary of the side-pocket and the 12-ring, has been considered as belonging to the former, but recent neutron diffraction results made evident that the proton on this site points rather to the 12-ring.[25] Sets of possible combinations of framework oxygens that can host protons were deduced by Alberti, using the Löwenstein rule and the hypothesis that at least one of the protonation sites sits in the 12-ring.[37] One of these combinations was later confirmed by neutron diffraction experiments:[25] there is a proton on O10 and O5 in the 12-ring, on O9 in the 8-ring and on O6 in the side-pocket. Embedded cluster ab initio calculations of Brändle and Sauer[38] found that the 14 proton sites are in a relatively narrow energy range of about 4.3 kcal/mol, with the most stable structure at Al4-O2-Si2. Only the five framework oxygens that point toward the 12-ring, a total of seven possible Brønsted sites, were considered in the ab initio PAW calculations of Demuth et al.[39] According to the GGA energy differences, four sites (Al1-O3, Al2-O3, Al2-O2 and Al2-O5) are in a relatively narrow range of 0 to 2.5 kcal/mol, but only one of them corresponds to the experimentally approved Brønsted site, Al2-O5.

There are a few ab initio computational studies which deal with the order of stability of acid sites in other zeolite frameworks, where precise experimental data are lacking and only indirect information is available about preferential protonation sites. CPMD/GGA simulation of the offretite structure[40] showed a net preference for protonation near Al-substituted T2 tetrahedra. Several pieces of experimental evidence lead to the convergent view that the Al-population on this T2 site, which is shared between two 6-rings, one belonging to the gmelinite, the other to the cancrinite cage, is higher than that of the T1 tetrahedron, which participates in the hexagonal prisms, connecting cancrinite cages.[41, 42]

PAW/GGA calculations on the zeolite gmelinite showed quite a small difference (less than 1.5 kcal/mol) among the four possible bridging oxygen sites, with a slight preference for O1 and O4.[43] Actually there are no experimental data to compare against for this rare framework structure.

Larin and Vercauteren published periodic Hartree-Fock ps-21G*/STO-3G results on the Brønsted-site ordering in the ABW, CAN, CHA, EDI and NAT frameworks.[44]

Recent neutron diffraction results are available for ferrierite.[24] However, to our knowledge, this system has not yet been the subject of theoretical modeling of the Brønsted sites. In two recent ab initio simulations of methanol in ferrierite,[45, 46] the acid OH group was placed intuitively at O6 between two T4 tetrahedral atoms, in agreement with the experimental findings.

The modeling of relative stabilities of Brønsted acid sites remains a difficult subject to treat, even by the most sophisticated theoretical tools. In order to establish a reliable correlation between proton siting and structural features of the framework, more experimental data would be required. From a practical point of view it would be advantageous to relate preferential protonation sites to some geometrical parameters of the siliceous framework. In fact, the basicity of a given framework oxygen can be related to the T–O–T angle: a smaller angle would mean an increase of the sp^n hybridization, and thus stronger O–H bonding, i.e. a higher probability of forming an acid site. Kramer and van Santen, on the basis of a systematic study of small ring-like clusters, proposed a correlation between the Si–O bond lengths of the all-silica lattices and the proton abstraction energies.[47] Empirical correlations exist between the Al-substitution and the framework geometry (bond lengths and angles), based on the different sizes of the AlO_4 and SiO_4 tetrahedra. Although this information can be helpful to check the plausibility of a given protonation site distribution, it seems to be insufficient to determine the locality of protons reliably.

It has been suggested in the case of offretite that minima of the electrostatic potential calculated for the Al-substituted framework, in the absence of the counter-ion, could indicate the most probable protonation sites.[40] Local reactivity indices like atomic softness and Fukui indices were also proposed to characterize the Brønsted acidity of zeolite active sites.[48, 49, 50] In our opinion, further studies are needed to verify the stability of the computation of such indices, in order to decide the usefulness of this conceptually attractive approach.

3.2 Local structural properties of Brønsted sites

Due to the substitutional disorder of the T-atoms, local geometrical parameters of the Brønsted sites are not accessible from diffraction studies, the diffraction pattern reflects only an average geometry. If the concentration of the T-atom substitution is small, the average geometry can be quite close to the structure of an ideal all-silica structure and the distortion becomes more and more pronounced with an increasing amount of Al (in zeolites) or of Si (in SAPOs). The most affected parameters are the T-O distances and the Si–O–T bond angles.

An average H–Al distance can be estimated by NMR techniques. The measurement of the second moment of the broad-line ^{1}H NMR signal of bridging OH groups and spin-echo double resonance experiments lead to values of r_{HAl}=2.38 Å and 2.43 Å, respectively.[26] A selective determination of the the individual H–Al distances on different acid sites is possible from an analysis of the spinning sideband patterns of the ^{1}H MAS NMR signals.[26] It has been shown that experimental H–Al distances, r_{HAl}, depend on the size of the ring where the proton is located: for 6-rings it is around 2.35 Å, for 8-rings it is in the range of 2.38 – 2.46 Å, for 10-rings it is between 2.46 and 2.50 Å and for 12-rings it is in the range of 2.46 – 2.52 Å.

The r_{HAl} distances calculated from the published data from the GGA geometry optimizations of the CHA zeotypes HSSZ-13 and HSAPO-34[34] are in the range of 2.38 – 2.44 Å for the oxygens in 8-rings, in good agreement with the experimental value (cf. Table 2). However, for the O3 acid sites the Al–H distances are too long: 2.45 Å for HSSZ-13 and 2.39 Å for HSAPO-34. Analogous behavior was found from the analysis of the optimized geometries of Shah et al.[33]

Similar analysis of the data published for the acid sites belonging to the 12-ring of mordenite[51] yield r_{HAl} distances in the range of 2.36 – 2.48 Å from the GGA optimizations and 2.33 – 2.42 Å from the LDA calculations, that seem to be too short for this ring size. LDA calculations on the offretite structure,[40] which has 12-, 8-, and 6-rings, yield also too short Al–H distances: they remain in the relatively narrow range of 2.36 – 2.40 Å, with one exception, (2.24 Å) corresponding to an unusually small Al–O–H angle.

It is important to underline that, even if the Brønsted site itself has the same constitution (Si–O(H)–Al) in alumino-silicates and silico-alumino-posphates, the typical geometrical parameters of the local structure are significantly dif-

	HSSZ-13				HSAPO-34			
	O1	O2	O3	O4	O1	O2	O3	O4
OH	0.974	0.976	0.975	0.975	0.973	0.975	0.976	0.974
AlO	1.904	1.876	1.938	1.876	1.828	1.817	1.828	1.812
SiO	1.904	1.876	1.938	1.876	1.828	1.817	1.828	1.812
HOAl	110.6	108.5	109.1	106.9	117.3	114.9	113.2	115.4
HAl	2.432	2.380	2.445	2.359	2.440	2.403	2.394	2.404

Table 2. Optimized local geometrical parameters of different acid sites.[34]

ferent. This is illustrated by the results derived from the comparative study of the HSSZ-13 and HSAPO-34, as shown on Fig. 3.

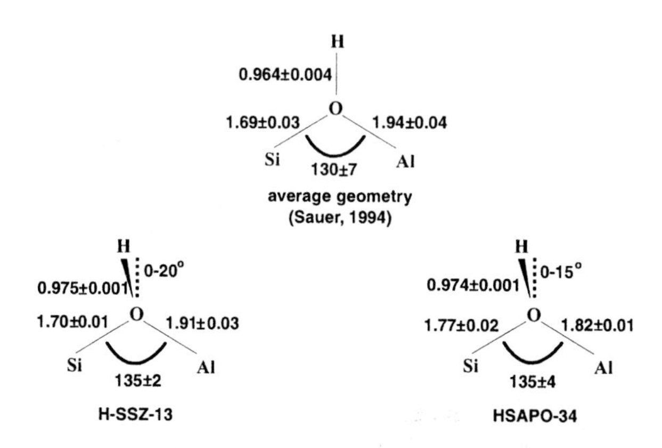

Figure 3. Average acid site geometry as proposed by Sauer et al.[2] on the basis of quantum chemical cluster calculations as compared to the HSSZ-13 and HSAPO-34 optimized acid site geometries.

The most significant difference is the simultaneous elongation of the Si–O bond, by about 0.7 Å accompanied by compression of the Al–O bond, by about 0.9 Å. These local geometrical parameters are in harmony with the conclusion drawn from the calculations of Shah et al.[52] that HSAPO-34 is less acidic that its aluminosilicate analogue, the HSSZ-13. Note however, that this conclusion does not seem to be generalizable to other frameworks. Experimental evidence seem to support the view that the relative acidity of the zeolite and the SAPO belonging to the same zeotype varies from one framework type to another.

The optimal proton position can be slightly out of the Si–O–Al plane. On the basis of the quadrupole coupling constant measurements and asymmetry values obtained from deuterium NMR experiments, Kobe et al.[53] suggested that the deuteron could be motionally averaged over two strongly out-of-plane positions. The only theoretical calculation which seems to support this hypothesis comes from the periodic Hartree-Fock calculations on sodalite, using Gaussian basis sets[54] which has predicted an out-of-plane angle of 25°, but ab initio pseudopotential plane wave calculations[34, 39, 55] indicate only a weak non-planarity. Recent periodic Hartree-Fock studies on the quadrupole coupling constants for H-form aluminosilicates confirmed also that the Si–O(H)–Al moiety is nearly planar.[56]

It is instructive to analyse the geometrical distortions of the $[TO_4]$ tetrahedra in the proximity of the substitution and protonation sites as compared to the geometry of the ideal SiO_2 or $AlPO_4$ framework structures. A thorough ab initio study of this problem in the CHA zeotypes HSSZ-13 and HSAPO-34 proved that the geometrical distortions remain localized on the tetrahedra directly connected to the substitution site.[34] The deformations, in spite of the differences of the reference structures, consisting in $[SiO_4]$ tetrahedra in one case and of alternating $[PO_4]$ and $[AlO_4]$ tetrahedra in the other, are remarkably similar. The $X^{(n-1)+} \rightarrow T^{n+}$ substitution, i.e. $Al^{+3} \rightarrow Si^{+4}$ for HSSZ-13 and $Si^{+4} \rightarrow P^{+5}$ for HSAPO-34, both, lead to a significant expansion of the substituted tetrahedron. The X–O(H) bond length increases by about 0.30 Å, and the other three X–O bond lengths are stretched by 0.07 Å. The T–O(H) bond length is also increased by about 0.09 Å. The considerable lengthening of the X–O(H) and T–O(H) bond lengths reflects their weakening, related to the formation of the O–H bond. The X–O(H)–T bond angle becomes smaller by 15°. The most significant bond length and bond angle deformations are schematically illustrated on Fig. 4. The tetrahedra themselves are distorted. The symmetrical deformation coordinate, formed by the three $\beta_n = \angle$ O(H)–T–O and the three $\alpha_n = \angle$ O–T–O bond angle changes as $\sum_{n=1}^{3}(\Delta\alpha_n - \Delta\beta_n)$, is about 15° for the substituted $[XO_4]$ tetrahedron and 10° for the $[TO_4]$ tetrahedron.[34] Similar trends can be observed from the analysis of other available structural studies.

3.3 Vibrational properties of acid sites

Infrared spectroscopy offers a powerful tool to characterize bridging OH groups, essentially via their O–H stretching vibrations.[27] For free OH groups, not interacting with adsorbed molecules, this band appears around 3600-3610 cm^{-1}. The O–H vibrational frequency depends on the framework topology and on the size of the ring towards which the proton is oriented. Accordingly, one can usually distinguish two sub-bands, the high- and low-frequency (HF and LF) ones, which broadly correspond to protons vibrating in large and small cages, respectively.

Most of the current quantum chemical models reproduce experimental frequencies with some systematic errors. In their respective basis set limit,

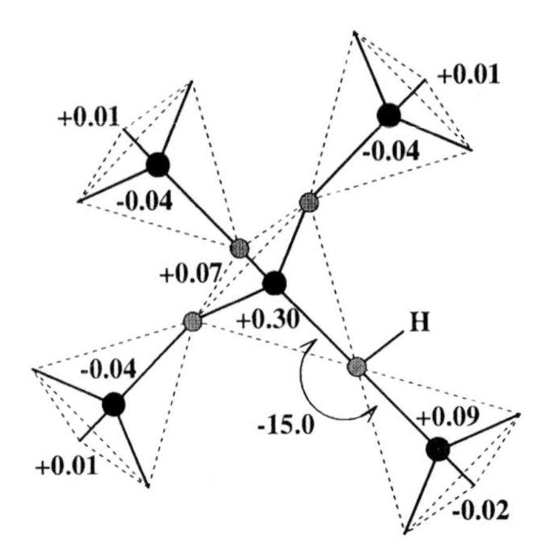

Figure 4. Deformation of an ideal framework after a T-atom substitution and protonation.

Hartree-Fock method considerably overestimates, DFT somewhat underestimates stretching frequencies, while correlated methods, like MP2 yield relatively reliable frequencies. The systematic error observed for the DFT calculations is not surprising in view of the overestimation of the equilibrium O–H bond lengths.[34]

The O–H stretching mode is a pure normal mode, with relatively strong anharmonicity, which can be deduced from the knowledge of the O–H potential curve at larger distances, obtained from the total energies computed at different O–H distances at the optimized structure of the total system. Note that, in view of the high frequency of the vibration, one should freeze the framework geometry, which is not supposed to be re-optimized for the various O–H separations.

The anharmonicity, $2\omega_e x_e$, which is defined experimentally as the difference between twice the fundamental frequency, ω_{01} and the overtone frequency, ω_{02},

$$2\omega_e x_e = 2\omega_{01} - \omega_{02} = \omega_e - \omega_{01} \tag{3.1}$$

corresponds also to the difference of the exact and harmonic, ω_e, frequencies. It can be determined by different methods. One possibility is to solve numerically the corresponding one-dimensional vibrational Schrödinger equation and use the calculated fundamental and overtone frequencies.[57] Alternatively, the anharmonicity constant can be obtained from the parameters of a Morse function, fitted to the potential curve,[58] or from the parameters of a third order polynomial approximation of the Morse function.[59] The polynomial approach seems to overestimate the anharmonicity ($2\omega_e x_e = 240 - 260\text{cm}^{-1}$) for the Brønsted site OH vibration,[33] while similar ab initio pseudopotential plane-wave calculations,[34] using the direct resolution of the one-dimensional Schrödinger equation, yield $2\omega_e x_e = 155 - 165\text{cm}^{-1}$ in close agreement with experimental values found in H-mordenite and HZSM-5.[60, 61] Similar differences were found between the two treatments of the anharmonicity in the same PAW calculations for gmelinite[43] and mordenite.[39] A difference of about 10 cm^{-1} was found between the LDA and GGA anharmonicities: the direct approach yields 172 and 180 cm^{-1}, while the polynomial approximation to the Morse potential gives 228 and 238 cm^{-1}, respectively.

The vibrational frequencies can also be estimated from molecular dynamics trajectories, using the Fourier transform of the velocity autocorrelation function. However, typical lengths of trajectories in ab initio molecular dynamics simulations (a few ps) do not allow one to draw quantitative conclusions from the these simulated vibrational spectra. The OH vibrational frequencies, calculated from the above "static" approach were compared to the "dynamical" ones, derived from the MD trajectory of gmelinite.[43] The "dynamical" frequencies were found about 200-300 cm^{-1} higher than the static ones, between 3740 and 3850 cm^{-1}, roughly 200 cm^{-1} above the experimental value, probably due to the errors related to the shortness of the trajectory and the relatively long integration timestep ($\Delta t = 1$ fs). A CPMD/LDA study on the offretite[40] lead, on the contrary, to frequencies which were 50-100 cm^{-1} too low, in the range of 3490 - 3525 cm^{-1}. This disagreement could have been reduced by using GGA, which seems to give systematically higher O–H frequencies by about 50 cm^{-1}, as has been observed in the case of mordenite acid sites.[39]

An elegant way to get O–H vibrational frequencies from short ab initio MD trajectories is based on the monitoring of $\mathrm{r_{OH}(t)}$, obtaining the instantaneous vibrational frequencies from the time delay, τ, between two consecutive maxima as $\omega = 2\pi/\tau$. A vibrational density of states can be constructed from the histogram of the instantaneous frequencies, giving some pieces of information about the expected width of the corresponding IR band.[55] The OH frequency, about 3550 cm^{-1}, obtained by this procedure from Car-Parrinello PAW trajectories for the sodalite framework[62] is very close to the experimentally expected value.

The case of HSSZ-13 and HSAPO-34, where we have both neutron diffraction and FTIR data,[22, 23] offers a particularly stringent experimental test for ab initio OH vibrational frequency calculations. Comparing the two systems, the HSAPO-34 OH frequencies are about 25 cm^{-1} higher that those of HSSZ-13. This trend is perfectly reproduced in the anharmonic frequencies obtained in ab initio pseudopotential plane wave calculations, even if the absolute frequencies are about 30 cm^{-1} lower than experimental values for both systems.[34] The anharmonic frequencies, obtained from a polynomial fit by Shah et al.[33] are less uniform in this respect: the HSAPO-34 frequencies coincide exactly with experiment, while for HSSZ-13 they are underestimated by 15 cm^{-1}. The experimental LF/HF splitting is in the order of 25 cm^{-1} for both cases. A comparison of experimental and theoretical splittings can be found in Table 3. In spite of the very good agreement with the experimental splitting for HSSZ-13, the calculations of Shah fail in predicting the splitting in HSAPO-34.[33] The calculations of Jeanvoine et al. are in reasonable agreement with experiment,[34] while the shell-model empirical potential results of Sastre et al.[29, 35] is in very good agreement with the experimental trends.

System		Jeanvoine[34]	Shah[33]	Sastre[29, 35]	exp.
HSSZ-13	$\Delta\omega(O2/O1)$	19	25	17	24
HSAPO-34	$\Delta\omega(O2/O4)$	10	0	27	30

Table 3. Calculated splittings of HF/LF OH vibrations in CHA zeotype systems, compared experiment.[22, 23]

Demuth et al.[39] limited their calculations to only those acid sites or H-mordenite which are located in the large 12-ring associated with the HF band, located experimentally[63] at 3612 cm^{-1}. The average frequency of the 7 possible sites obtained with GGA functional theory calcculations was at 3593 cm^{-1}, about 20 cm^{-1} below the experimental value. No calculations were performed for the OH groups vibrating in the 8-ring and in the side pocket, appearing experimentally at about 3586 cm^{-1}.

As mentioned before, the experimental assignment is usually based on the observation that the HF band is associated with a proton vibrating inside a larger pore, while the LF band is associated to OH groups located in small pores. Secondary H-bond formation with other oxygen atoms in the smaller ring is evoked as an explanation for the weakening of these O–H bonds, leading to a red-shift of the vibration.[3, 40] A critical test of this hypothesis showed that in the case of the HY zeolite the H$\cdots$O secondary contacts in the 6-ring are longer than typical H-bond distances, therefore it cannot be considered as a satisfactory rationale for the HF/LF splitting.[36]

It was suggested also that the local geometry, in particular the Al–O–Si angle could be correlated to the OH frequency; a hypothesis supported by ab initio calculations on small clusters, where the regression coefficient is practically 1.0.[36] Several authors attempted to verify this correlation, which was found in most cases moderately good.[34, 39] Even if the general trend (smaller bond angle – higher frequency) was confirmed, the regression coefficients were not high enough to consider the correlation truly convincing. From these observations it has been concluded that local environment of the acid site cannot be the main factor determining acidities.[35]

Among the non-local criteria, the possible role of the long range electrostatic interactions was recognized quite early on. On the basis of a simple energy decomposition scheme, where the OH bond energy is considered as the sum of an intrinsic bond energy and an electrostatic term, it was proposed that the difference of the OH force constants at different acid sites could be related to the difference of the electric field gradient along the OH stretch coordinate.[64] Recent empirical force field studies on a dozen different zeolites and SAPOs illustrated that the OH frequency shifts do not depend on the cavity size, but the norm of the electric field calculated at the proton is very well correlated with it.[65]

4. Protonation of water by Brønsted acid sites

The interaction of the Brønsted sites with various probe molecules constitutes a valuable tool for characterizing their acid strengths. Strong bases, like ammonia, induce an immediate proton transfer from the OH group and the formation of an NH_4^+ cation. Weak bases, like CO, form weak H-bonded complexes and the characteristic spectroscopic changes, such as the shift of the OH stretching vibration, are considered as a sensitive measure of the acid strength. While in these cases the chemical nature of the probe molecule after interaction with the acid site is well established, the nature of the water molecule interacting with an acid proton in zeolitic materials is much less clear. In simple terms, the main question is whether the water molecule is *physisorbed*, without altering its identity as H_2O, or the proton of the acid site is transferred to the water to form hydronium ions, H_3O^+, a situation which is referred to as a *chemisorbed* water molecule. The debate around this question illustrates well the complementary role that theory and experiment can play to understand a seemingly simple, nevertheless quite complex physicochemical situation.

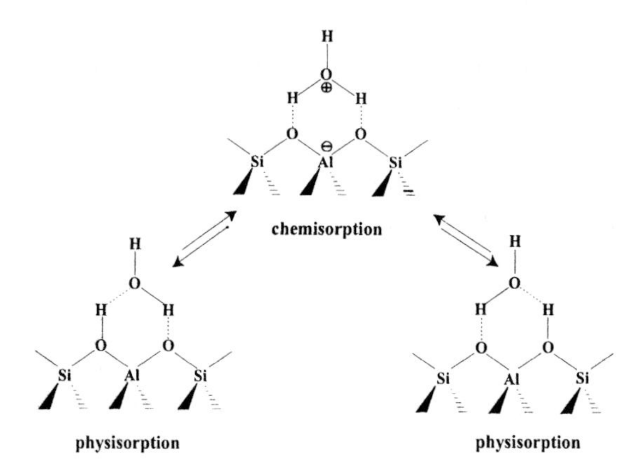

Figure 5. Schematic illustration of physisorbed and chemisorbed water on an acid site.

It has been observed by infrared spectroscopy that water adsorption is associated with the disappearance of the zeolitic OH vibration at about 3600 cm^{-1} and four new characteristic bands appear: a sharper band at 3660 cm^{-1}, associated with the OH stretching mode of the adsorbed water molecule, and the so-called (A,B,C) trio at $\approx$2900 cm^{-1}, at $\approx$2400 cm^{-1} and at $\approx$1600 cm^{-1}, constituted from much broader bands. This spectrum was attributed to the formation of adsorbed hydronium species, H_3O^+, with the A and B bands associated with the antisymmetric and symmetric vibrations of the hydronium OH groups.[66, 67] The plausibility of this model was questioned on the basis of IR spectra of H_2O and D_2O adsorbed on HZSM-5 and DZSM-5, arguing for only a partial proton transfer towards the water.[68] Another interpretation of these spectral features, relying on the hypothesis of a physisorbed water molecule, was proposed on the basis quantum chemical calculations,[69] and confirmed by further FTIR studies on a $H_2^{18}O$.HZSM-5 system.[70] Accordingly, these bands would result from Fermi resonance between the inhomogeneously broadened OH stretching mode with the overtones of the OH bending modes of the zeolite. An alternative interpretation of the IR bands, without invoking the contribution of Fermi resonances, has been suggested from the analysis of CPMD simulations on water in sodalite.[71]

High-quality ab initio calculations of various cluster models,[72, 73] were in favor of the view that water is only physisorbed on the zeolite and found that the proton-transferred hydronium ion structure is a transition state between two minima corresponding to the water H-bonded to the acid site (cf. Fig. 5).

Further support for the absence of hydronium species was provided by inelastic neutron scattering measures on HZSM-5.[74] This vibrational technique allows one to calculate the intensities with a reasonable accuracy from atomic displacements. Comparing the intensities obtained from quantum chemical cluster model calculations[72] for the two possible adsorption mechanisms, it was found that a water molecule attached to the acid site by two hydrogen bonds is in good agreement with the experimental intensities, while the chemisorbed model with the formation of the hydronium ion is out of question, at least for this particular framework structure. The interpretation of various NMR data obtained for water-loaded zeolites seems to be more ambiguous. ^{1}H MAS NMR experiments were interpreted by a fast proton exchange between water molecules, bridging OH groups and hydronium ions,[75] and a measure of the concentration of the H_3O^+ ions was proposed from the analysis of the wide line ^{1}H NMR spectra of hydrated samples.[76, 77]

Quantum chemical cluster calculations have pointed out also that, although the proton transfer to a single water molecule was not favored, it was possible towards water dimers,[72] suggesting that at low water concentrations the H_2O must be physisorbed, while at a higher water loading the formation of hydronium ions is not excluded.

The validity of these quantum chemical results were questioned seriously by low-temperature neutron diffraction studies on HSAPO-34 loaded with water in an equivalent amount to the number of the acid protons.[78] The structure obtained by the Rietveld analysis of the powder diffraction data put into evidence the existence of H_3O^+ ions, H-bonded to the acid site O1, co-existing with intact water molecules physisorbed on the O2 acid site. Obviously, these experimental data were in sharp contrast with quantum chemical calculations. In addition to the cluster model calculations,[72] results on embedded cluster models,[79] corrected for long range electrostatic interactions, as well as CPMD/PAW plane wave DFT calculations on analogous zeolitic systems[62] lead to conclusion that the proton cannot be transferred from the acid site to a single water molecule. After the publication of the experimental structure of the hydrated HSAPO-34, provocative accounts about the apparent failure of the theoretical approach appeared in the chemical press. One of these reports, entitled *Quantum mechanics proved wrong*,[80] had at least the merit to stimulate theoretical chemists to answer to this challenge by attempting to improve models that are the closest possible to the experimental situation.

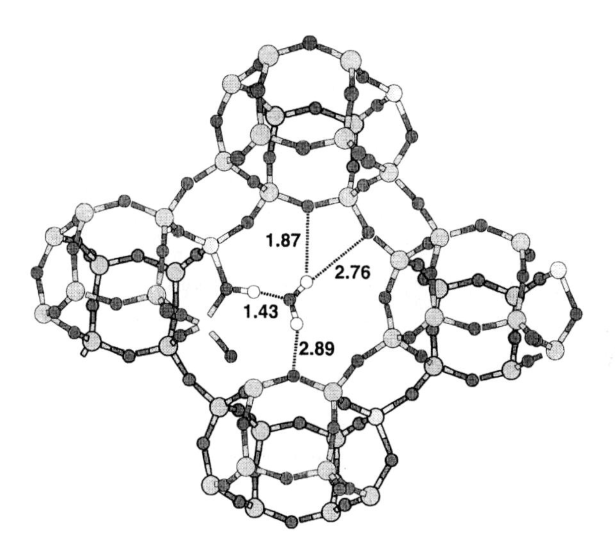

Figure 6. Optimized structure of a single water molecule in the HSAPO-34 framework.

The most obvious improvement of the model was to include the full zeolite structure in ab initio pseudopotential plane wave DFT calculations. However, the attempts to localize proton transferred HSAPO-34.H_2O complexes have proved to be unsuccessful: the water molecule remained physisorbed.[81] A typical structure, selected from the numerous local minima, is shown in Fig. 6, illustrating that topologically distant framework oxygens participate also in the stabilization of such a complex. A correct account of this feature would necessitate quite extended cluster models, which are usually limited to 2 or 3 oxygen atoms in the vicinity of the acid site.

Ab initio molecular dynamics studies on the monohydrated HSAPO-34 model confirmed the impossibility of proton transfer in this case: after equilibration the acid proton remains on the framework oxygen and vibrates with an amplitude of about 0.2 Å around a mean distance of 1.05–1.06 Å. The center of the water molecule fluctuates between 2.4 and 2.8 Å from the acid oxygen atom and even in the moments of its closest approach the proton does not get nearer than 1.4 Å to the water oxygen. The situation changes dramatically in similar simulations performed for the di-hydrated HSAPO-34. The mean OH distance becomes only slightly longer, 1.11 Å, but it fluctuates with an amplitude, which is doubled with respect to the mono-hydrated case. At several points of the trajectory (see Fig. 7) the proton gets closer to the water than to the framework oxygen, indicating the possibility of forming an hydronium ion.[81]

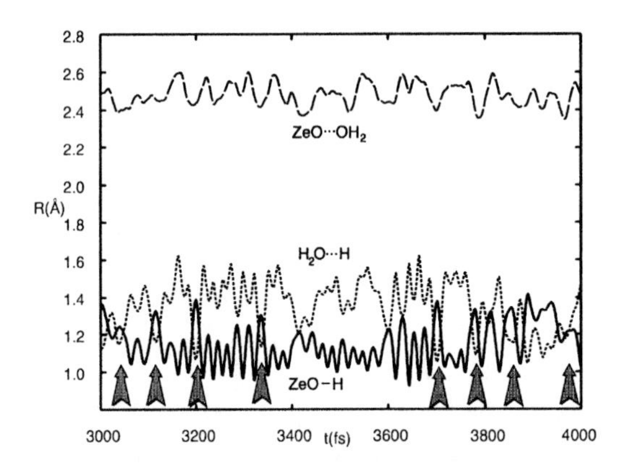

Figure 7. A piece of the HSAPO-34.2H_2O molecular dynamics trajectory.[81] The arrows indicate the proton transfer events.

In the light of these molecular dynamics simulations one could attempt to reinterpret the experimental observations of Smith et al. The stoichiometric proportion of H_2O molecules and acid protons should be understood globally and it does not mean necessarily that each acid proton has its own water partner. A local increase of the water concentration cannot be excluded and these configurations favor the formation of $H_5O_2^+$ species. While the protonated water molecule has a relatively fixed, locked position near the acid site, giving rise to a sufficiently high diffraction intensity, it is not surprising that a second, much more fluxional water would be not visible in neutron diffraction experiments.[81]

CPMD simulations were performed on the same hydrated HSAPO-34 system using a larger, doubled unit cell.[82] This larger simulation cell allowed for the examination of various arrangements of the acid sites and construct models which are as close as possible to the experimental acid site and water concentrations. In particular, the stabilization of the proton transfer complex by one and two additional water molecules was also examined.[82] It was concluded that the presence of two stabilizing water molecules is mandatory to obtain a complete proton transfer towards the water oxygen. Finally, both of these studies confirmed that quantum mechanics was not proven wrong, on the contrary, it is an indispensable tool for obtaining the correct physical picture behind the experimental data.

The factors that govern the protonation equilibrium have been summarized[82] as: (i) proton affinity of the water molecule or of the $(H_2O)_n$ cluster, (ii) acid strength of the zeolitic material, (iii) relative stabilization of the neutral and protonated water cluster by H-bond interactions with the framework oxygens. Factors (ii) and (iii) depend on the framework constitution (alumino-silicate or SAPO) and on the framework geometry (size of the cavities). Therefore it is not surprising that the conditions of hydronium ion formation may depend on the nature of the zeolite. In AIMD simulations, using GGA functional and PAW techniques, spontaneous proton transfer has been observed for a single water molecule in the gmelinite framework,[51] and even water-mediated proton jumps between different acid sites could be observed during the picosecond long trajectory. Proton exchange reaction has been observed in the ab initio simulations of methanol in sodalite, too.[71] CPMD calculations using GGA and PAW technique on the sodalite and chabazite structure found that a minimum of two water molecules was necessary to observe proton transfer from the acid site to the water molecule.[62, 71]

The conditions for protonation of the methanol molecule show strong similarities to the case of water. Most of the ab initio pseudopotential or PAW plane wave geometry optimizations and dynamical simulations agree that the proton transfer occurs only at higher methanol coverages. Various framework structures were explored, like SOD,[55, 62, 71] CHA,[45, 46, 52, 83, 84, 85, 86] FER,[45, 46, 85], MFI[45, 46, 85, 86] and TON.[46]

5. Protolytic cracking of alkanes

Cracking of small saturated hydrocarbons, catalyzed by zeolites, can proceed via two mechanisms,[87, 88] both involving carbocations: the bimolecular chain reaction, which involves carbenium ions that are further transformed by β-scission, and the unimolecular protolytic mechanism, involving alkanium ions that are formed by the direct protonation of the alkane by the Brønsted acid OH groups of the catalyst. This latter mechanism, originally proposed by Haag and Dassau, is the predominant one at about 800 K in medium-pore zeolites, like HZSM-5, which favor monomolecular reactions.[89] While relatively few theoretical studies have been done on the bimolecular, β-scission mechanism,[90, 91] the protolytic mechanism has been the subject of numerous quantum chemical studies. "Theory has advanced to the point that it is now having a significant impact on the development of the understanding of alkane cracking mechanisms" wrote Jentoft and Bates in their 1997 review paper on the subject.[87]

The overwhelming majority of the theoretical studies were performed on cluster models of the catalytic site. In spite of the fact that the role of space confinement and the secondary interactions with the framework atoms is well-known, there are only a few electronic structure calculations on lattice models involving hydrocarbons, using either periodic DFT calculations,[92, 93] or embedding methods.[94, 95] In this brief account of the subject we attempt to overview some of the recent computational results of the literature and present some new data obtained from ab initio DFT pseudopotential plane wave calculations on C_1 - C_4 alkanes in the chabazite framework.

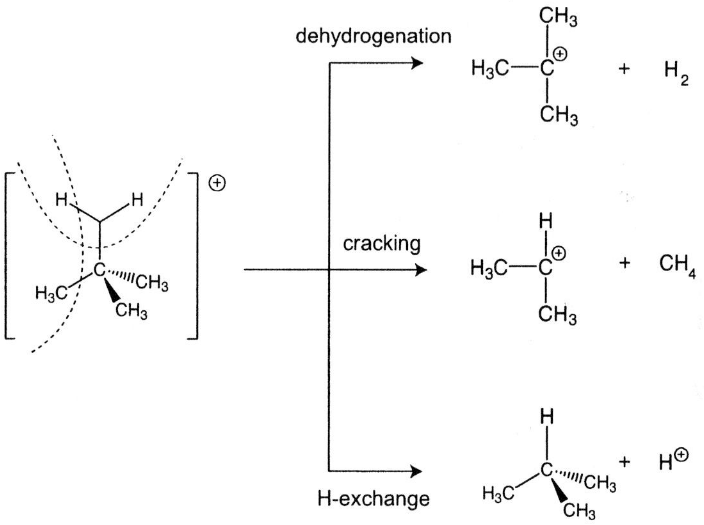

Figure 8. Reaction routes of a protonated alkane molecule.

The key step of the Haag-Dassau protolytic cracking mechanism is the formation of a pentacoordinated carbocation, the alkanium ion. As illustrated on Fig. 8, this ion can undergo different transformations: (i) H-exchange, by releasing a proton to another framework oxygen, (ii) dehydrogenation, leading to alkenes of the same chain length, (iii) cracking, leading to smaller saturated and unsaturated hydrocarbons. The common precursor for all of these reaction channels is the physisorbed, H-bonded alkane molecule. The energy difference between the protonated transition state and the physisorbed complex is the *intrinsic* activation energy. It should be distinguished from the *apparent* activation energy, which takes into account the energy of physisorption, according to the scheme on Fig. 9.

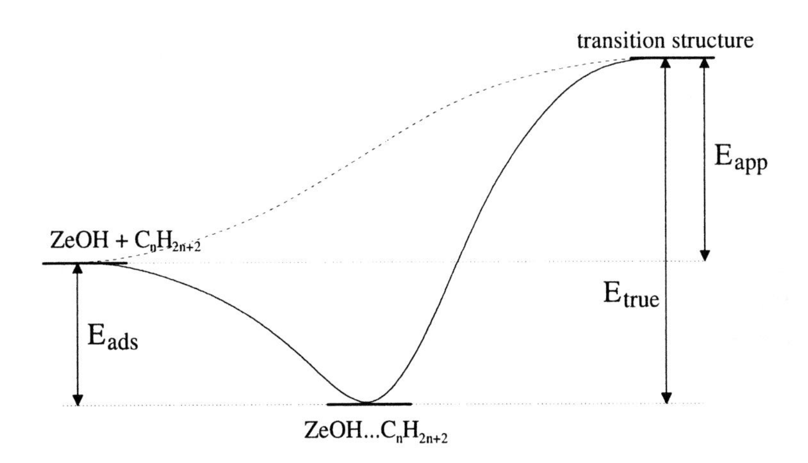

Figure 9. Schematic illustration of the apparent and intrinsic (true) activation energy and their relationship with the adsorption energy.

Lercher et al.[96] have shown that the intrinsic activation energy of the protolytic cracking of light n-alkanes over HZSM-5 is practically independent on the carbon chain length and the increase in the rate of reaction (decrease of the apparent activation energy) is essentially due to the increase of the adsorption energy. A recent study of the monomolecular cracking of n-hexane over three different zeolites, H-Y, HZSM-5 and H-MOR have shown that the differences in the apparent activation energy can be attributed essentially to the differences in the heat of adsorption of n-hexane on the zeolite frameworks.[97] It follows from these observation that a complete theoretical study of hydrocarbon cracking should correctly account for both energetic components: the physisorption energy and the intrinsic activation energy. Furthermore, if one wishes to model transition-state shape selectivity, the use of complete zeolite models, including steric interactions with the zeolite cavities, is mandatory.

5.1 Physisorption of alkanes

The physisorption of alkanes in zeolites is essentially due to van der Waals forces. The main contributions to the stabilization are the induction energy, roughly speaking the electrostatic interaction of the dipole moment induced by the electric field of zeolite with the framework charge distribution and the dispersion energy, the universal quantum mechanical attracting force due to the fluctuating charge distributions of both the framework and the alkane molecules. Both energy components are proportional to the polarizability of the alkane, which increases linearly with the chain length.

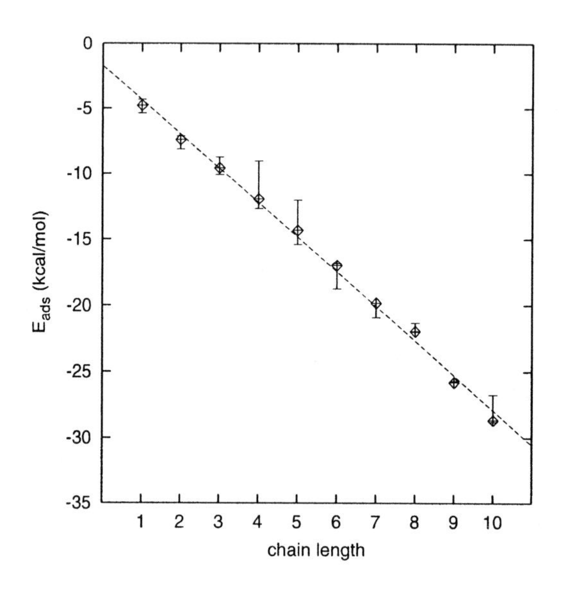

Figure 10. Experimental heats of adsorption of n-alkanes after Vlugt at al.[98]

For instance, the experimental heats of adsorption of $C_4 - C_9$ alkanes, obtained by temperature-programmed desorption techniques in silicalite/ZSM-5, increases linearly with the chain length.[99] On the basis of a compilation of available experimental data on silicalite/ZSM-5, Vlugt et al.[98] proposed a set of "best" heats of adsorption for the $C_1 - C_{10}$ n-alkanes. From these data an average increment of about 2.62±0.06 kcal/mol per carbon atom could be deduced (cf. Fig. 10). The effect of the framework structure has been demonstrated by calorimetric measurements of isosteric heats of adsorption of methane, ethane and propane in high-silica zeolites.[100] In a homologous series of frameworks with one-dimensional channels of different effective diameter, the heat of adsorption of methane varied from 6.5 (small pore TON framework) to 3.4 (large-pore UTD-1 zeolite).[100] A chromatographic study of the adsorption of n-alkanes confirmed the trend that a smaller pore size leads to a strong increase of the adsorption enthalpy with the chain length.[101] Using empirical potential energy functions in configurational-bias Monte Carlo simulations, very good agreement could be obtained for heats of adsorption[6] and for adsorption isotherms.[98] Furthermore, these simulations offer the possibility to predict the preferred adsorption modes as a function of the alkane chain length.[102] Molecular mechanics optimization, molecular dynamics and Monte Carlo techniques were compared for the adsorption energies of light alkanes in silicalite,[103] showing the near equivalence of these simulation approaches, provided that an appropriate force field is used.

In contrast to the quantitatively correct adsorption energies obtained from parameterized potential energy functions, quantum chemical approaches are not always able to describe appropriately van der Waals forces. Dispersion forces are related to intermolecular electron correlation, which is completely missing from Hartree-Fock theory. Therefore, correlation effects should be included by some means, such as MP2 theory using at least triple-zeta + polarization basis, in order to obtain acceptable adsorption energies between a hydrocarbon and a cluster model of the zeolite.[95] Note that the size of the zeolite cluster model has a noticeable effect on the calculated complexation energies. Present-day density functionals fail to account for dispersion effects. Even if some authors claim that specific choices of the functionals are able to describe the van der Waals minima of weakly bound complexes, it seems to be clear that the *long-range behavior* of the intermolecular potential energy curve is always incorrect. The proper asymptotic limit (R^{-6}) cannot be reproduced without including specific van der Waals terms in the functional, which depends on the frequency-dependent charge density susceptibility function.

5.2 Transition state

The alkanium ions, formed by the direct attack of the Brønsted acid proton on the C–C or C–H bond of the alkane, although detected as a highly energetic species in gas phase, can exist only as transition structures, stabilized by the interaction with the zeolite surface. The carbocation component of the transition structure can be either an alkanium type, RH_2^+, like for the H-exchange reaction, or rather a carbenium type, R^+, like for the dehydrogenation and cracking reaction channels.[104] Following the proposal of Kazansky and Senchenya,[105] it is now widely accepted that only alkoxide structures constitute stable intermediates and carbocations appear exclusively in unstable transition structures. Experimental support for the existence of carbenium ions, covalently bound to the framework oxygen to form alkoxy-like structures, was provided by ^{13}C NMR spectroscopy.[106] This postulate has been questioned quite recently by DFT calculations on cluster zeolite models reacting with isobutane, demonstrating the possibility of a dehydrogenation reaction path which proceeds towards isobutene without alkoxide intermediate.[107]

Ab initio DFT pseudopotential plane wave calculations were performed to study the transition structures and the corresponding activation energies for methane, ethane, propane, n-butane and isobutane in the chabazite framework, which can be considered as a model of a small-pore zeolite catalyst.[108]

In contrast to the cluster calculations, where the active site model is usually symmetric with respect to the central Al-atom, the framework oxygens are not equivalent in zeolite structures. In the case of the simplest reaction, the H-exchange reaction of methane, all possible oxygen neighbors of a given acid site may be candidate for the proton-exchange. Among the possible arrangements, only one of the adsorption complexes, involving the O4 and O2 framework oxygens, is exothermic, the other combinations are several kcal/mol less stable. The intrinsic activation energy was found to be 26.0 kcal/mol. The structures of the transition state and the left and right adsorption complexes are illustrated on Fig. 11.

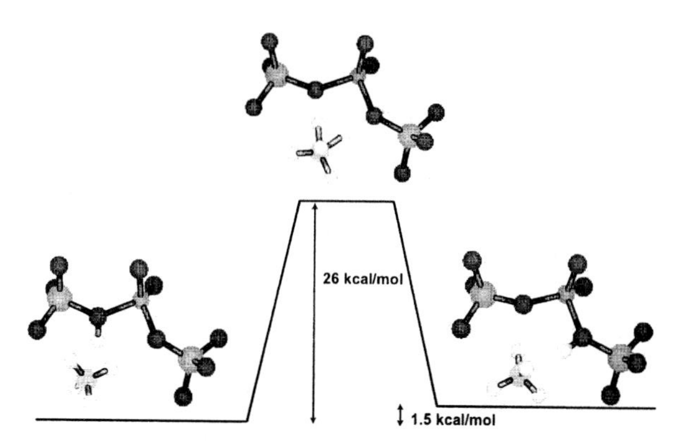

Figure 11. H-exchange reaction of the methane between the O2 and O4 oxygens of the framework.

In the case of cracking reactions the proton should attack a C–C bond. The main question is whether the alkanium ion is a short-lived transition state or whether it is a stable intermediate corresponding to a minimum of the potential energy surface. A quite exhaustive ab initio study on a cluster model, involving *a posteriori* long-range electrostatic corrections, has been performed on the ethane cracking in HZSM-5.[95] The transition state is an alkanium ion, formed by the protonation of the C–C bond and the calculated barrier, taking into account zero-point energy (-2 kcal/mol), thermal correction for a reaction temperature of 773 K (-1 kcal) and long-range electrostatic correction (-15 kcal/mol), was found to be 54.1 kcal/mol. This is only slightly higher than the experimental value of 47$\pm$3 kcal/mol, measured for propane and n-butane in HZSM-5.[96] As mentioned before, the intrinsic reaction barrier seems to be constant for the n-alkane series, therefore the ethane activation energy should also be close to this value.

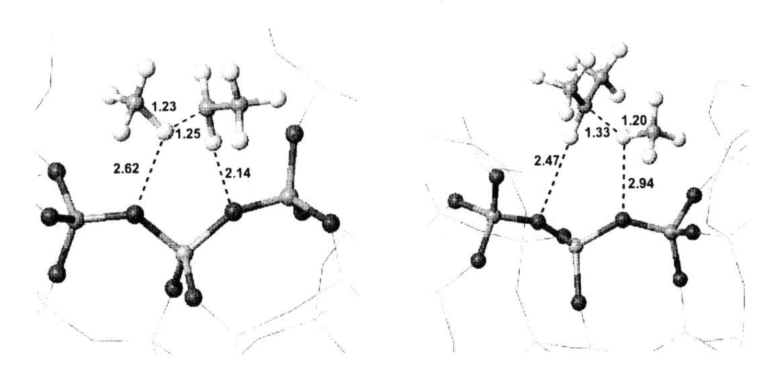

Figure 12. Transition structures of propane and n-butane cracking in the chabazite framework.

The transition structures, shown in Fig. 12, found in the ab initio DFT-GGA pseudopotential plane wave calculations for ethane, propane and n-butane[108] are is qualitative agreement with the above discussed results of Zygmunt et al.[95] The intrinsic reaction barriers (without vibrational corrections) are between 42.9 and 51.5 kcal/mol, in quite good agreement with the experimental value (cf. Table 4). The transition state formed by the attack of the proton on the central C–C bond is slightly smaller (37.3 kcal/mol) in harmony with the expected higher stability of the alkanium ion. It is worth remarking that the n-butane could not fit to the small pore of the chabazite framework, therefore it had to adopt a gauche conformation in order to be able to be exposed to the attack of the Brønsted acid site.

bond	C_2H_6	C_3H_8	n-$C_4H_{10}^a$	n-$C_4H_{10}^b$	i-C_4H_{10}
H1–O	3.67	2.62	3.01	3.25	2.94
H1–C1	1.23	1.20	1.20	1.25	1.14
H1–C2	1.25	1.32	1.33	1.27	1.58
H2–C2	1.09	1.11	1.10	1.10	1.10
H2–O'	2.23	2.14	3.44	2.31	2.79
C1-C2	1.96	2.08	2.12	2.47	2.47
Al–O	1.74	1.74	1.74	1.73	1.73
Al–O'	1.73	1.74	1.72	1.73	1.73
Barrier	51.49	42.89	44.41	37.26	35.93

Table 4. Some geometrical features of the alkane/acid site transition states for protolytic cracking. (a) cleavage on the primary C–C bond; (b) cleavage on the secondary C–C bond.

H-abstraction reactions occur preferably on tertiary carbon centers. Therefore isobutane is a good candidate for studying this reaction channel. The geometry of the adsorbed complex, as well as that of the transition structure (Fig. 13) is qualitatively similar to the structures found by Milan and

Nascimento[107] in DFT cluster calculations. The computed intrinsic activation energy (45.16 kcal/mol) is in reasonable agreement with the value of 42.4 kcal/mol that can be deduced from recent adsorption heat measurements of isobutane in silicalite[109] (13.0 kcal/mol) and apparent activation energies for dehydrogenation[110] (29.4 kcal/mol). Again, the cluster model calculations yield a slightly high activation energy[107] of 53 kcal/mol.

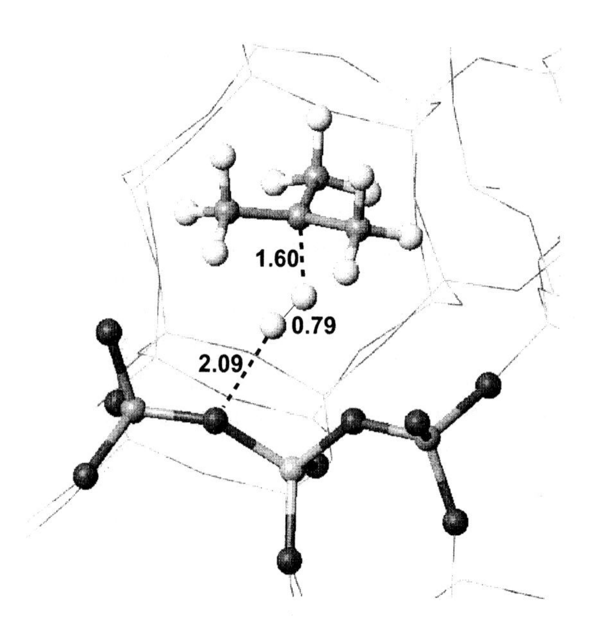

Figure 13. Transition structure of the isobutane dehydrogenation reaction in chabazite.

6. Conclusions

Progress in ab initio methodology and in computer technology during the last 5 years made it possible to do large-scale ab initio simulations on chemically meaningful models of relatively complex reaction systems, such as the hydrocarbon transformation on industrially interesting zeolite catalysts. Although some methodological improvements are necessary in order to correctly take into account weak intermolecular forces, which contribute significantly to the physisorption of alkanes in zeolite pores, these calculations performed on periodic models of the catalyst have the great advantage of automatically including practically all physical interactions on the same footing. This is a considerable advantage with respect to the popular QM/MM embedding methods, where the choice of the interaction potential and the quantum and classical subsystems may be biased.

It is most probable that in the next few years we are going to witness an exponential increase of these kinds of ab initio molecular dynamics simulations on more and more complicated catalytic systems, exploring complex reaction mechanisms. One of the challenges in this respect is the use of special molecular dynamics simulation techniques adapted to activated processes, which could be the appropriate tool for the theoretical prediction of rate constants.

In the mean time, with the accumulating theoretical data, one will certainly need to rethink simple conceptual models, which are able to establish structure/reactivity, structure/catalytic activity relationship without performing lengthy and still costly dynamical simulations on each and every new variant of a given reaction type.

Acknowledgments

The authors thank the Research Group CNRS G1209 *Ab Initio Molecular Dynamics Applied to Catalysis* for supplying a stimulating scientific environment to accomplish this work. The collaboration on the use of ab initio pseudopotential methods was particularly fruitful with Profs Jürgen Hafner and Georg Kresse (Universität Wien). Discussions with Dr. L. Benco (Wien) and Prof. M. A. C. Nascimento (Rio de Janeiro) are gratefully acknowledged.

References

1. Sauer, J. *Chem. Rev*. **1989,** *89,* 199.
2. Sauer, J.; Ugliengo, P.; Garrone, E.; Saunders, V. R. *Chem. Rev*. **1994,** *94,* 2095.
3. van Santen, R. A.; Kramer, G. J. *Chem. Rev*. **1995,** *95,* 637.
4. Frenkel, D.; Smit, B. *Understanding molecular simulations: from algorithms to applications;* Academic Press: Boston, 1996.
5. Chandler, D. *J. Chem. Phys.* **1978,** *68,* 2959.
6. Bates, S. P.; v. Well, W. J. M.; v. Santen, R. A.; Smit, B. *J. Am. Chem. Soc.* **1996,** *118,* 6753.
7. Smit, B.; Loyens, L. D. J. C.; Verbist, G. L. M. M. *Faraday Discuss.* **1997,** *106,* 93.
8. Catlow, C. R. A.; Ackermann, L.; Bell, R. G.; Gay, D. H.; Holt, S.; Lewis, D. W.; Nygren, M. A.; Sastre, G.; Sayle, D. C.; Sinclair, P. E. *J. Mol. Catal. A* **1997,** *115,* 431.
9. Sastre, G.; Fornes, V.; Corma, A. *J. Chem. Soc., Chem. Commun.* **1999,** 2163.
10. Sierka, M.; Sauer, J. *Faraday Discuss.* **1997,** *106,* 41.
11. Car, R.; Parrinello, M. *Phys. Rev. Lett.* **1985,** *55,* 2471.

12. Remler, D. K.; Madden, P. A. *Mol. Phys.* **1990,** *70,* 921.
13. Payne, M. C.; Teter, M. P.; Allan, D. C.; Arias, T. A.; Joannopoulos, J. D. *Rev. Mod. Phys.* **1992,** *64,* 1045.
14. Vanderbilt, D. *Phys. Rev. B.* **1990,** *41,* 7892.
15. Blöchl, P. E. *Phys. Rev. B.* **1994,** *50,* 17953.
16. Kresse, G.; Joubert, D. *Phys. Rev. B.* **1999,** *59,* 1758.
17. Kresse, G.; Furthmüller, J. *Comp. Materials Science* **1996,** *6,* 15.
18. *Structure Commission of the International Zeolite Association: Database of Zeolite Structures;* http://www.iza-structure.org/databases/: URL, 2001.
19. Cheetham, A. K.; Eddy, M. M.; Thomas, J. M. *J. Chem. Soc., Chem. Commun.* **1984,** 1337.
20. Czjek, M.; Jobic, H.; Fitch, A. N.; Vogt, T. *J. Phys. Chem.* **1992,** *96,* 1535.
21. Bull, L. M.; Cheetham, A. K.; Hopkins, P. D.; Powell, B. M. *J. Chem. Soc., Chem. Commun.* **1993,** 1196.
22. Smith, L. J.; Cheetham, A. K.; Marchese, L.; Thomas, J. M.; Wright, P. A.; Chen, J.; Gianotti, E. *Catal. Lett.* **1996,** *41,* 13.
23. Smith, L. J.; Davidson, A.; Cheetham, A. K. *Catal. Lett.* **1997,** *49,* 143.
24. Martucci, A.; Alberti, A.; Cruciani, G.; Radaelli, P.; Ciambelli, P.; Rapacciuolo, M. *Microporous and Mesoporous Materials* **1999,** *30,* 95.
25. Martucci, A.; Cruciani, G.; Alberti, A.; Ritter, C.; Ciambelli, P.; Rapacciuolo, M. *Microporous and Mesoporous Materials* **2000,** *35-36,* 405.
26. Hunger, M. *Catal. rev. - Sci. Eng.* **1997,** *39,* 345.
27. Brunner, E. *Catal. Today* **1997,** *38,* 361.
28. Ito, M.; Shimoyama, Y.; Saito, Y.; Tsurita, Y.; Otake, M. *Acta Crystallogr., Sect. C* **1985,** *41,* 1898.
29. Sastre, G.; Lewis, D. W.; Catlow, C. R. A. *J. Phys. Chem. B* **1997,** *101,* 5249.
30. Brändle, M.; Sauer, J.; Dovesi, R.; Harrison, N. M. *J. Chem. Phys.* **1998,** *109,* 10379.
31. Dovesi, R.; Saunders, V. R.; Roetti, C.; Causà, M.; Harrison, N. M.; Orlando, R.; Aprà, E. "CRYSTAL95 User's Manual", Technical Report, University of Torino, 1996.
32. Sierka, M.; Sauer, J. *J. Chem. Phys.* **2000,** *112,* 6983.
33. Shah, R.; Gale, J. D.; Payne, M. C. *Phase Transitions* **1997,** *61,* 67.
34. Jeanvoine, Y.; Ángyán, J. G.; Kresse, G.; Hafner, J. *J. Phys. Chem. B* **1998,** *102,* 5573.

35. Sastre, G.; Lewis, D. W. *J. Chem. Soc., Faraday Trans.* **1998,** *94,* 3049.
36. Eichler, U.; Brändle, M.; Sauer, J. *J. Phys. Chem. B* **1997,** *101,* 10035.
37. Alberti, A. *Zeolites* **1997,** *19,* 411.
38. Brändle, M.; Sauer, J. *J. Am. Chem. Soc.* **1998,** *120,* 1556.
39. Demuth, T.; Hafner, J.; Benco, L.; Toulhoat, H. *J. Phys. Chem. B* **2000,** *104,* 4593.
40. Campana, L.; Selloni, A.; Weber, J.; Goursot, A. *J. Phys. Chem. B* **1997,** *101,* 9932.
41. Chen, H. T.; Wang, X. K.; Luo, L. W.; Yuan, Y. Z.; Wang, Z. J.; Ding, T. D.; Li, X. H.; Hu, C. *Chem. Phys. Lett.* **1996,** *252,* 375.
42. Alberti, A.; Cruciani, G.; Galli, E.; Vezzalini, G. *Zeolites* **1996,** *17,* 457.
43. Benco, L.; Demuth, T.; Hafner, J.; Hutschka, F. *J. Chem. Phys.* **1999,** *111,* 7537.
44. Larin, A. V.; Vercauteren, D. P. *J. Mol. Catal. A* **2001,** *168,* 123.
45. Stich, I.; Gale, J. D.; Terakura, K.; Payne, M. C. *J. Am. Chem. Soc.* **1999,** *121,* 3292.
46. Haase, F.; Sauer, J. *Microporous and Mesoporous Materials* **2000,** *35-36,* 379.
47. Kramer, G. J.; van Santen, R. A. *J. Am. Chem. Soc.* **1993,** *115,* 2887.
48. Deka, R. C.; Vetrivel, R.; Pal, S. *J. Phys. Chem. A* **1999,** *103,* 5978.
49. Deka, R. C.; Roy, R. K.; Hirao, K. *Chem. Phys. Lett.* **2000,** *332,* 576.
50. Pal, S.; Chandrakumar, K. R. S. *J. Am. Chem. Soc.* **2000,** *122,* 4145.
51. Benco, L.; Demuth, T.; Hafner, J.; Hutschka, F. *Chem. Phys. Lett.* **2000,** *324,* 373.
52. Shah, R.; Gale, J. D.; Payne, M. C. *J. Chem. Soc., Chem. Commun.* **1997,** 131.
53. Kobe, J. M.; Gluszak, T. J.; Dumesic, J. A.; Root, T. W. *J. Phys. Chem.* **1995,** *99,* 5485.
54. Nicholas, J. B.; Hess, A. C. *J. Am. Chem. Soc.* **1994,** *116,* 5428.
55. Schwarz, K.; Nusterer, E.; Margl, P.; Blöchl, P. E. *Int. J. Quantum Chem.* **1997,** *61,* 369.
56. Larin, A. V.; Vercauteren, D. P. *Int. J. Quantum Chem.* **2001,** *82,* 182.
57. Senchenya, I. N.; Garrone, E.; Ugliengo, P. *J. Mol. Struct. (Theochem)* **1996,** *368,* 93.
58. Bates, S. P.; Dwyer, J. *Chem. Phys. Lett.* **1994,** *225,* 427.
59. Shah, R.; Gale, J. D.; Payne, M. C. *J. Phys. Chem.* **1996,** *100,* 11688.
60. Kazansky, V. B.; Kustov, L. M.; Borokov, V. Y. *Zeolites* **1983,** *3,* 77.

61. Garrone, E.; Kazansky, V. B.; Kustov, L. M.; Sauer, J.; Senchenya, I. N.; Ugliengo, P. *J. Phys. Chem.* **1992,** *96,* 1040.
62. Nusterer, E.; Blöchl, P. E.; Schwarz, K. *Chem. Phys. Lett.* **1996,** *253,* 448.
63. Makarova, M. A.; Wilson, A. E.; v. Liemt, B. J.; Mesters, C. M. A. M.; d. Winter, A. W.; Williams, C. *J. Catal.* **1997,** *172,* 170.
64. Ferenczy, G. G.; Ángyán, J. G. *J. Chem. Soc. Faraday Trans.* **1990,** *86,* 3461.
65. Lewis, D. W.; Sastre, G. *J. Chem. Soc., Chem. Commun.* **1999,** 349.
66. Jentys, A.; Warecka, G.; Derewinski, M.; Lercher, J. A. *J. Phys. Chem.* **1989,** *93,* 4837.
67. Marchese, L.; Chen, J.; Wright, P. A.; Thomas, J. M. *J. Phys. Chem.* **1993,** *97,* 8109.
68. Parker, L. M.; Bibby, D. M.; Burns, G. R. *Zeolites* **1993,** *13,* 107.
69. Pelmenshikov, A. G.; van Santen, R. A. *J. Phys. Chem.* **1993,** *97,* 10678.
70. Wakabayashi, F.; Kondo, J. N.; Domen, K.; Hirose, C. *J. Phys. Chem.* **1996,** *100,* 1442.
71. Schwarz, K.; Nusterer, E.; Blöchl, P. E. *Catal. Today* **1999,** *50,* 501.
72. Krossner, M.; Sauer, J. *J. Phys. Chem.* **1996,** *100,* 6199.
73. Zygmunt, S. A.; Curtiss, L. A.; Iton, L. E.; Erhardt, M. K. *J. Phys. Chem.* **1996,** *100,* 6663.
74. Jobic, H.; Tuel, A.; Krossner, M.; Sauer, J. *J. Phys. Chem.* **1996,** *100,* 19545.
75. Hunger, M.; Freude, D.; Pfeifer, H. *J. Chem. Soc., Faraday Trans.* **1991,** *87,* 657.
76. Batamack, P.; Dorémieux-Morin, C.; Fraissard, J.; Freude, D. *J. Phys. Chem.* **1991,** *95,* 3790.
77. Semmer-Herlédan, V.; Heeribout, L.; Batamack, P.; Dorémieux-Morin, C.; Fraissard, J.; Gola, A.; Benazzi, E. *Microporous and Mesoporous Materials* **2000,** *34,* 157.
78. Smith, L.; Cheetham, A. K.; Morris, R. E.; Marchese, L.; Thomas, J. M.; Wright, P. A.; Chen, J. *Science* **1996,** *271,* 799.
79. Greatbanks, S. P.; Hillier, I. H.; Burton, N. A. *J. Chem. Phys.* **1996,** *105,* 3770.
80. *Chem. Ind.* **1996,** *4,* 117.
81. Jeanvoine, Y.; Ángyán, J. G.; Kresse, G.; Hafner, J. *J. Phys. Chem. B* **1998,** *102,* 7307.

82. Termath, V.; Haase, F.; Sauer, J.; Hutter, J.; Parrinello, M. *J. Am. Chem. Soc.* **1998,** *120,* 8512.
83. Shah, R.; Gale, J. D.; Payne, M. C. *Int. J. Quantum Chem. Symp.* **1997,** *61,* 393.
84. Haase, F.; Sauer, J.; Hutter, J. *Chem. Phys. Lett.* **1997,** *266,* 397.
85. Stich, I.; Gale, J. D.; Terakura, K.; Payne, M. C. *Chem. Phys. Lett.* **1998,** *283,* 402.
86. Gale, J. D.; Shah, R.; Payne, M. C.; Stich, I.; Terakura, K. *Catal. Today* **1999,** *50,* 525.
87. Jentoft, F. C.; Gates, B. C. *Topics in Catalysis* **1997,** *4,* 1.
88. Corma, A.; Orchillés, A. V. *Microporous and Mesoporous Materials* **2000,** *35-36,* 21.
89. Kotrel, S.; Knözinger, H.; Gates, B. C. *Microporous and Mesoporous Materials* **2000,** *35-36,* 11.
90. Frash, M. V.; Kazansky, V. B.; Rigby, A. M.; van Santen, R. A. *J. Phys. Chem. B* **1998,** *102,* 2232.
91. Hay, P. J.; Redondo, A.; Guo, Y. *Catal. Today* **1999,** *50,* 517.
92. Boronat, M.; Zicovich-Wilson, C. M.; Corma, A.; Viruela, P. *Phys. Chem. Chem. Phys.* **1999,** *1,* 537.
93. Ugliengo, P.; Civalleri, B.; Dovesi, R.; Zicovich-Wilson, C. M. *Phys. Chem. Chem. Phys.* **1999,** *1,* 545.
94. Sinclair, P. E.; de Vries, A. H.; Sherwood, P.; Catlow, C. R. A.; van Santen, R. A. *J. Chem. Soc., Faraday Trans.* **1998,** *94,* 3401.
95. Zygmunt, S. A.; Curtiss, L. A.; Zapol, P.; Iton, L. E. *J. Phys. Chem. B* **2000,** *104,* 1944.
96. Narbeshuber, T. F.; Vinek, H.; Lercher, J. A. *J. Catal.* **1995,** *157,* 388.
97. Babitz, S. M.; Williams, B. A.; Miller, J. T.; Snurr, R. Q.; Haag, W. O.; Kung, H. H. *Appl. Catal. A* **1999,** *179,* 71.
98. Vlugt, T. J. H.; Krishna, R.; Smit, B. *J. Phys. Chem. B* **1999,** *103,* 1102.
99. Millot, B.; Methivier, A.; Jobic, H. *J. Phys. Chem. B* **1998,** *102,* 3210.
100. Savitz, S.; Siperstein, F.; Gorte, R. J.; Myers, A. L. *J. Phys. Chem. B* **1998,** *102,* 6865.
101. Denayer, J. F.; Baron, G. V.; Martens, J. A.; Jacobs, P. A. *J. Phys. Chem. B* **1998,** *102,* 3077.
102. van Well, W. J. M.; Cottin, X.; Smit, B.; van Hooff, J. H. C.; van Santen, R. A. *J. Phys. Chem. B* **1998,** *102,* 3952.
103. Nascimento, M. A. C. *J. Mol. Struct. (Theochem)* **1999,** *464,* 239.

104. Blaszkowski, S. R.; Nascimento, M. A. C.; van Santen, R. A. *J. Phys. Chem.* **1996,** *100,* 3463.

105. Kazansky, V. B.; Senchenya, I. N. *J. Catal.* **1989,** *119,* 108.

106. Haw, J. F.; Nicholas, J. B.; Xu, T.; Beck, L. W.; Ferguson, D. B. *Acc. Chem. Res.* **1996,** *29,* 259.

107. Milas, I.; Nascimento, M. A. C. *Chem. Phys. Lett.* **2001,** *338,* 67.

108. Parsons, D.; Ángyán, J. G. *to be published* **2001,** .

109. Sun, M. S.; Shah, D. B.; Xu, H. H.; Talu, O. *J. Phys. Chem. B* **1998,** *102,* 1466.

110. Yanping, S.; Brown, T. C. *J. Catal.* **2000,** *194,* 301.

MODELLING OF OXIDE-SUPPORTED METALS

M. ALFREDSSON*, S.T. BROMLEY† AND C.R.A. CATLOW
Davy Faraday Research Laboratory
The Royal Institution of Great Britain, 21 Albemarle Street,
London W1S 4BS, UK

1. Introduction

Supported metal catalysts generally show an increase in catalytic activity compared to the pure oxide or metal. Yet these systems are not well characterised, owing to the fact that such catalysts typically consist of a range of different supported metal sites, from small clusters to monolayer islands, all with non-uniform distributions in size and shape. One way to begin to understand such complex systems is to attempt to capture some essential part of the full system by developing model catalysts experimentally or using computer modelling techniques. This chapter concentrates on the latter but in the context of the relevant experimental data.

The experimental approach needed to understand supported metal cluster catalysts, is to prepare new catalysts based on systems of high uniformity e.g. supported clusters with identical size and composition. There are a number of ways of attaining this goal, using:

1. Size selected clusters from molecular beams deposited on 'planar' 2-dimensional (2-D) supports [1]
2. Ship-in-a-bottle synthesis of organometallic clusters (usually within porous hosts) followed by denuding of the organic ligands [2].
3. Adsorption of organometallic clusters on various supports followed by denuding of the organic ligands [3].

In the present paper we will discuss the metal-support interactions, using computational techniques. However, at a first sight even 'idealised' catalysts are difficult to study computationally due to the complex dynamical processes in the catalyst's preparation (e.g. cluster-surface impacts, constrained chemical synthesis, thermal ligand-denuding). To avoid some of these problems, one can use, as a guide, the results of experiment on the preparation and final state of such catalysts to make informed assumptions about the effects of these processes. Although from experiment it is known that these processes do lead to catalytic systems which are relatively well-defined, there are still a large number of potential degrees of freedom

† Present address: TOCK, DelftChemTech, TU Delft, Julianalaan 136, 2628 BL Delft, The Netherlands

M.A. Chaer Nascimento (ed.), Theoretical Aspects of Heterogeneous Catalysis, 109–147.

(e.g. variable cluster and support morphologies, different types of cluster-support interactions, variable electronic states of the clusters) to consider in any modelling study.

Our first example concerns bimetallic clusters supported on mesoporous SiO_2-based materials (see ***Fig. 1***), which provide a large internal area of contact with the metal owing to their large pore size. This property offers the possibility of creating isolated, catalytically active sites in a structurally well defined manner at the inner walls of the mesopores. Large collections of such bimetallic clusters can be attached to substrates (such as silica) and are used, for example, in petroleum refining for the dehydrogenation of hydrocarbons.

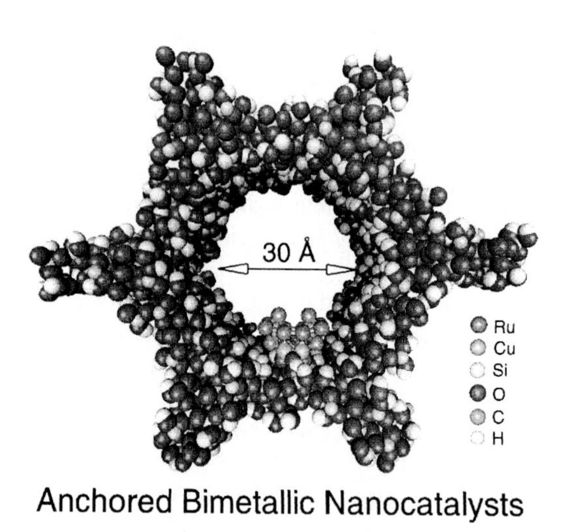

Fig. 1 Molecular graphic representation of a bimetallic cluster supported on MCM41 by R.G. Bell

For these systems, we investigate the technique of deriving clusters from organometallic precursors. Of the three preparative techniques listed above, the latter two are the most developed and have also proved to give well-defined catalysts with high activity and selectivity [4]. In addition to their catalytic performance such systems also have the following advantages for study via computational methods:

(i) As the final denuded clusters derive from identical precursor anions, the final distribution of cluster chemical composition and structure, after the removal of the carbonyl ligands may also assumed to be narrow. This *narrow distribution* allows us to concentrate on a few clusters with a chemical composition representative of the whole ensemble.

(ii) The clusters, after removal of their ligand sheaths, are very stable to sintering (especially when one of the constituents is either silver or copper [3,5]), resulting in discrete clusters thus validating the idea that the experimental data should pertain to *an ensemble of separate clusters* and not conglomerates.
(iii) The degree of separation between clusters remains large (approx. 30 Å) due to the spacing by organic counterions [6]. This ensures that *no significant cluster-cluster interactions* occur; modelling isolated clusters is therefore clearly of value.

In this investigation we concentrate on two systems of bimetallic clusters derived from well-defined carbonylated precursors: $Ru_{12}Cu_4C_2Cl_2(CO)_{32}$ and $Pd_6Ru_6(CO)_{24}$, for which detailed experimental data are available in the literature [3,4].

Secondly, we shall discuss noble metals (Pd and Pt) supported on 2-D Zirconia and Ceria surfaces. The industrial relevance of these systems arises from their use in the three-way automobile catalyst, which employ noble metals supported on CeO_2 or on a solid solution of CeO_2/ZrO_2 [7]. Pure ZrO_2 is used as the support material in solid oxide fuel cellls (SOFC) [8] and sensors [9]. Zirconia shows three crystal modifications, of which we concentrate our discussions on the fluorite structure, often referred to as cubic Zirconia (c-ZrO_2). Although the cubic phase of bulk Zirconia is not thermodynamically stable, except at very high temperatures (>2600 K) [10], cubic ZrO_2 is stabilised by impurities such as Ca^{2+} and Y^{3+}, and is used in many industrial applications. In addition the dopants introduce oxygen vacancies into the structure. Thus for this particular metal/oxide interfaces we will need to pay special attention to the rôle of defects on the surface and interfacial properties.

The metals supported on a 2-D surface are investigated by periodic models, with the aim of characterising the properties of the supported metal particles as function of their cluster size: from isolated supported metal atoms to regular overlayers. We will discuss two different noble metals: Pd and Pt. The properties of these metals are similar, but when supported on an oxide surface they show in many cases different behaviour. One important distinction between them concerns the shape of the supported clusters, which are known to depend on the strength of the metal-support interaction, and therefore on parameters such as the choice of the substrate, working temperature and loading [11]. Pt is reported to form 2-dimensional particles followed by transformation into 3-dimensional particles [12], while Pd tends to form 3-dimensional but flat (raft-like) particles when supported on a ZrO_2 support [13].

For the study of catalytic processes on the metal/oxide interfaces, there are several properties that are of particular interest, of which the most significant are:
The geometric structure of the interface. Many catalytic reactions are surface sensitive and it is well known that the metal-support interactions are strongly textural and stochiometrically dependent [11,14]. Moreover, it is of particular interest to understand if the metal clusters introduce new active sites, where chemical reactions can occur on the oxide support.
The *electronic structure of the metal/oxide system*, in particular the way in which the electronic properties of both the clusters and the oxide are perturbed by the

interfaces. These properties are directly correlated with the catalytic activity of the surface.
The rôle of defects and dopants on the metal-support interactions. It is well known that defects and dopants change both the electronic and the geometric structure of the interface; hence an understanding of these effects is of particular interest.

We start by giving a short report of the computational methods employed in our calculations, including a geometrical description of the systems investigated. We describe first the cluster calculations on silica based materials, followed by periodic calculations on noble metals supported on oxides. The results for both systems yield detailed information on structure and bonding on these complexes.

2. Method

A crucial feature of the metal/oxide interface is that it combines the extended nature of the support with the limited size of the supported metal particles – a feature that is common to the study of defects in solids. Such systems pose special problems for realistic modelling. The requirements of defining computationally tractable, as well as accurate models are of particular importance. Three different approaches are common, namely bare clusters, embedded clusters and periodic slab models. All three are associated with approximations, and the best choice must be defined by the correct compromise between cost and accuracy.

In the cluster model, the surface may be represented by a finite number of atoms in the immediate surroundings of the active site. Cluster models may include important edge effects if they are too small, but are computationally expensive if large clusters are chosen. For covalent/semi-covalent supports this cluster-model approach has proved to be both adequate and viable [15]. For supports for which long-range electrostatic interactions and polarisation effects are important, such as ionic oxides, approaches which embed clusters in a point-charge (PC) environment, or in more sophisticated representations of the crystal surroundings via embedding potentials may be used [16]. These methods have the advantage that the cluster size may be limited but it can be difficult to derive the correct embedding potential. Finally, in the slab model a layer of material is created, periodic in two dimensions and of finite thickness along the third. Slab calculations may include significant interactions between periodic images of the active site. It is also important to model the surface with a sufficiently thick slab. To study metal-support interactions we have in the current paper employed both cluster models and periodic slab calculations where appropriate.

The well-proven computational technique of choice for dealing with transition metal clusters is Density-Functional Theory (DFT) [17]. The main advantage of DFT over other quantum mechanical methods is that it allows for an integrated treatment of electron correlation effects combined with relatively high computational speed. Both of these are important for transition metal clusters with their high number of energetically similar atomic and electronic configurations. The calculations performed are described in further detail below.

2.1 CLUSTER CALCULATIONS

2.1.1 Bare cluster calculations

It is not currently possible to model the full thermal decomposition of a large number of carbonyl ligands from relatively large bimetallic clusters followed by the subsequent electronic and atomic relaxation using only DFT. To attempt to study metal clusters derived from organometallic precursors therefore, we need a rapid approximate method for generating plausible models of the final denuded state of the clusters.

One frequently used technique employed for quickly generating reasonably accurate chemical models is based on the use of interatomic potentials. For a wide variety of organic molecules and inorganic materials such potentials, used with modelling methods such as Molecular Dynamics (MD) or Monte Carlo, and structural optimisation algorithms, can lead to valuable structural and dynamic information. For metallic and metal-containing systems, however, there has been less success in the use of potentials due to the delocalised nature of the metal bonding and the inadequately of the pair and other sample potential models in describing bonding in metals. In the present case, however, the problem is simplified as we need to treat only the metal cluster and can ignore the carbonyl ligands, which are removed by thermolysis.

Treatment of bare monometallic [18], and bimetallic [19] clusters by interatomic potentials is usually achieved using simple pairwise potentials (e.g. Lennard-Jones, Morse) which being essentially non-bonded and non-specific are flexible enough to allow one, employing various optimisation schemes, to search for global energy minima structures of large (of the order of 100 atoms) clusters. Such a methodology generally results in highly symmetric clusters, and often, in the case of bimetallic clusters, an even spatial distribution of the two metal atom types. For the cases considered here, which often involve low symmetry and/or the inclusion of a silica support, the aim is not necessarily to find the global energy minimum structure of a cluster by randomly mixing two metal types, but to characterise the cluster structure which results from a particular dynamic process starting from a small specific metal carbonyl precursor. That is, there are specific known structural constraints, which are not only computationally advantageous to use, but physically direct the evolution of the cluster's structure.

For our problem, due to the lack of specific metal-metal potentials for our clusters, general-purpose potentials were used. Specifically, the Extensible Systematic Forcefield (ESFF) [20,21] and the MM+ forcefield [22,23] were used. For these potentials a number of parameters, (not only non-bonded), are used such as bonded and torsional terms, which are calculated from rule-based algorithms. The resulting parameterisation is not expected to be particularly accurate, but allows choice of specifically bonded candidate structures for which the connectivities can be constrained, based on the precursor skeleton, in the energy minimisations. This method has the advantage of retaining chosen structural units and symmetries of the precursor but does not allow for changes in coordination in the minimisations. By using two interatomic potential sets and by varying manually the internal metal metal connectivities, together with MD simulated annealing, we aim to sample the potential energy surface of cluster configurations densely enough to be sufficiently close to true energy minima to allow DFT optimisation to be employed.

Once a number of candidate clusters have been formed by this process the results may be assessed by two different methods. Firstly the total energies can be calculated for each structure, via single-point DFT calculation, and compared to find the lower energy cluster structures. Secondly the candidate structures can be compared with experimental data (e.g. EXAFS) to establish how physically realistic the results are. Using both these methods of assessment the best candidates can then be fully optimised at an *ab initio* level via DFT, which allows for any bond breaking/forming and charge transfer to occur.

All-electron DFT calculations were performed using the $DMOL^3$ [24] code. These incorporated scalar relativistic corrections and employed the non-local exchange and correlation functional Perdew-Wang91 [25] denoted GGA in the rest of the paper, which is generally found to be superior to the local density approximation (LDA) [26] for cluster modelling [15] (see also *section 2.2.3*). Numerical atomic basis functions were used throughout, allowing for greater accuracy than a Gaussian type basis set of the same size. In particular, we used the double numerical basis set (DNP) [24] employing two functions for each occupied valence orbital in a free atom plus polarisation functions. As the basis functions used are given numerically, the number of grid points used for the numerical integration can be significant [27]. For most of the geometry optimisations, the default 'medium' grid mesh was used with the higher density 'fine' grid mesh being employed for the later optimisation steps and for calculation of the total energies. Typical convergence to within 0.005 Å and 0.00001 Ha was obtained. All energies reported in this section are uncorrected for Basis Set Superposition Error (BSSE) due the difficulty this poses for multiply connected systems with large relaxations [28] as exemplified in the cluster calculations reported. The absolute BSSE errors in surface absorption energies are highlighted in the section on periodic calculations for atoms and layers, which also show that generally the relative energy scales are unaffected. We should note that BSSE is expected to be larger for Gaussian type than numerical basis sets [29]. Thus, for large cluster calculations, using the same high-level calculations including large basis set and fine grid mesh, it is reasonable to suppose that the BSSE error will be relatively small (typically 5-10% of the binding energy) and of constant magnitude. This allows the calculated energies to at least be taken as mirroring accurately the relative energetic trends and stabilities. All total and binding energies for the bimetallic clusters are thus given as relative energies.

2.1.2 Silica-bound clusters

For the silica-bound clusters (see ***Fig. 1***), hydrogen-terminated fragments of silica were used to model the interaction between the clusters and the silica surface. Due to the amorphous character of the pore walls of MCM41 silica, various fully-relaxed, fragments represented possible internal pore surface sites. The use of hydrogen-terminated fragments of silica to represent extended silica supports, as well as being computationally expedient, is validated by (*i*) the semi-covalent nature of the purely siliceous support, requiring no large long-range electrostatic/polarisation effects to be accounted for, (*ii*) quantum mechanical and forcefield calculations on silica clusters and extended systems showing the local nature of the support relaxation [30], and (*iii*) Pure and 'embedded' DFT studies of metal atoms and monometallic clusters on hydrogen-terminated silica fragments showing the local nature of the metal-silica interaction [15]. The results of the application of this methodology to two bimetallic cluster catalysts are reported *section 3*.

2.2 PERIODIC CALCULATIONS

Calculations using this technique are applied to the systems comprising metal atoms supported on the surfaces of ZrO_2 and CeO_2. Since many of the results for these systems are still not published elsewhere, we give a more detailed description of the calculations on these systems.

2.2.1 Geometrical details

In the current study we compare two different substrates; c-ZrO_2 and CeO_2 which both crystallise in the fluorite structure. We have investigated the {111} surface for both the substrates (see ***Fig. 2***), while for c-ZrO_2 we also included the {011} surface in our study (see ***Fig. 3***). The {111} surface, as modelled in our calculations, is oxygen terminated and built up of 9 atomic layers (see ***Fig. 2***). The choice of surface model is taken from Gennard *et al.* [31], who investigated models containg 9, 12 and 18 atomic layers, employing the Hartree-Fock (HF) Hamiltonian, and found the surface energy for the 9-layer slab to be within 0.003 J/m^2 of that of the 18-layer slab for c-ZrO_2; the comparable figure is 0.068 J/m^2 for the {111} surface of CeO_2.

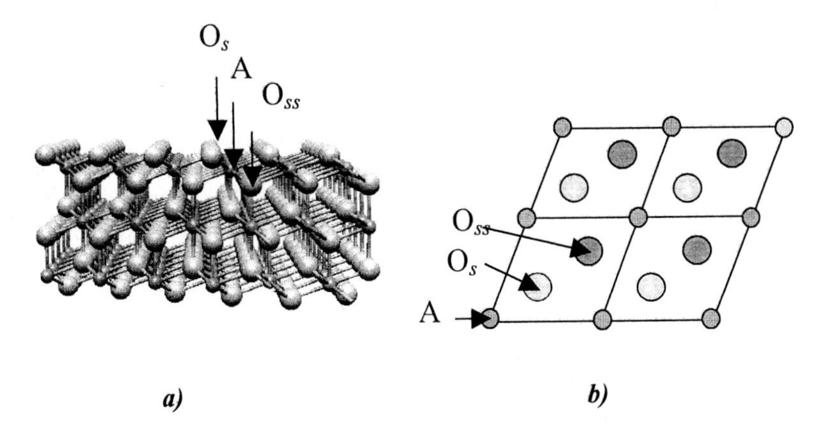

Fig. 2 a) side view of the {111} surface of c-ZrO_2 and CeO_2. Larger and smaller circles represent oxygen and zirconium, respectively. ***b)*** Schematic top view of the same geometry. The shading of the circles indicates in which atomic layer the atoms are located, where light grey represents the surface layer. The three different adsorption sites discussed in the text are marked in the figure, where A denotes the cation (Zr/Ce) in the substrate. Fig. ***b*** represents the 2x2 supercell, created from the primitive 1x1 surface cell .

Gennard and co-workers [31] also performed HF calculations on the {011} surface for *c*-ZrO_2 and CeO_2 (see ***Fig. 3***). This surface may be considered as built up of kinks and ions, instead of forming a flat unreconstructed surface as the {111} surface, with repeating O-Zr-O terraces separated by empty interstices in the oxygen sub-lattice. These empty rows (denoted 'channels' in the rest of the paper) may provide suitable surface sites, where active noble metals can be accommodated. Mulliken charges presented in *ref. 31* shows that ionic charges are largely unaffected by increasing the slab thickness above 4-atomic layers, which is important when studying chemical properties (especially the catalytic activity) of the {011} surface. The {011} surface in our model is built up of 5 atomic layers.

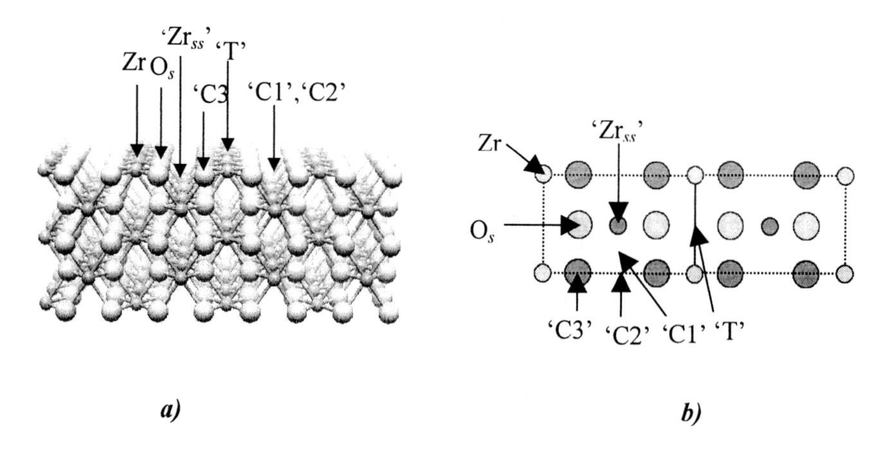

Fig. 3 a) side view of the {011} surface of *c*-ZrO_2 and CeO_2, and ***b)*** Schematic top view of the same geometry. The shading of the balls indicates in which atomic layer the atoms are located, where light grey represents the surface layer. The different adsorption sites discussed in the text are marked in the figure.

When depositing the ad-layer on the oxide surface the following procedure was adopted:

1) First we optimised the bulk structure for the current number of k-points and Hamiltonian.
2) Next from the optimised bulk structure we cut out the clean surfaces.
3) The ideal surface was then optimised, keeping cell shape and axes fixed at the optimised bulk parameters.
4) Subsequently, the metal ad-layer was deposited on the optimised oxide surface. The adsorption sites were chosen such that the metal adlayers were commensurate with the underlaying substrate.
5) Finally, we optimised the coordinates of the slab + metal atoms again keeping the cell volume and shape fixed.

For the {111} surface, three different surface sites were investigated. First, we deposited the metal adlayer on top of the outermost surface oxygen (denoted O_s in ***Fig. 2***). In this position the noble metal is in a 1-fold oxygen coordinated position. In the second and third sites; *i.e.* on top of Zr/Ce and O_{ss}, the noble metal is in a 3-fold oxygen coordinated position with respect to the outermost oxygens. Deposition of a 'monolayer' of the metal, results in an overlayer structure corresponding to an hexagonal lattice with an *a*-axis of 3.66 Å for *c*-ZrO_2 and 3.92 Å on top of CeO_2, *i.e.* the optimised cell parameters from our bulk calculations. The surface coverage(θ) on the {111} surface corresponds to θ=1.00 monolayers (ML) *i.e.* all equivalent adsorption sites are occupied in the primitive surface cell.

For the lower metal coverage considered (θ=0.25 ML) (only reported for *c*-ZrO_2) we designed a 2x2 supercell, resulting in an intra-metal distance of 7.32 Å (see ***Fig. 2***). At this metal-metal distance there are neither Pd nor Pt interactions and the metal adlayer may be considered as composed of isolated atoms. As for θ=1 (ML) three adsorption sites were investigated: O_s, Zr and O_{ss}.

In addition we studied a defective {111} surface of *c*-ZrO_2 built up from the 2x2 supercell. Owing to the mirror/glide plane in the centre of the *c*-ZrO_2(111) slab (see *section 2.2.2*) a defect was constructed on both surfaces of the slab, resulting in a defect density of θ=0.25 ML per surface. In the current paper, we consider the neutral oxygen defect, created according to:

$$c\text{-}ZrO_2(111) \leftrightarrow c\text{-}ZrO_{2-x} + x/2\ O_2 \quad (1)$$

The defect was introduced into the outermost oxygen layer of the {111} surface. In our investigations, we have found the energetically most favourable adsorption site, for the Pd atom, to be on top of the oxygen defect. As a consequence, we only consider the latter adsorption site in the current paper.

On the {011} surface, the noble metals were deposited first on top of the outermost oxygen (denoted O_s in ***Fig. 3***), representing a 1-fold oxygen coordinated site on the step; secondly on top of the surface Zr ion; and thirdly on the terrace site denoted 'T' in ***Fig 2*** (at coordinates (0.5, 0.0, z)). In addition we investigated four different sites in the 'channel', starting with the site on top of the sub-surface Zr ion (denoted 'Zr_{ss}') with coordinates (0.5,, 0.5, z), as well as the sites 'C1', 'C2', 'C3' with coordinates (0.25, 0.25, z), (0.0, 0.5, z) and (0.0, 0.25, z), respectively. The structure of the overlayer at the {011} surface is orthorhombic with cell-axes: *a*=3.66 Å and *c*=5.16 Å for *c*-ZrO_2.

In addition, we have employed MgO as a model system in *section 2.2.4*. MgO crystallises in the rocksalt structure and is well characterised both experimentally [32] and computationally [33]. For MgO we have concentrated our studies on the {001} surfaces (see ***Fig. 4***), the structure of which is well known to undergo small surface relaxation and reconstruction [32], making MgO the ideal prototype system for theoretical studies of the adsorption process. In the calculations presented here we used a 3-layer atomic slab model on which we deposited the metal adlayer on top of three different sites (see ***Fig. 4***); first on top of the oxygen ion (O_s), second on top of the Mg ion (Mg); and third on top of the bridge site (denoted 'Bridge'). The resulting overlayer structure is the $p(1x1)$ structure with the cell parameter a=2.84 Å.

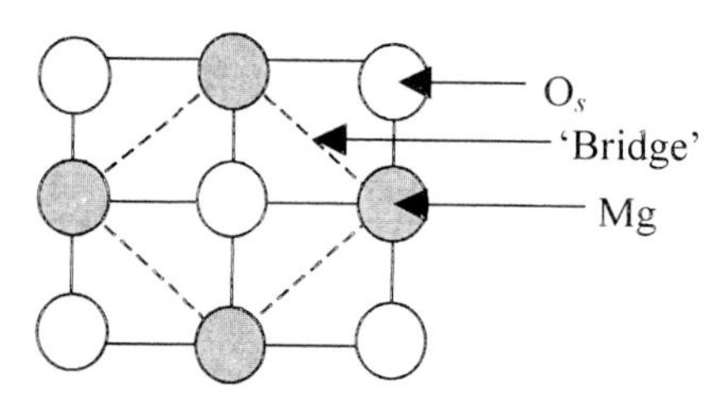

Fig. 4 Top view of the MgO(001) surface. Cations marked as dark circles and anions open circles. The three adsorption sites discussed in the text are shown in the figure. The surface cell corresponding to θ=1 ML is indicated as dashed lines.

2.2.2 Computational details

In our periodic quantum mechanical (PQM) calculations, we have employed two different techniques to expand the electron charge density: a plane-wave basis set (employing the VASP [34] and CASTEP codes [35]); and a local basis set of Gaussian type functions (employing the CRYSTAL98 (*CR98*) program [36]). A detailed description of the basis sets employed for Zr (HayWSC-311d31), Mg(8-511g*) and O(8-411g*) is given in *ref. 37*. For the Pd and Pt metals, we have found that it is important to employ a basis set, which takes into account the diffuse properties exhibited in the bulk phase. We therefore used the basis sets optimised for the bulk structures by Kokalj *et al.* (given in *ref. 37*) - (Pd: HayWSC-2111d31) and (Pt: HayWSC-2111d31) for the two metals, respectively. For calculations performed with the VASP code, non-local ultrasoft pseudo-potentials were employed as described in *ref. 34*, while a description of the ultrasoft pseudopotentials employed in the CASTEP calculations is found in *ref. 35*. For both the VASP and CASTEP calculations we have used an energy cut-off (E_{cut}) of ~380 eV. Owing to the open shell nature of many of the noble metals investigated, all calculations have been performed as spin polarised, employing a k-point grid of 6x6x1.

The 3-D periodic boundary conditions employed by VASP and CASTEP imply that the surfaces are modelled as repeated slabs, separated by a 'pseudo-vacuum'. In the current investigation a pseudo vacuum of ~10-15 Å has been applied, depending on the system under investigation. All slabs (including our *CR98* calculations) are modelled with two equivalent surfaces, due to symmetry; *i.e.* each slab contains a mirror/glide plane in the middle. To keep the symmetry restrictions introduced for the clean surfaces, the metal adlayers were deposited on both surfaces of the slab.

Geometry optimisation was performed with respect to atomic positions and cell parameters, within the conjugate gradient scheme. To confirm that the optimised geometries were in the 'global' minimum, the geometry optimisations were started from a number of different structures, such as the ideal and optimised clean surfaces.

In all cases the same minimum was found. To prevent phase transformations for the Zirconia surfaces, the cell volume and shape were kept fixed and only the ionic coordinates were optimised. In this way, we introduced symmetry restrictions to prevent further phase transformations. No such transformations were observed.

Adsorption energies were calculated as the energy difference between the optimised geometry of the total metal/oxide system and the optimised clean surface plus the energy of the metal-layers, (divided by two as the slab contains two equivalent surfaces), *i.e.*:

$$\Delta E_{ads} = \frac{1}{2}\left[E_{tot}^{opt}(M/AO_n) - E_{tot}^{opt}(AO_n) - 2\times E_{tot}(M-layer)\right], \quad (2)$$

where M denotes the noble metals and A the cation in the substrate.

We should note that the adsorption energies obtained experimentally are more correctly compared with formation energies ($\Delta E_{ads}^{formation}$) calculated according to equation (3):

$$\Delta E_{ads}^{formation} = \frac{1}{2}\left[E_{tot}^{opt}(M/AO_n) - E_{tot}^{opt}(AO_n) - 2xE_{tot}(M-atom)\right], \quad (3)$$

in which the metal ad-layer is considered as built up from the isolated metal atoms. $\Delta E_{ads}^{formation}$ calculated according to equation (3), therefore, also includes adsorbate-adsorbate interactions, and not exclusively the adsorbate-substrate interactions obtained from equation (*2*).

All adsorption energies presented in the rest of *section 2* and in *section 4* are, however, calculated with equation (2), unless otherwise reported, as this approach is sufficient for comparative purpose, which is our main concern in this study.

Electron charge densities (ECHD) presented in the paper are calculated as the difference between the optimised Pd/Pt-ZrO_2 interface and the free Zirconia surface and Pd/Pt adlayer, respectively:

$$\Delta\rho = \rho_{tot}^{opt}(M/ZrO_2) - \rho_{tot}(ZrO_2) - \rho_{tot}(M-layer), \quad (4)$$

where ρ denotes the electron charge density, and M represents the metal; Pd or Pt.

2.2.3 *Choice of Hamiltonian*

For comparison we have employed two different Hamiltonians: the local density approximation (LDA) and Perdew-Wang91 (GGA). Cluster and embedded cluster calculations by Pacchioni *et al.* [38] and Illas *et al.* [15,39] for a number of different functionals have shown that the gradient corrected methods, such as Perdew-Wang91, and the hybrid methods (eg. B3LYP [40,41]) show better agreement with experimental data than the LDA method. However, the latter functional has often successfully been used for metals. Nevertheless, the behaviour of the different DFT functionals have yet not been examined for the metal/oxide interface applying PQM calculations, which allow us to study a higher metal coverage than used in previous cluster calculations. Owing to the fact that we cannot employ the Hartree-Fock Hamiltonian in our plane-wave calculations the hybrid methods are excluded in the current investigation. We concentrate our discussion on geometric and energetic properties.

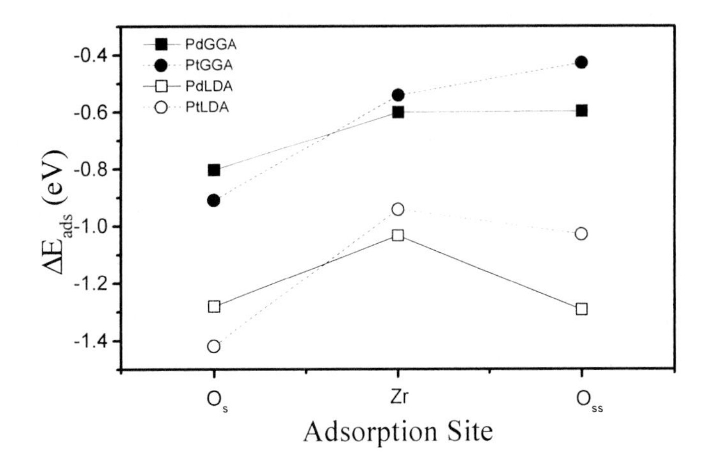

Fig. 5 Adsorption energies (ΔE_{ads}) calculated for the three different surface sites on the {111} interface discussed in the text. Pd is marked as squares and Pt as circles. Two Hamiltonians are compared. ΔE_{ads} given in eV.

In ***Fig. 5*** we show the adsorption energies for Pd and Pt deposited on the {111} surface of c-ZrO_2 (metal coverage θ=1.00 ML). As expected the LDA Hamiltonian overestimates the adsorption energy compared to the GGA Hamiltonian. The energy difference between the two functionals is ~ 0.6 eV, for both the Pd- and Pt-layers. This overbinding property of the LDA Hamiltonian is also reflected in the optimised bond distances. For all adsorption sites investigated the 'equivalent' metal-oxygen distances are shorter for the LDA than the GGA Hamiltonian (see **Table 1**). As a particular example, we discuss the situation in which the Pd and Pt-layers have been deposited on top of O_s (see ***Fig. 2***), for which it is found that the optimised Pd-O distances are 2.09 and 2.03 Å for GGA and LDA, respectively. The corresponding distances for the Pt-layer are 2.02 and 1.98 Å, concluding that for both the metal ad-layers, the metal-oxygen distances are shorter employing the LDA than the GGA method. The experimentally reported Pt-O distance, which was obtained from EXAFS measurements, is 2.03 Å [42], indicating that our GGA value is in good agreement with experiment, while the LDA value is underestimated.

Table 1 Shortest optimised Pd-O bond lengths on the metal/ZrO_2(111) interfaces (θ=1 ML). Pd-O distances are given in Å.

	Pd		Pt	
Adsorption Site	LDA	GGA	LDA	GGA
O_s	2.03	2.09	1.98	2.02
Zr	2.44	2.79	2.65	2.83
O_{ss}	2.43	2.69	2.64	2.82

Another feature observed in ***Fig. 5*** concerns the relative stability of the various adsorption sites applying the different Hamiltonians. When Pd is deposited on top of O_s ΔE_{ads} is favoured by more than 0.3 eV compared to the next most stable surface sites (Zr and O_{ss}), while employing the LDA Hamiltonian we cannot make any distinctions in ΔE_{ads} between the O_s and O_{ss} sites. We suggest that this behaviour of the LDA Hamiltonian is related to the surface relaxations involved when the metal is deposited on the oxide surface. In fact, we find that when Pd is deposited on top of O_s, the outermost oxygen is outwardly displaced by >0.05 Å, while the same value is more than 0.2 Å when Pd is deposited on top of O_{ss}. For the GGA Hamiltonian, the corresponding values are 0.05 and 0.1 Å. Hence, the larger surface relaxations associated with the LDA Hamiltonian give rise to an enhanced Pd-O overlap for the O_{ss} site, which is reflected in the favourable adsorption energy.

The GGA Hamiltonian almost certainly yields the more reliable results and is employed throughout the remainder of the paper.

2.2.4 Plane-wave versus local basis set calculations

The calculated adsorption energies vary over a wide range, depending not only on the Hamiltonian employed but on the description of the electron charge density (*i.e.* plane-wave versus local basis set calculations). As an example, we refer to the case in which a monolayer of Ag is deposited on top of the {001} surface of MgO, for which $\Delta E_{ads}^{formation}$ varies in the range of 0.2 to 0.9 eV (see **Table 2**) [43], to be compared with the experimental value of 0.25 eV [44]. This large variation in calculated $\Delta E_{ads}^{formation}$ gave rise to the following investigation, in which we have compared $\Delta E_{ads}^{formation}$ for two different noble metals (Pd, and Ag) supported on the {001} surface of MgO.

Table 2 $\Delta E_{ads}^{formation}$ for a monolayer of Ag adsorbed on the (001) surface of MgO. ΔE_{ads} given in eV. HF-CC denotes Hartree-Fock calculations with *a posteriori* correlation correction within the GGA scheme.

Method	$\Delta E_{ads}^{formation}$
HF-CC[43]	0.20
FP LMTO[45]	0.88
FLAPW[46]	0.30
Image Model[47]	0.32
Expt.[44]	0.25

Fortuitously, the adsorption energies obtained using the VASP code (calculated according to equation (*3*)) are in excellent agreement with experimentally reported values for the Ag monolayer deposited on top the MgO support (see ***Fig. 6***). But for the two noble metals studied the adsorption energies obtained with *CR98* are overestimated by approx. 100% compared to adsorption energies calculated with the

VASP code. Correcting for the BSSE, via the counterpoise method [48], it is nicely shown in ***Fig. 6*** that $\Delta E_{ads}^{formation}$ are in agreement with adsorption energies calculated with the VASP code.

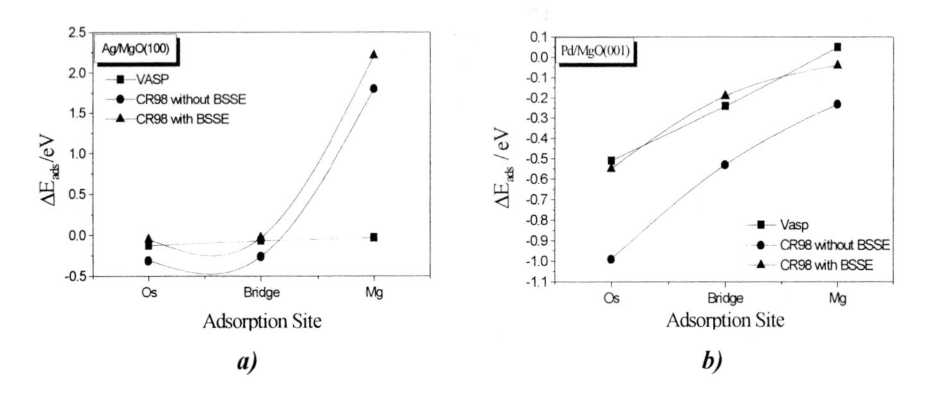

Fig. 6 $\Delta E_{ads}^{formation}$ calculated according to equation (*3*) for the metal adlayers of ***a)*** Ag and ***b)*** Pd adlayers on MgO(001). ΔE_{ads} obtained with the VASP code are denoted as■; ΔE_{ads} from CR98 without BSSE correction as ●; and CR98 with BSSE correction as ▲. Adsorption sites are presented in ***Fig. 4***.

Despite the improvement in $\Delta E_{ads}^{formation}$,when including the BSSE correction, we find that the Ag-layer deposited on top of the Mg surface ion, gives rise to a non-favourable adsorption energy (see ***Fig. 6***), while $\Delta E_{ads}^{formation}$ calculated using the VASP code shows a very small energy difference between the different adsorption sites. We propose that this discrepancy between the two programs, again, is caused by basis set effects in the *CR98* calculations. A conclusion based on results on earlier calculations by Heifets *et al.* [42], who performed CRYSTAL calculations employing a slightly more 'ionic' basis set than we did. The latter authors find that on all adsorption sites discussed (O_s, Mg and 'Bridge' site), the adsorption energies are favourable, in agreement with our results obtained from the VASP code.

We conclude that calculations on metal/oxide interfaces have critical basis set dependence when we employ a basis set of Gaussian type to calculate adsorption energies. This feature indeed shows the need to re-optimise the Gaussian type basis set employed in our calculations, and to take into account BSSE corrections in the investigations of the interface if we are to reproduce correctly adsorption energies and electronic properties.

3. Results for the bimetallic clusters on mesoporous Si-based materials

3.1 FREE SPACE CLUSTERS

Although in the heterogeneous catalysts described in *sections 1 and 2.1* the clusters are bound to a substrate it is useful to first explore the low energy cluster conformations in free space. This approach is justified, at least as a first step in an investigation, by the following factors:

- Under conditions of weak cluster-support interaction the cluster may adopt free-space-like conformations.
- With methods based upon organometallic precursors the cluster is involved in a number of processes (e.g. thermal, chemical) in free-space or solution or weakly-bound to the support before finally strongly anchoring to the support.
- Where the cluster-support interaction is strong, the fluid internal relaxation of the cluster may allow it to adopt a free-space-like structure.
- The calculation of free-space clusters while giving insight into supported clusters is also computationally easier (no support atoms, higher cluster symmetries).

In the following example of Pd_6Ru_6 clusters, the MCM41 supported system formed from thermolysis of $Pd_6Ru_6(CO)_{24}$ is found to be an exceptional hydrogenation catalyst [4]. The clusters though are also found to be fairly weakly bound to the silica support and thus are quite likely (in the absence of sintering) to adopt free-space-like conformations. Following the methodology outlined in *sections 1 and 2.1.1* for such systems we can thus explore some of the energetically low-lying free-space cluster structures.

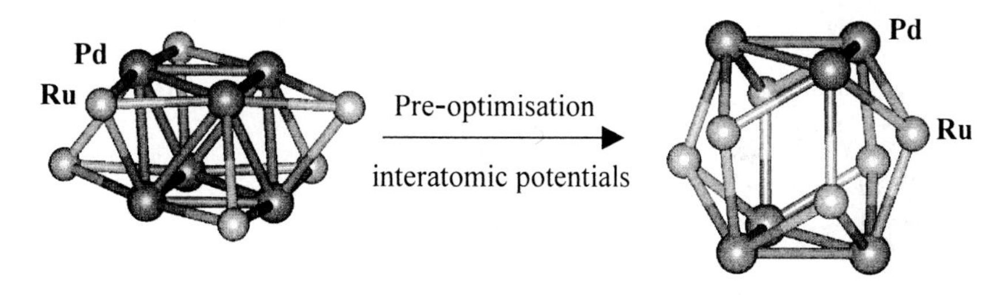

Fig. *7* Pre-optimisation of the denuded bimetallic skeleton of the Ru_6Pd_6 carbonyl cluster using interatomic potentials.

Firstly we take the purely metallic skeletal framework as our precursor cluster, assuming the complete decarbonylation by thermolysis. Using this bimetallic skeleton we can then perform a pre-optimisation using general purpose interatomic potentials guided by experimental data and chemical intuition. In ***Fig.*** *7* we show the denuded skeletal framework and a pre-optimised structure. For the final state of the catalyst we know from Extended X-ray Fine Structure (EXAFS) that both the Ru and the Pd atoms have an average coordination number of approximately 5.3 [4]. From direct inspection of the precursor metallic framework structure, see ***Fig.*** *7*, the Ru atoms have a coordination of 3 and the Pd atoms of 7. Although the interpretation of the EXAFS spectrum is somewhat complicated by the fact that Ru and Pd are essentially indistinguishable by EXAFS, and thus the average total coordination for this precursor structure is close to 5.3, the spectrum favours an interpretation in which all atoms have a more similar coordination environment. From chemical intuition it is reasonable to suppose that the Pd-Pd bonds in the skeletal framework structure will be the weakest, Pd being closed shell in the atomic state. Under this assumption we can manually break these bonds and optimise the structure using interatomic potentials (MM+, ESFF). The interatomic potential pre-optimised structure is shown in ***Fig.*** *7*. Here we see that the structure takes on a more icosahedral form but retains the symmetry of the precursor skeleton – known to be D3d from X-ray crystallography of the precursor molecular crystal. Also the Pd coordination is reduced to 5, in accord with the EXAFS, and the Ru coordination has stayed constant at 3. On closer inspection the Ru-Ru distances have now become so much shorter as to be really Ru-Ru bonds. This bond formation can not be anticipated by the simple interatomic potential based energy minimisation. Using the atomic coordinates of this pre-optimised structure as an input to a DFT calculation the emergence of Ru-Ru bonding immediately becomes apparent. This this is shown in **Table 3** where the relevant Ru-Ru bond lengths (Ru2-Ru1, Ru2-Ru3 – see ***Fig. 8***) for the DFT-optimised D3d Ru_6Pd_6 clusters are below 2.35 Å.

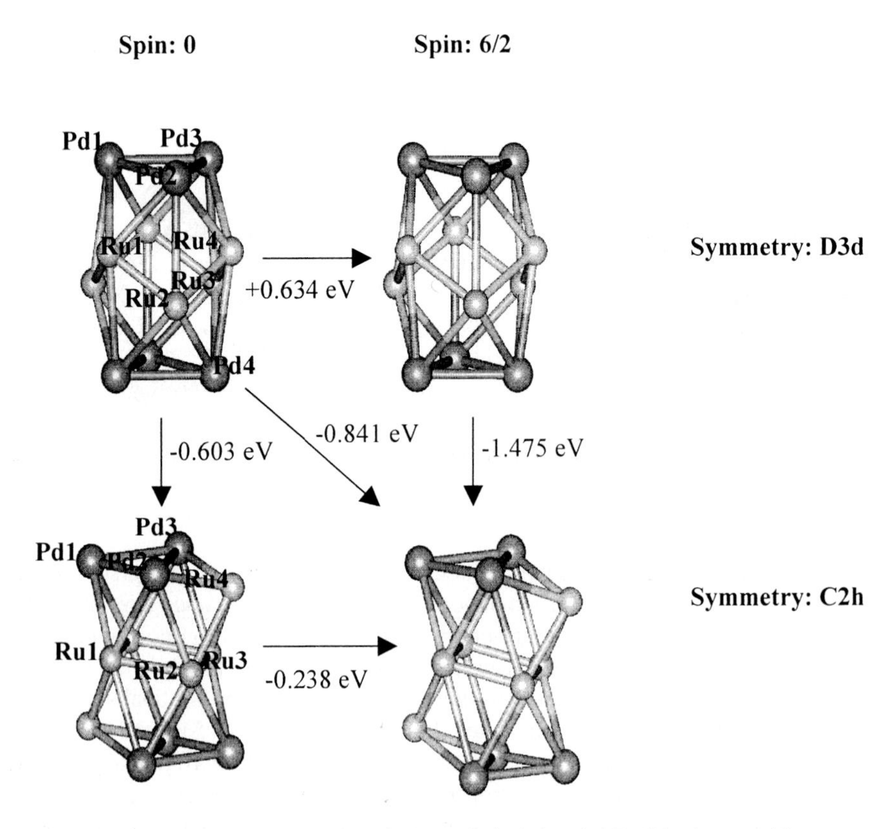

Fig. 8 Changes in relative total energies of DFT-optimised denuded Ru_6Pd_6 clusters with respect to variation in spin state and symmetry.

Table 3 Variation in interatomic distances (Å) of DFT-optimised Ru_6Pd_6 clusters with respect to changes in spin state and symmetry.

Rel. Total Energy (eV)	**0.000**	**-0.634**	**-1.237**	**-1.475**
Spin	6/2	0	0	6/2
Sym	D_{3d}	D_{3d}	C_{2h}	C_{2h}
Pd2-Pd1	2.813	2.813	2.740	2.738
Pd2-Pd3	2.813	2.813	3.114	2.757
Pd2-Ru1	2.613	2.631	2.662	2.666
Pd2-Ru2	2.963	2.994	2.711	2.879
Pd2-Ru4	2.613	2.631	2.548	2.604
Ru2-Ru1	2.325	2.293	2.383	2.357
Ru2-Ru3	3.540	3.488	2.397	2.378
Ru2-Ru4	2.325	2.293	2.480	2.486
Ru2-Pd4	2.613	2.631	2.614	2.660
Ru4-Pd4	2.613	2.994	4.105	4.050

Optimisation of the structure derived from the interatomic potentials pre-optimisation yields a bonded bimetallic cluster with D3d symmetry and is shown in the upper part of ***Fig. 8***. The high D3d symmetry of this structure can actually be found also optimising directly the precursor bimetallic skeletal framework using DFT and employing the D3d symmetry to speed up the calculation. This finding confirms the validity of our approach of using pre-optimisation using interatomic potentials to rapidly get close to the same minimum energy structure. For this cluster structure, each atomic centre has a coordination number of 5 matching well the experimental EXAFS coordination of 5.3 (a 10 percent margin of experimental error in EXAFS derived coordination numbers is commonly accepted). Unfortunately though the clusters bond lengths do not correspond well to the EXAFS spectra (indicating longer Ru-Ru bonds) and thus we are forced to conclude that this cluster is not representative of the clusters in the actual catalyst. To further investigate the Pd_6Ru_6 cluster, both its electronic state and its geometric state can be now varied.

Varying the electronic state of the cluster and consistently optimising using DFT (retaining the D3d symmetry) yields a number of clusters with different stable spin states, and geometries which closely correspond to the initial structure but with an interesting variance of the metal-metal bond lengths [49]. It is found that the closed shell (spin 0) cluster is the most energetically stable cluster for this symmetry and conformation. The relative energies of the spin 0 and spin 6/2 states are shown in ***Fig. 8***.

Looking at the structure of the D3d cluster we can see that it is has a very open form and is likely to be a structure in a local minimum. One possible, and chemically sensible, way to form a more compact structure is to allow more Ru-Ru (d7s1 atomic state) bonding in the cluster. One simple way to do this is to condense the Ru puckered ring configuration of the D3d structure, breaking the longest Pd-Ru bonds. Looking at **Table 3**, and ***Fig. 8*** we can directly see this by the formation of new Ru-

Ru bonds (Ru2-Ru3 distance going from ~3.5Å to ~2.4Å) and the breaking of Pd-Ru bonds (Ru4-Pd4 distance going from ~2.6Å to ~4.1Å). Going through the process of pre-optimisation using interatomic potentials and final DFT optimisation again for such a conformation results in a cluster with C2h symmetry as shown in the lower part of ***Fig. 8***. Such a cluster takes much longer to optimise using DFT and could not simply be found by further DFT optimisation of the local minimum D3d structure and fixing a lower symmetry. The resulting cluster is found to be lower in energy that the D3d cluster and again confirms the methodology involving pre-optimisation. In this case, in contrast to the D3d structure, exploring the spin states of the cluster (each time re-optimising the structure) it is found that an open shell spin 6/2 electronic spin state is lowest in energy. This new lower energy, and lower symmetry cluster structure shows an improvement in the fit to the EXAFS data but still is not perfect. In particular, although for both clusters the coordination environment is consistent with the average value of 5.3, the Ru-Ru distances are too low (*cf* ~2.3Å for the D3d clusters, ~2.4 Å for the C2h clusters and 2.65Å average EXAFS value). Such small calculated metal-metal bond lengths of the assumed totally-denuded clusters has been addressed by other authors [50] who suggest that residual carbon atoms may still reside on the cluster's surface increasing the average experimental intermetallic bond lengths. To accurately reproduce experimental data, it might also be necessary to include the effect of cluster-support interactions in our calculations, but the results discussed above have, we considered, identified the essential features of the cluster structures. The methodology used shows us the importance of the use of different computational techniques combined with the guiding influence of experimental data and chemical intuition to find the most favourable catalytic cluster structures. Further details of the full geometric and electronic structures and energetics of other tested Pd_6Ru_6 cluster conformations will be reported elsewhere [51].

3.2 SILICA-BOUND CLUSTERS

For supported clusters that are strongly bound to a substrate, the exploration of the energetically favorable free-space clusters may seem to be unnecessary and even misleading considering that one should really find the lowest energy cluster-support state. In fact this is found not always to be true and especially not for strongly bound clusters which undergo pre-treatment processes (e.g. thermolysis, chemical reactions) before binding to a support. In the case of $Ru_{12}Cu_4C_2$ clusters there also seems to be the added process of large internal cluster relaxation after the thermolysis of the carbonyl ligand sheath. By checking the relative energies of various free-space and silica bound clusters and investigating how their structures correspond to the experimental EXAFS data we can begin to deduce the relative probabilities and importance of these processes.

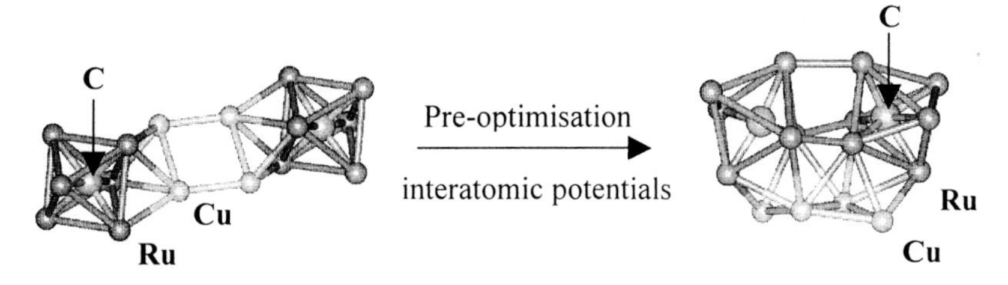

Fig. 9 Pre-optimisation of the denuded bimetallic skeleton of the $Ru_{12}Cu_4Cu_2$ carbonyl cluster using interatomic potentials.

Following the methodology demonstrated above we start under the assumption of total decarbonylation (supported by lack of C-O stretching modes in the IR spectrum of the final supported catalyst) and take the denuded metallic skeletal framework of the precursor salt, known from XRD on the molecular crystal, as our initial structure. This structure is shown on the left in ***Fig. 9***. From EXAFS on the final array of supported clusters, in the catalyst the Cu-Ru coordination number and the Cu-Cu coordination number is know to be higher than in the precursor structure [3], which implies that the copper atoms condense together and that the ruthenium atoms also bond with the copper atoms. From analysis of the EXAFS spectrum it is strongly suggested that the clusters are bound to the silica support via Cu-O-Si bonds. This mode of silica-cluster bonding is also supported, and shown to be relatively energetically favourable over binding interactions, by DFT calculations on small copper clusters interacting with silica surface defects [52]. Here we also consider the likely possibility that this interaction can occur via reaction of the precursor carbonyl with the numerous dangling silanol groups found on the inner surface of the MCM41 pores. One likely way for the cluster to condense, whilst allowing the copper atoms access for bonding with the silica surface, is to suppose that the two Ru_6C octahedra coalesce (allowing for energetically favourable Ru-Ru bonding) above a condensed copper base. Assuming this general structural change, followed by geometry optimisation using interatomic potentials, gives the cluster structures similar to that on the right in ***Fig. 9***. Due to the large number of ways in which the cluster can fold-up, a number of pre-optimised structures were formed. The best candidate clusters were selected for the DFT optimisations. Our choices were based upon DFT single-point energy calculations of the structures and their approximate match to the EXAFS data by inspection of metal coordination environments (metal-metal bond lengths are expected to be too inaccurate for reasonable comparison from the interatomic potentials pre-optimisations).

Using DFT to optimise fully these pre-optimised cluster structures is computationally expensive, even using the rather low C_2 symmetry. For this particular cluster, in addition to its relatively large size and complexity, another reason for the high computational expense in the DFT optimisations came from an unexpected large structural relaxation of the ruthenium-encapsulated carbon atoms, which escape to the surface of the cluster upon optimisation. This structural change could not be simply anticipated by conventional fixed-coordination interatomic potentials in the pre-optimisation step, as much bond-breaking and forming was required. This structural change was found to be very energetically favourable, reducing the total cluster energy by 5-10 eV depending upon exactly how the carbon atom escapes. Four examples of the phenomenon are shown in ***Fig. 10*** for two DFT optimised free space clusters (FS1,FS2), and two silica bound clusters (B1,B2). For the silica-bound clusters we represent the silica surface by a hydrogen-terminated four-membered silica ring and the cluster bonding to occur via two oxygen atoms each bound to two copper atoms of the cluster. This cluster-silica binding mode was found to be most favourable with respect to the EXAFS data in [53].

The DFT optimised cluster FS1 is found to have the most energetically favourable free-space structure compared to other clusters investigated and is closest in morphology to silica-bound cluster B1. Cluster B1 is also found to be the best representative silica-bound cluster with respect to the experimental data found thus far and is compared favourably with the experimental EXAFS data for both coordinations and bond-lengths in *ref. 54*. Conversely cluster B1 is not found to be the lowest energy cluster structure compared with other possible clusters bound to a silica four-ring in the same manner.

The minimum energy silica-bound cluster was found to be B2, which is closest in morphology to the free-space cluster FS2, which is higher in energy than FS1. Using the relative total energies given in Fig.9 and using the idealised 'reaction' scheme

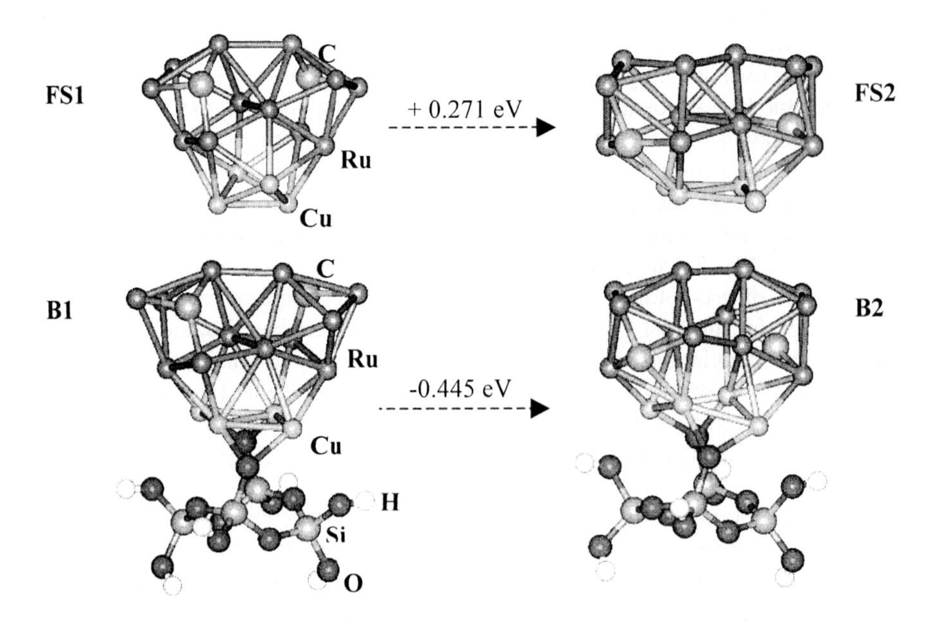

Fig. 10 Relative total energies of DFT-optimised $Ru_{12}Cu_4Cu_2$ clusters in free-space and bound to a cluster representation of the silica substrate.

$$Ru_{12}Cu_4C_2 + Si_4(OH)_8 \rightarrow Ru_{12}Cu_4C_2/Si_4O_8H_6 + H_2$$

we can simply derive relative silica binding energies for B1 and B2. This turns out to be (0.445+0.271= 0.716 eV) *i.e.* B2 is more strongly bound to silica w.r.t B1 by 0.716eV. In other words cluster FS2 is a higher energy free-space cluster which nevertheless binds much more strongly to the silica support than does FS1. This can be qualitatively understood from the proximity of the relatively electronegative carbon atoms to the clusters base drawing charge from the copper atoms and increasing the O-Cu bond strength. This effect is examined in more detail in [55]. Due to its low relative total energy and silica-binding energy it may be imagined that B2 would be the most representative of the actual catalytic clusters. That this is not apparently the case requires explanation.

This result can be explained by remembering that the manner in which the clusters become bound to the surface is a complex dynamical process, the final state of which we have attempted to estimate by the methodology described above. That the morphology of cluster B1 is very similar to the lowest energy free-space cluster and that silica-bound cluster B1 fits the experimental data the best, indicates that the condensing of the cluster framework occurs before it is bound to the silica support. If the cluster first binds to the silica support, then condenses, other possible lower energy silica-bound structures would be expected which are found not to fit the EXAFS data as well. From work in progress, the transitions from FS1>FS2 and B1>B2 appear to have large energy barriers (indicated by the dashed line in the ***Fig. 10***) which is commensurate with the large structural differences in both. Investigations, in fact, suggest that the flexibility of the silica support can determine whether certain such transitions can occur. The full computational/EXAFS comparison of numerous other free-space and silica-bound clusters will be given in [55].

In summary, our study shows the power and the necessity of computational methods in unravelling the structural chemistry of silica-supported metal clusters. It has been shown that the structure of such clusters is highly sensitive to the processes by which the cluster-support interaction occurs and not just dependent on the strength of the binding interaction. Further studies are aimed at also investigating the electronic structure of supported clusters and the effects of ligands on cluster structure.

4. Results for Pd and Pt supported on Zirconia and Ceria 2-D surfaces

4.1 PD AND PT SUPPORTED ON C-ZRO$_2$

4.1.1 Cluster growth

The growth mode of the metal clusters; *i.e.* the morphology of the films in the initial stage of the growth is of particular interest to understand if the metal clusters grow as 3-D islands (Weber mode), layer+layer (Frank van der Merve mode) or layer+island

(Stranski-Krastanov mode). For many catalytic reactions the metal cluster's size and shape controls the turn-over rate of the reaction, resulting in a particular interest for the understanding of the cluster growth. Three processes are defined for the initial growth process; the adsorption step, the diffusion step and the nucleation step. We will pay special attention to these three processes.

Calculating the adsorption energies (ΔE_{ads}) for Pd and Pt, with respect to different metal coverages on the {111} surface of c-ZrO_2, we have the tool to discuss the adsorption and diffusion processes involved in the initial stages of the metal growth. The results presented below concern the ideal system with no defects, and no interactions with small molecules present at the surface. Nevertheless, we believe the results presented can guide the understanding of the initial growth process for the Pd and Pt metals on the Zirconia substrate. Further theoretical and experimental data are, however, needed to establish the correct adsorption and diffusion processes on the surface.

Analysing ΔE_{ads} for a Pd coverage of 0.25 ML (see ***Fig. 11***) we find that Pd prefers the 3-folded oxygen site denoted O_{ss} by more than 0.3 eV in comparison to the next most stable adsorption site (Zr). However, the energy difference between the two sites is small, suggesting appreciable mobility of the Pd-atom on the c-ZrO_2{111} surface. From the fact that the 3-fold oxygen sites are the most stable, we conclude that the Pd-atom will adsorb primarily on terrace sites. Increasing the metal coverage to θ=1 ML, the metal overlayer instead shows a higher preference for the 1-fold oxygen site (O_s). The energy difference between O_s and O_{ss} is again small (0.3-0.4 eV), which indicates that the Pd-layer should be observed on both terraces and steps on the ZrO_2 surface. Further more we would expect Pd to show significant mobility for both higher and lower metal coverage, suggesting that the migration of the metal clusters are size independent, assisting the formation of larger metal clusters.

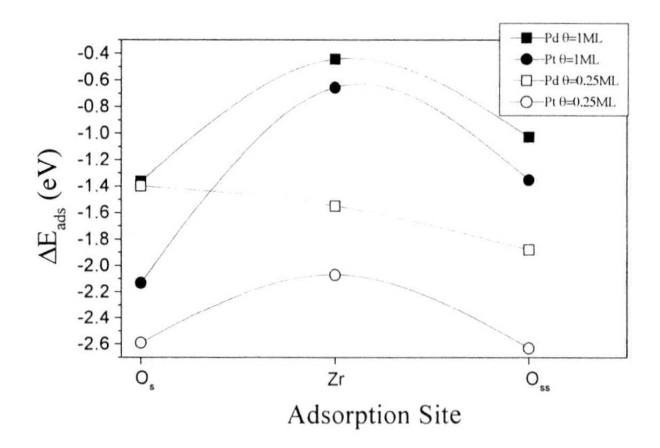

Fig. 11 Adsorption energies (ΔE_{ads}) calculated for the three different surface sites on the {111} interface discussed in the text. Pd is marked as squares and Pt as circles. Two metal coverges are compared (θ=1 and 0.25 ML, represented by filled and open symbols, respectively). ΔE_{ads} given in eV.

The Pt-atom (represented by θ=0.25 ML) shows no difference in adsorption energy between the O_s and the O_{ss} sites (***Fig. 11***), suggesting both that there will be high mobility of the Pt-atom on the ZrO_2\{111\} surface, and that the adsorption process will occur on both terrace and step sites. However, as the metal coverage increases the surface mobility decreases, and for θ=1 ML, Pt favours the O_s site by more than 0.7 eV before the O_{ss} site. Consequently, our calculations suggest that the diffusion process is cluster size dependent, and we would expect to observe smaller metal clusters of Pt than of Pd. We suggest therefore that Pt will predominantly decorate surface steps at higher metal loading, owing to its preference for lower oxygen co-ordination sites; *i.e.* we expect to observe Pt on steps and defects, which exhibit a lower co-ordination number than for the non-defect \{111\} surface.

Moreover, again, as for the \{111\} surface Pd does not differentiate between the O_s and the terrace site, demonstrated by calculations performed on the \{011\} surface of *c*-ZrO_2. In ***Fig. 12*** adsorption energies for the higher metal coverage (θ=1 ML) of Pd and Pt layers are presented. As for the \{111\} surface, we see that Pt clearly prefers the 1-fold oxygen site before the terrace site, *i.e.* the energy difference between the site denoted O_s and 'T' in the figure is more than 0.4 eV. From ***Fig. 12*** it is also seen that the energy difference between the various sites in the channel is smaller for the Pd-layer than the Pt-layer suggesting a higher mobility for the Pd than the Pt-layer, which is in accordance with the results obtained for the higher metal coverage on the \{111\} surface.

In addition, ***Fig. 12*** shows that the Pt/ZrO_2 interface is the most thermodynamically stable, in agreement with findings for the \{111\} interfaces, and that adsorption energies on the \{011\} surface are greater than on the \{111\} surface, indicating that the clean \{011\} surface of Zirconia is catalytically more active than the \{111\} surface.

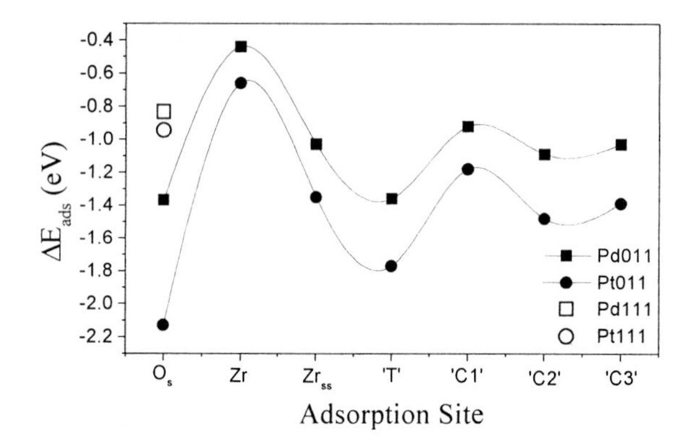

Fig. 12 Adsorption energies (ΔE_{ads}) calculated for seven different surface sites on the {011} interface discussed in the text. Pd is marked as squares and Pt as circles. ΔE_{ads} for the Pd/Pt adsorbed on top of O_s on the {111} interface is marked with open squares and circles, respectively. ΔE_{ads} given in eV.

The higher mobility of the Pd than the Pt-layer agrees with indications from experimental data [56,57]. Employing the STM technique, Bowker and co-workers found that the cluster size distribution of Pd on the $TiO_2(111)$ surface readily moves towards larger cluster sizes after annealing at 700 K, while the size distribution of Pt is unchanged after annealing, with smaller clusters being stable also at higher temperatures. This result indicates a higher surface mobility of the Pd than the Pt clusters, but further experimental data is needed to verify this conclusion. Furthermore, it was found that Pd shows a higher probability to adsorb on both terrace and step sites, while Pt is essentially observed at steps when supported on the $TiO_2(111)$ surface, an observation that is in line with our calculations albeit for a different oxide support.

The calculations discussed above show the importance of low coordination sites on the substrate in the first adsorption and diffusion step. We have, therefore, as a complement to the ideal {111} surface investigated the energetics of the defective {111} surface. As discussed in *section 2.2.1*, the defect considered is a neutral oxygen defect created according to equation (*1*). The calculated adsorption energy for a Pd metal atom on the neutral oxygen defect site, and found that sorption at this site is ~5 times more favourable (ΔE_{ads} ~5.5 eV) than on the non-defective {111} surface of *c*-ZrO_2. This result clearly shows the importance of surface defects in the metal adsoprtion. We propose that the defects create trapping sites where the Pd-atom can be easily adsorbed, and function as a nucleation site for larger clusters. Again, further work, both experimentally and theoretically, is needed on this important problem.

4.1.2 Geometric Structure

Deposition of a metal ad-layer (θ=1ML) results in significant surface relaxations, especially when the metal is deposited on a 3-fold oxygen co-ordination site, which gives rise to an outward relaxation of more of ~0.1 Å of the outermost surface oxygens. We note that the Pd-layer results in slightly larger displacements than those caused by the Pt adlayer (see **Table 4**).

Table 4 Surface relaxations in the outermost atomic layers of the {111} surface (for M on top of O_s site), reported for two different metal coverages: θ=1 and 0.25 ML. Surface displacements are calculated as the difference of the ideal {111} surface and the relaxed geometry of the Pd and Pt/ZrO_2 interfaces. Negative and positive values indicates inwardly and outwardly displacements, respectively. For θ=0.25 ML, O_s denotes the surface ion to which an metal atom is bound, while O_s' represent the non-bound surface oxygens; equivalent notation for the other surface layers. Displacements are given in Å.

Surface layer	Pd/*c*-ZrO_2(111)		Pt/*c*-ZrO_2(111)	
	θ=1 ML	θ=0.25 ML	θ=1 ML	θ=0.25 ML
O_s	+0.04	+0.29	+0.04	+0.51
O_s'		-0.09		-0.10
Zr	- 0.09	+0.03	- 0.07	+0.03
Zr'		-0.10		-0.13
O_{ss}	-0.11	-0.02	- 0.03	-0.01
O_{ss}'		-0.07		-0.11

Also for the lower metal coverage (θ=0.25 ML), the surface relaxations are of the order of ~0.1 Å for the outermost surface oxygen ions (see **Table 4**). However, in the case when Pd or Pt is supported on top of the 1-fold oxygen site, O_s, the surface oxygen ion to which the metal atom is adsorbed, is inwardly relaxed by 0.3 and 0.5 Å for the Pd and Pt atoms, respectively. The general observation for θ=0.25 ML is that surface relaxations are smaller than those observed for the higher metal coverage.

Surface relaxation is also important for the {011} surface (see **Table 5**). As for the clean surface, the relaxations are associated with a columnar displacement of the Zr ions [31]: *i.e.* the surface Zr ions are inwardly relaxed, while the subsurface Zr ions relax outwardly along the z-direction. The oxygens show a negligible surface relaxation. The largest displacement (> 0.2 Å) is observed when the metals are deposited on top of Zr, while when adsorbed on top of the sites denoted O_s and Zr_{SS}, the outermost Zr ions are inwardly relaxed by ~0.1 Å.

Table 5 Surface relaxation obtained as the difference of the ideal {011} surface and the 'relaxed' geometry of the Pd and Pt/ZrO_2 interface (θ=1ML). Negative and positive values indicate inward and outward displacements, respectively; displacements given in Å.

Surface layer	Pd/ZrO_2			Pt/ZrO_2		
	O_s	Zr	Zr_{ss}	O_s	Zr	Zr_{ss}
Zr_1	-0.12	-0.25	-0.12	-0.07	-0.26	-0.10
Zr_2	0.1	0.01	0.11	0.07	0.01	0.10
Zr_3	0.	0.	0.	0.	0.	0.
Zr_4	-0.1	-0.01	-0.11	-0.07	-0.01	0.10
Zr_5	0.12	0.25	0.12	0.07	0.26	-0.10

A bond length analysis for the Pd/*c*-ZrO_2(011) interface (see **Table 6**) reveals that the r(PdO) distance on top of O_s is ~2.0 Å (comparable with the 1-fold site at the {111} surface discussed in *section 2.2.3*), while Pd supported on top of Zr, gives a r(PdO) distance of more than 3.0 Å, explaining the unfavourable situation for the metal on top of Zr (ΔE_{ads} is unfavourable by more than 1 eV compared to the terrace site 'T' (see ***Fig. 12***)). In the 'Zr_{ss}-site', r(PdO) distances have again decreased (~2.20 Å), and the r(PdZr) distances have increased (~3.0 Å), which justify the small energy difference between the two sites, O_s and Zr_{ss} sites. As for the {111} surface, Pt-O distances are slightly shorter (r(PtO)=1.99 Å) when the Pt ad-layer is deposited on top of O_s (the energetically most favourable adsorption site). These results are in good agreement with EXAFS measurements on the Pt/ZrO_2 interface, which report a Pt-O distance of 2.03 Å [42].

Table 6 Bond lengths optimised for Pd and Pt/ZrO_2 {011} interfaces (θ=1ML). The notation of the distances in the table is presented in the top figures for the three different adsorption sites discussed in the text. Bond lengths are given in Å.

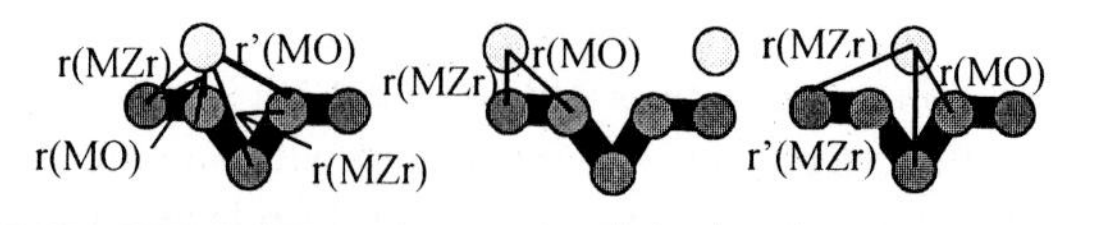

	Pd/*c*-ZrO_2(011)			Pt/*c*-ZrO_2(011)		
	O_s	Zr	Zr_{ss}	O_s	Zr	Zr_{ss}
r(MO)	2.06	3.13	2.32	1.99	3.08	2.21
r'(MO)	3.33	-	-	3.26	-	-
r(MZr)	3.12	2.46	3.04	3.04	2.40	3.54
r'(MZr)	3.99	-	3.77	3.92	-	3.67

Optimising the Pt-O distance for the isolated Pt atom on the c-ZrO_2 interface we obtain a bond length of 1.94 Å on top of O_s (see **Table 7**). Thus, the metal-oxygen distance is shortened by ~0.1 Å compared to the Pt monolayer. Also the Pd-O distance is shorten compared to the bond lengths optimised for θ=1 ML (rPdO)=2.02 Å for θ=0.25 ML. As for the higher coverage, the Pt-O and Pd-O distances in the 3-fold oxygen sites are in the range of 2.2-2.6 and 2.6-2.8 Å, respectively.

Table 7 Bond lengths optimised for Pd and Pt/ZrO_2 {111} interfaces. The notations of the distances in the table are presented in the top figures for the three different adsorption sites discussed in the text. Two metal coverages are presented (θ=1 and 0.25 ML) Bond lengths are given in Å.

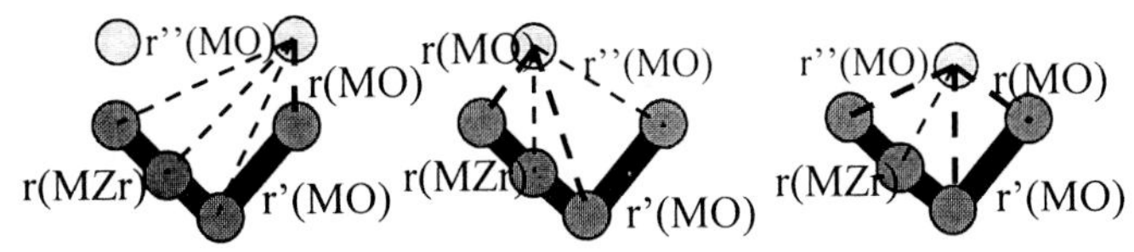

	θ = 1 ML					
	Pd/c-ZrO_2(111)			Pt/c-ZrO_2(111)		
	O_s	Zr	O_{ss}	O_s	Zr	O_{ss}
r(MO)	2.07	2.79	2.69	2.02	2.82	2.57
r'(MO)	4.11	4.04	3.31	4.02	4.02	3.01
r''(MO)	4.20	5.44	4.53	4.10	5.46	4.45
r(MZr)	3.49	2.68	3.26	3.41	2.68	3.07
	θ = 0.25 ML					
	Pd/c-ZrO_2(111)			Pt/c-ZrO_2(111)		
	O_s	Zr	O_{ss}	O_s	Zr	O_{ss}
r(MO)	2.02	2.69	2.41	1.94	2.64	2.42
r'(MO)	3.93	3.70	2.67	3.70	3.68	2.67
r''(MO)	4.00	5.26	4.35	3.88	5.62	4.47
r(MZr)	3.30	2.45	2.87	3.10	2.44	2.82

4.1.3 Bonding character of the metal oxide interface

The nature of the metal-support interactions (*i.e.* as covalent or electrostatic) for the metal/ZrO_2 interface is directly correlated to the catalytic activity of the surface, and in particular the way in which the electronic properties of both the clusters and the oxide are perturbed by the interfaces. We know from previous investigations that the bonding character varies between different metals, for example, the Ag/MgO bond is best described by electrostatic interactions [43], while Ni supported on MgO shows significant hybridisation [58]. The bonding of Pd supported on the {001} surface of MgO(001) is described as being mainly of an electrostatic nature in *ref. 59* (*i.e.* no hybridisation between the Pd and MgO support), while the interactions between Pd and the {111} surface of MgO show a substantial hybridisation [60].

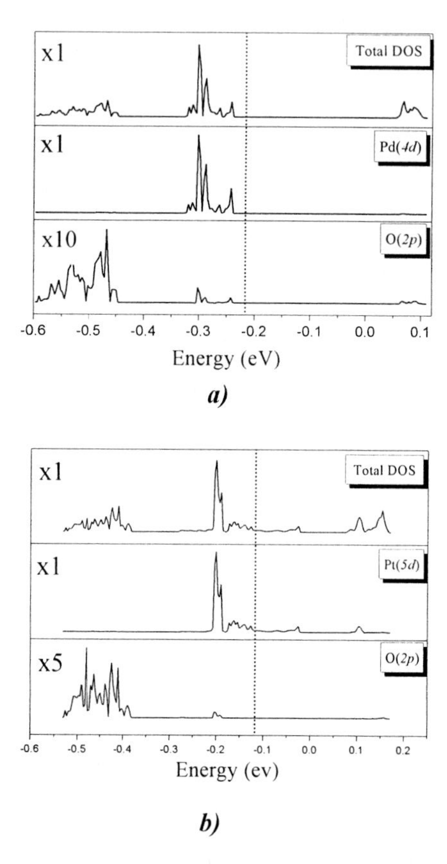

Fig. 13 Projected Density of States (PDOS) for the optimised {111} surface, for the ***a)*** for the Pd/ZrO_2, and in ***b)*** for the Pt/ZrO_2 interfaces when the metal is adsorbed on top of O_s. The Fermi level is marked as a dashed line, and energies are given in eV. Note the scaling of the PDOS in comparison to the total DOS. Calculations performed with the Hartree-Fock Hamiltonian.

Inspection of the partial density-of-states (PDOS) for the higher Pd and Pt coverage (θ=1 ML) on the {111} surface show no evident differences between the two metals, and no hybridisation between the outermost oxygens and the metal (neither Pd nor Pt) (see ***Fig. 13***). It is further established, by performing a Mulliken Population analysis, that there is no substantial electron charge transfer from the metal to the support (see **Table 8**). However, electron re-distribution maps (see ***Fig. 14***) demonstrate a localised perturbation at the interface region, which gives rise to a significant polarisation of the metal ad-layers. The features presented in ***Fig. 14*** are similar for both the metal ad-layers discussed in this paper; the main difference is that the adsorbate-adsorbate interactions are larger for Pt than for Pd. This is illustrated by the more pronounced features of the *d*-orbitals for the Pt-metal, and the quadrupole moments reported in **Table 8**: *i.e.* for Pt, the quadrupole moment is more than 4 a.u., while for Pd it is less than 0.3 a.u. when the metal is deposited on top of O_s, demonstrating a larger deformation of the electron cloud for Pt than for Pd. Owing to the large dipole moment created in the Pd-layer and the quadrupole moment in the Pt-layer the bond-character is best described by electrostatic forces caused by to the polarisation of the metal ad-layer. Moreover, the lack off electron charge transfer from the metal to the support suggests the metal ad-layer be best described in terms of oxidation state 0.

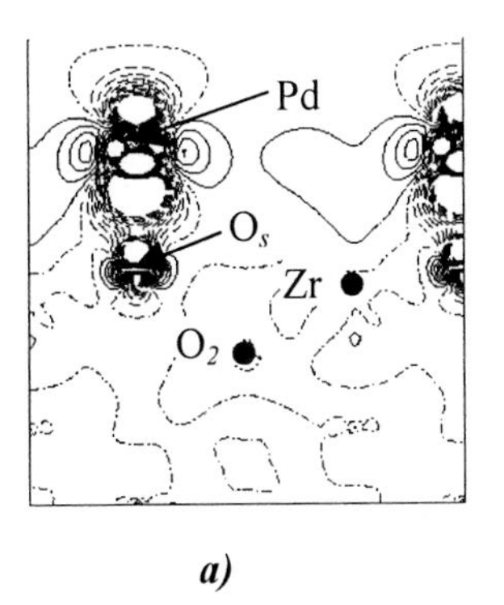

a)

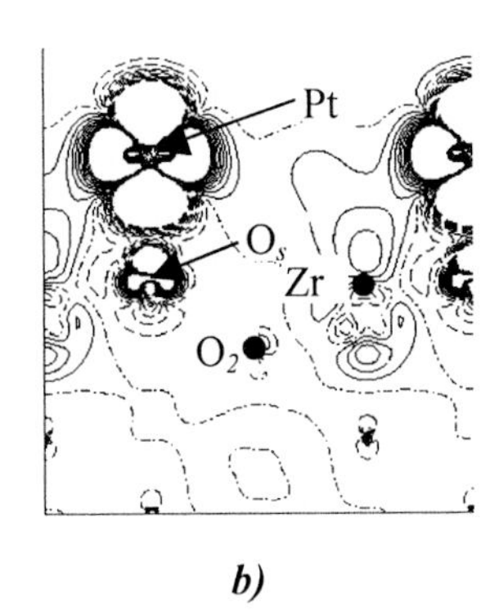

b)

Fig. 14 Electron difference density maps (defined according to equation 4 in the text) for ***a)*** the Pd/ZrO_2 and ***b)*** the Pt/ZrO_2 interfaces. For both interfaces the adsorption site O_s is presented. Contour levels at $\pm 0.001\ e/a_0^3$. Electron excess is marked as solid contours, electron deficiency as dashed contours and the zero line as dotted lines.

For a metal coverage of 0.25 ML (see ***Fig. 15***) we instead observe a significant hybridisation between the metal and the oxygen, here demonstrated for the adsorption sites O_{ss}. Analysing Mulliken charges for the Pd atom on top of O_s, however, reveals that the electron charge transfer is negligible also for this metal coverage (see **Table 9** - which is true also for the O_{ss} site), suggesting the metal atom stays in the oxidation state Pd^0. This feature can be explained as a balance between an electron donation from the metal (Pd(*3d*), and Pt(*4d*)) to the substrate and a backdonation from the oxygen (*2p*) to the metal atom.

4.2 REDOX PROPERTIES OF THE Pd/*c*-ZrO_2 INTERFACE

For the study of the catalytic activity we are particularly interested in the redox properties of the metal/oxide interfaces. From X-ray Photoelectron spectroscopy (XPS) it has been verified that the Pd^{2+} [61] and the Pt^{4+} [62] species are present at the interface before reduction with H_2. However, our calculations on the non-defect surface presented above suggest that the Pd and Pt species are best described in

Table 8 Mulliken charges, dipole moments, and quadrupole moments calculated for the Pd and $Pt/ZrO_2\{011\}$ interfaces, employing the Hartree-Fock Hamiltonian. Values refer to the geometry optimised structure at the GGA level. Mulliken charges for the relaxed clean surface are: O_s (-1.37 *e*); Zr (+2.76 *e*); and Zr_{sub} (+2.95 *e*). The quadrupole moment corresponds to the operator z^2-$x^2/2$-$y^2/2$. Zr' refer to the Zr ion in the subsurface, while O_s' represents the surface oxygen on which no metal atom is adsorbed.

Ads. Site	Pd/ZrO_2														
	Mulliken (*e*)					Dipole moment (*a.u.*)					Quadrupole moment (*a.u.*)				
	Pd	O_s	Zr	O_s'	Zr'	Pd	O_s	Zr	O_s'	Zr'	Pd	O_s	Zr	O_s'	Zr'
O_s	0.03	-1.40	2.75	-1.37	2.97	0.03	0.11	0.19	0.19	0.11	-2.48	0.36	-0.24	0.08	0.30
Zr	0.07	-1.35	2.65	-	2.96	-0.29	0.20	0.27	-	0.13	0.38	0.43	-0.38	-	0.38
O_2	0.21	-1.43	2.65	-	2.97	1.26	-0.08	-0.08	-	0.	-3.23	-0.03	0.73	-	-0.07

Ads. Site	Pt/ZrO_2														
	Mulliken (*e*)					Dipole moment (*a.u.*)					Quadrupole moment (*a.u.*)				
	Pt	O_s	Zr	O_s'	Zr'	Pt	O_s	Zr	O_s'	Zr'	Pt	O_s	Zr	O_s'	Zr'
O_s	0.17	-1.43	2.63	-1.37	2.98	0.67	0.10	0.51	0.14	0.08	-3.83	-0.02	0.46	0.29	0.32
Zr	0.43	-1.36	2.30	-	2.97	0.53	0.21	1.03	-	0.10	1.44	0.46	1.40	-	0.31
O_2	+0.09	-1.34	2.62	-	2.94	1.48	-0.15	0.43	-	0.04	-5.36	-0.09	0.63	-	-0.03

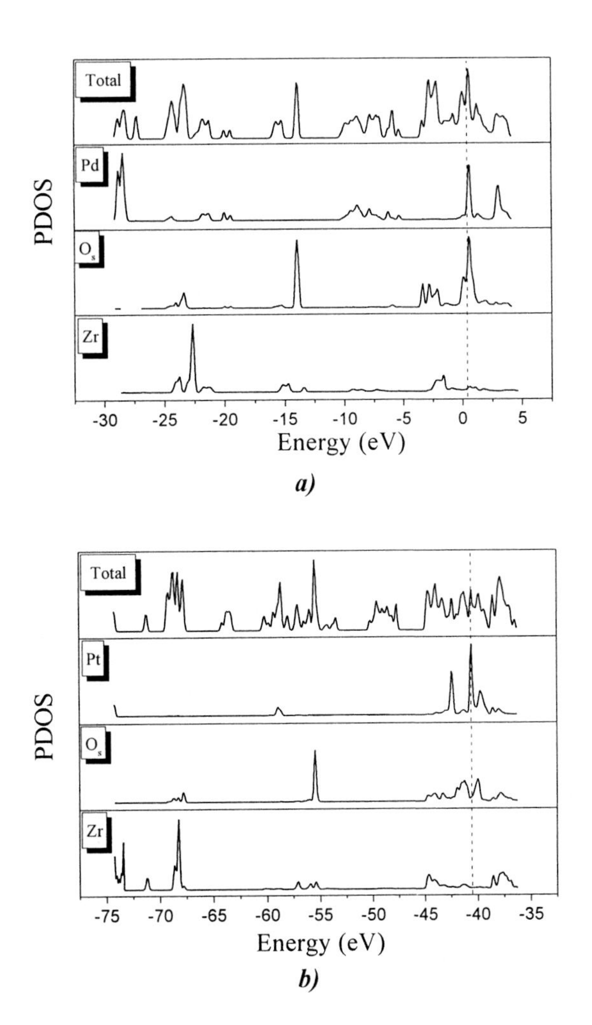

Fig. 15 Projected Density of States (PDOS) for the optimised {111} surface ***a)*** for the Pd/ZrO_2, and in ***b)*** for the Pt/ZrO_2 interfaces. PDOS are presented for the adsorption site O_{ss}. The Fermi level is marked as a dashed line, and energies are given in eV.

oxidation state 0 (Pd^0 and Pt^0, respectively). We will here discuss two different possibilities for the formation of the ionic species of Pd on the interface: firstly the formation of a PdO species on the surface by oxidation of the Pd-atom with oxygen according to the process:

$$\tfrac{1}{2}O_2 + Pd/ZrO_2\{111\} \rightarrow PdO/ZrO_2\{111\}. \qquad (A)$$

Secondly, we consider the formation of a PdO species on the defective {111} surface of *c*-Zirconia:

$$\tfrac{1}{2}O_2 + Pd/ZrO_{2-x}\{111\} \rightarrow PdO/ZrO_{2-x}\{111\}. \qquad (B)$$

The reaction energy ($\Delta E^{O_2}_{reaction}$) of the $O_2(g)$ molecule, involved in the formation of the PdO species on the perfect and defective surface reveals that $\Delta E^{O_2}_{reaction}$, according to equation (*A*) presented above, is –0.29 eV (see ***Fig. 16***), while the same reaction on the defective surface (*B*) is unfavourable (0.27 eV) (see ***Fig. 16***). But if we change the geometry, such that the atomic oxygen adsorbed on top of Pd is pointing towards the Zr ion in the defect site, we find that the $O_2(g)$ molecule indeed dissociates also on the defective surface. This result indicates that the formation of the PdO species, *i.e.* the oxidation step, would not occur on the defective surface unless the geometry is favourable. We note that $O_2(g)$ does not dissociates on the clean non-defective $ZrO_2(111)$ surface, while on the defective {111} surface our calculations predict the $O_2(g)$ molecule to dissociates creating a non-defective {111} surface, with similar geometrical and energetical properties as the {111} surface optimised in *section 4.1.1* and *4.1.2*.

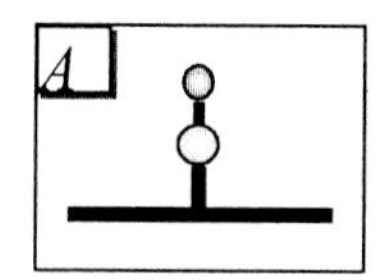

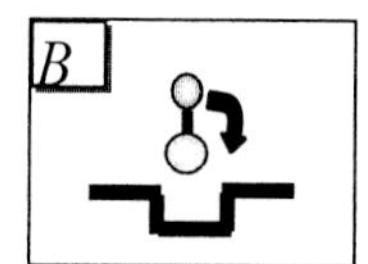

Fig. 16 Schematic picture of the PdO entity adsorbed on the non-defective (*A*) and defective (*B*) $ZrO_2(111)$ surface. Defect created according to equation (*1*). The arrow in (*B*) indicates the oxygen movement discussed in the text.

Mulliken charges are well known to be very basis set and Hamiltonian dependent [63], and from **Table 9** it is seen that the HF Hamiltonian, as expected, shows a more localised solution than the GGA Hamiltonian, resulting in lower ionic charges for the latter Hamiltonian; yet the trends for the two methods considered are consistent, and unless otherwise stated the results presented below are obtained employing the GGA Hamiltonian. As discussed previously, we find that depositing the Pd-atom on the non-defective surface, results in a negligible electron charge transfer from the metal to the support, and Pd is better described as metallic (oxidation state Pd^0). However, if the Pd atom is deposited on top of the defective site Pd is reduced ($Pd^{-0.5}$), and only when a PdO species is formed on top of the perfect surface do we find that Pd is oxidised ($Pd^{0.3}$).

To understand better the stability of the PdO species formed on the ideal and defective {111} surfaces of *c*-ZrO_2, we investigated the adsorption energies of the PdO entity in the different cases. From ΔE_{ads} presented in **Table 9** it is seen that the PdO entity is less stable than the Pd atom adsorbed in the identical adsorption sites; *i.e.* the adsorption energy for the isolated Pd atom on the non-defective {111} surface is –1.40 eV, while the adsorption energy for the isolated PdO entity is –0.23 eV. We find the same results for the defective {111} surface of *c*-ZrO_2.

Table 9 ΔE_{ads} and $\Delta E^{O_2}_{reaction}$ for a PdO entity and $O_2(g)$ molecule on the {111} surface of c-ZrO_2. ΔE^{dis} for the PdO/ZrO_{2-x} is given for two different geometries. Mulliken charges (q) for atomic Pd and O, after interaction with the substrate. Mulliken charges are presented for two different Hamiltonians (GGA and Hartree-Fock). Energies are obtained with the GGA Hamiltonian given in eV; Mulliken charges in e.

System	ΔE_{ads}	$\Delta E^{O_2}_{reaction}$	q(Pd) (GGA)	q(O) (GGA)	q(Pd) (HF)	q(O) (HF)
Pd/ZrO_2	-1.40	-	+0.04	-	+0.04	-
PdO/ZrO_2	-0.23	-0.21	+0.27	-0.33	+0.49	-0.47
Pd/ZrO_{2-x}	-5.50	-	-0.47	-	-0.59	-
PdO/ZrO_{2-x}	-4.12	+0.27/ -0.71	-0.24	-0.35	-0.26	-0.46

Analysing the optimised geometries for the PdO/c-ZrO_2 non-defective {111} surface reveals that the Pd-O_s distances decrease by more than 0.15 Å when a ½$O_2(g)$ molecule is adsorbed on the Pd atom. We associate this decrease in Pd-O distance with the electron re-distribution of the PdO entity. The Pd-O bond length within the PdO entity is ~1.80 Å (observed on both the perfect and defective {111} surface), which is more than 0.1 Å shorter than the Pd-O bond length calculated between the isolated Pd atom and O_s.

4.3 SUBSTRATE EFFECTS

In the current section we compare the c-ZrO_2 and CeO_2 interfaces, owing to their similarity in structure. The main difference in geometry is the larger metal-metal distances in the deposited metal adlayer for the CeO_2 interface, commensurate with the underlaying oxide substrate. We note that all calculations presented in this section were performed with the CASTEP code, while geometry optimisations presented in the previous part of this section were done with the VASP program. As pointed out in *section 2.2.2* the pseudo-potentials employed in the CASTEP code is different from those employed with the VASP program, explaining the energy differences for the Zirconia interface presented in ***Fig 16*** compared to energies shown in ***Fig. 11***.

In *section 4.1.1*, it was determined that the Pt/c-ZrO_2 interface for both the {111} and the {011} surfaces were thermodynamically more stable than the corresponding Pd interface. Comparing the Pd and Pt interfaces for the {111} surfaces of CeO_2, we instead find (see ***Fig. 17***) the Pd/CeO_2 interface to be energetically more favourable, by more than 0.2 eV, than the corresponding Pt interfaces. Moreover, the Pt/CeO_2 interface shows a remarkable affinity for the 1-fold oxygen coordinated site (O_s) that is not observed for Pd/CeO_2. In fact for all three adsorption sites investigated the shortest Pt-O obtained is ~2.00 Å, owing to a displacement of the Pt-layer towards the O_s surface site. Only in the case of the O_{ss} adsorption site the shorter Pt-O distance is a combined motion of the Pt-layer and an outwardly relaxation of the surface oxygens, while the latter motion was the dominating contribution in the discussions for the ZrO_2 substrate earlier. The energy difference in ΔE_{ads} between the two sites O_s and O_{ss} for the Pt/CeO_2 interface is ~0.3 eV.

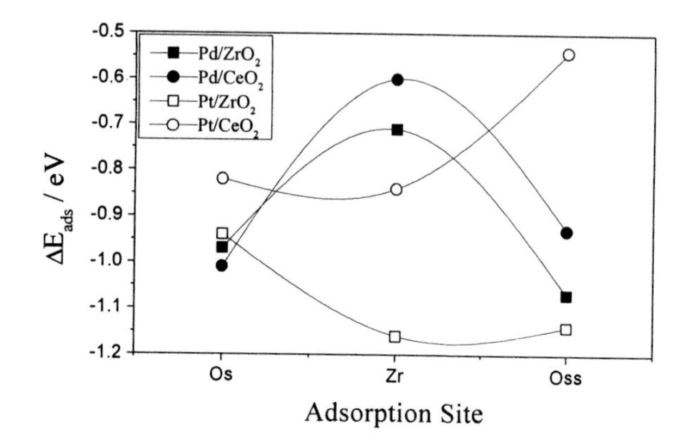

Fig. 17. Adsorption energies (ΔE_{ads}) calculated for three different surface sites on the {111} interfaces of c-ZrO_2 and CeO_2. Pd is marked as filled and opened squares on the ZrO_2 and CeO_2 substrate, respectively, while Pt is marked as filled and opened circles. ΔE_{ads} given in eV.

For the Pd/CeO_2 interface the Pd-layer shows a negligible energy difference (0.1 eV) in ΔE_{ads} between the 1-fold O_s and 3-fold O_{ss} oxygen sites. The latter result is in agreement with findings for the Pd/ZrO_2\{111\} interface. But the Pd/ZrO_2 interface is again 0.1 eV energetically more favourable than the Pd/CeO_2 interface. The small energy differences between the different adsorption sites for the Pd/CeO_2 interface, suggest a high mobility of Pd also on CeO_2 interfaces. We find that calculated Pd-O distances for the Pd/CeO_2 interface varies in the range of 2.0 to 2.6 Å, depending on the adsorption site. The shortest Pd-O distance is obtained on top of O_s, also found for the Pd/ZrO_2 interface, while on top of the 3-fold oxygen sites Pd-O varies between 2.4-2.6 Å. We note that M-O distances are similar for the two interfaces discussed.

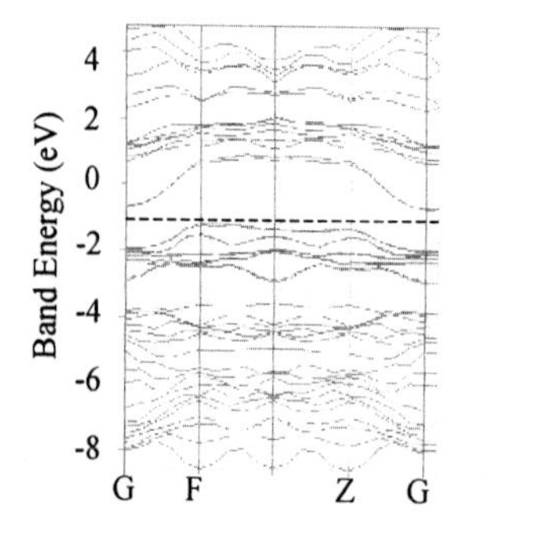

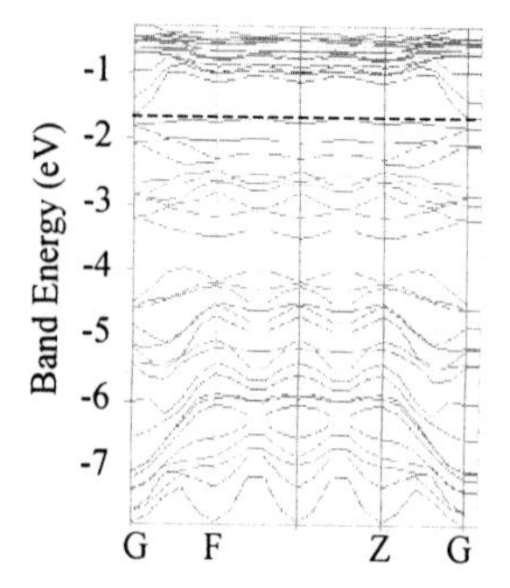

Fig. 18 The electronic band structures obtained for the ***a)*** Pd/c-ZrO_2 and ***b)*** CeO_2 interfaces. Adsorption site is O_{ss} (***Fig. 2***). Energies are given in eV. Standard notations for the primitive cell in reciprocal space are used.

To gain insight into the chemical properties of the noble metal/CeO_2 interfaces we present the band structures for the Pd/CeO_2 and Pd/ZrO_2 (see ***Fig. 18***). A remarkable feature is the small band gap observed for the CeO_2 in comparison to the Zirconia interface: the Pd/ZrO_2 interface shows a significant band gap of ~1.7 eV, while the band gap for Pd/CeO_2 is less than 0.1 eV. Another observation is the larger perturbation to the valence band of the ZrO_2 interface compared to CeO_2, indicating a stronger metal support interaction in the ZrO_2 than the CeO_2 system.

5. Summary and Conclusions

Our discussion started with the free-space bimetallic Pd_6Ru_6 cluster, derived from a carbonyl precursor. Such studies can tell us much about internal relaxation of catalytic clusters and is most valid for systems where such internal relaxation can occur before binding with the substrate and/or where the cluster-support interactions are weak. Both these conditions apply to the Ru_6Pd_6 cluster. We found that the electronic and the geometric structures of the denuded free clusters are closely interdependent, with the lowest energy states for different cluster geometries having different spin-states. In particular, the lower energy clusters were found to have a lower symmetry than that of the precursor carbonyl cluster (D3d > C2h). Work is under way to establish the energetic barriers for such structural transitions, which appear to be very small from initial results. The final state energetics reported show the structural destabilising effect of the carbonyl denuding procedure and show that internal relaxations alone invalidate attempts to base catalytic cluster models solely on the unrelaxed symmetry/geometry of the precursor skeletal framework.

Taking into account the inner walls of the mesoporous silica support, used for such clusters, via hydrogen-terminated silica cluster representations, the effect of internal relaxation effects versus the effect of cluster-support interactions on the structure of the catalytic clusters could be assessed. For these studies the $Ru_{12}Cu_4C_2$ cluster was studied for which it is known from experiment that there are strong interactions with the silica support. This study shows that although the support can alter the form of the cluster morphology, it is not necessarily the clusters which bind most strongly to the support which are favoured in the catalyst, which is explained by showing that the lowest energy free-space clusters are not those which bind most strongly to the silica support, *but* are those which closely match the form of the best silica-bound clusters as judged by experimental EXAFS. In other words these results tend to confirm the idea that the internal relaxation is the strongest influence on the morphology of the catalytic clusters. Experimentally this finding suggests that the denuding process (via thermolysis) of carbonyl clusters results firstly in the condensing and internal relaxation of the cluster which then subsequently binds to the silica support.

From analysis of the adsorption energies for different metal coverages on *c*-ZrO_2, we propose that the cluster growth for the Pt metal shows a considerable size dependence, which is not observed for Pd. Instead, the differences in ΔE_{ads} for Pd are less than 0.4 eV for the various surface coverages discussed. Our calculations also suggest Pd to be less textural dependent than Pt, *i.e.* Pd shows a significant migration on both the {111} and {011} surface, while Pt will hence show a negligible diffusion on both surfaces. Moreover, we emphasise that for both θ=1 and 0.25 ML, the Pt interface is more thermodynamically stable than the corresponding Pd interfaces, and that the Pd and Pt/ZrO_2(011) interfaces are energetically more favourable than the {111} interfaces, suggesting the {011} surface of *c*-ZrO_2 is catalytically more active than the {111} surface. Furthermore our calculations find that neutral oxygen defects on the oxide support act as possible nucleation sites for the cluster growth.

Analysing the bonding character of the {111} interface reveals that, owing to a significant polarisation of the metal ad-layer induced by the oxide support, the bonding character is best described as electrostatic: for the Pd-layer dominated by a dipole-dipole interaction, while the Pt-support interaction originates from an induced quadrupole moment. Decreasing the metal coverage to 0.25 ML introduces an appreciable hybridisation between the surface oxygens and the adsorbed metal atom. However, for neither metal coverages investigated did we observe any electron charge transfer, suggesting an oxidation state of the metal ad-layer of 0. Oxidation of Pd with O_2(*g*) results in an electron charge transfer to the 'atomic' oxygen from the Pd atom of ~0.3 e. However, the sorption of Pd atoms on the defective {111} surface results in a reduction of the Pd atom, confirming that defects are important for the chemical properties of the adsorbed metal.

It has been discussed previously that the three-way exhaust catalyst shows higher catalytic efficiency when a solid solution of ZrO_2 and CeO_2 is used as the support material than the pure CeO_2 support. A simple explanation for this finding could be, as predicted by our calculations, that the Pd and Pt/ZrO_2 systems are more thermodynamically stable than the corresponding CeO_2 interfaces. Another reason may be associated with the high affinity to the 1-fold oxygen site for the Pt-layer on the CeO_2 surface, also justifying the importance of defects on highly reducible CeO_2 surfaces. Band structure analysis indicates that the higher catalytic activity of the Zr doped CeO_2 surface cannot be simply connected with a structural stability of the system, but in addition it introduces important changes in the interactions of the system.

Acknowledgement

First we would like to acknowledge Dr. R. G. Bell, who kindly provided us with ***Fig. 1***. Financial support from The Swedish Foundation for International Cooperation in Research and Higher Education (STINT) and EPSRC (Grant No. GR/L72527) are gratefully acknowledged by M.A, and S.T.B., respectively. EPSRC are thanked for the computer facilities provided at the Royal Institution of Great Britain and the High Performance Computing facilities of the universities of Edinburgh and Manchester. We are also grateful to Dr. A. Sokol and Dr. D. Shephard for useful discussions.

References

[1] Heiz U., Sanchez A., Abbet A., Schneide W-D, (1999) *J. Am. Chem. Soc.*, **121**, 3214
[2] Ichikawa M., (1992) *Adv. Catal.*, **38**, 238
[3] Shephard D. S., Mashmeyer T., Johnson B. F. G., Thomas J. M., Sankar G., Ozakaya D., Zhou W., Oldroyd R. D., Bell R. G., (1998) *Chem. Eur. J.*, **4**, 1214
[4] Raja R., Sankar G., Hermans S., Shephard D. S., Bromley S. T., Thomas J. M., Johnson B. F. G., Mashmeyer T., (1999) *Chem. Commun.*, 1571
[5] Shephard D. S., Mashmeyer T., Sankar G., Thomas J. M., Ozakaya D., Johnson B. F. G., Raja R., Oldroyd R. D., Bell R. G., (1997) *Angew. Chem.*, **36,** 2242
[6] Zhou W., Thomas J. M., Shephard D. S., Johnson B. F. G., Ozkaya D., Mashmeyer T., Bell R. G., Ge Q., (1998) *Science*, **280**, 705
[7] Kašpar J., Fornasiero P. and Graziani M., (1999) *Catalysis Today*, **50**, 285
[8] Singhal S.C., (2000) *Solid State Ionics*, **135**, 305
[9] Farrauto R.J. and Heck R.M., (2000) *Catalysis Today*, **55**, 179
[10] Yoshimura M., (1998) *Am. Ceram. Soc. Bull.*, **67**, 1950
[11] Stakheev A.Yu., Kustov L.M., (1999) *Applied Catalysis A: General*, **188**, 3
[12] Roberts S. and Gorte R. J., (1991) *J. Phys. Chem.*, **95**, 5600
[13] Neergaard Waltenburg H. and Møller P.J., (1999) *Applied Surface Science*, **142**, 305
[14] The Surface Science of Metal Oxides, (1999) *Faraday Discussions*, **No. 114**
[15] Lopez N., Francesc I., Pacchioni G., (1999) *J. Phys. Chem B* **103**, 1712
[16] Bromley S.T., Sokol A. A., French S., Catlow C. R. A., Rogers S., Kendrick J., Watson M., King F., Sherwood P., in preparation (QUASI – Quantum Simulation in Industry, EU Esprit IV 25047, www.ri.ac.uk/DFRL/QUASI)
[17] Grönbeck H., (1996) 'On the structure of and bonding in metal clusters' thesis Gothenburg University and Chalmers University of Technolology, Gothenburg, Sweden; and references therein
[18] Wales D. J., Doye J. P. K., (1997) *J. Phys. Chem.* **A 101**, 5111
[19] Zhu L., DePristo A. E., (1997) *J. Catal.*, **167**, 400
[20] Barlow S., Rohl A. L., Shi S., Freeman C. M., O'Hare D., (1996) *J. Am. Chem. Soc.*, **118**, 7578
[21] Discover 3.0, (1997) MSI, San Diego,
[22] Allinger N. L., *J. Am. Chem. Soc.*, (1977) **99** 8127; Hyperchem Computational Chemistry Manual, (1996) Ch11, Hypercube
[23] Hyperchem Release 6.0, (1996) Hypercube Inc.,Ontario
[24] Delley B. J., (1990) *J. Phys. Chem.* **92**, 508; Dmol^3 User Guide, (1997) MSI, San Diego,
[25] Perdew J.P., Chevary J.A., Vosko S.H., Jackson K.A., Pederson M.R., Singh D.J. and Fiolhais V., (1992) *Phys. Rev. B*, **46**, 6671
[26] Perdew J.P. and Zunger A., (1981) *Phys. Rev. B*, **23**, 5048
[27] Goursot A., Pápai I., Daul C., (1994) *Int. J. Quant. Chem.*, **52**, 799
[28] Fuentealba P., Simon-Manso Y., (1999) *Chem. Phys. Lett.* **314**, 108
[29] Delley B.. (1990) *J. Chem. Phys.* **92**, 508
[30] Kramer G. J., de Man A., van Santen R. A., (1991) *J. Am. Chem. Soc.*, **113**,

6435
[31] Gennard S., Corà F. and Catlow C.R.A., (1999) *J. Phys. Chem. B*, **103**, 10158
[32] Heinrich V.E. and Cox P.A., (1994) 'The Surface Science of Metal Oxides', Cambridge University Press, Cambridge, UK
[33] Causà M., Dovesi R., Pisani C. and Roetti C., (1986) *Surf. Sci.* **175**, 551
[34] Kresse G. and Furthmüller J., (1999) *VASP 4.4 Vasp the Guide*, Tecnische Universität Wien, Wien,
[35] Lindan P., (1999) *The Guide 1.1 to CASTEP 3.9*, Daresbury Laboratory, Warrington, UK
[36] Saunders V.R., Dovesi R., Roetti C., Causà M., Harrison N.M., Orlando R. and Zicovich- Wilson C.M., (1999) *CRYSTAL 98 User's Manual*, University of Turin, Turin,
[37] http://www.ch.unito.it/ifm/teorica/Basis_Sets/mendel.html
[38] Neyman K.M., Vent S., Rösch N. and Pacchioni G., (1999) *Topics in Catalysis*, **9**, 153
[39] López N. and Illas F., (1998) *J. Phys. Chem. B*, **102**, 1430
[40] Becke A. D., (1993) *J. Chem. Phys.*, **98**, 5648
[41] Lee C., Yang W. and Parr R.G., (1988) *Phys. Rev.* **B37**, 785
[42] Bitter J.H., Seshan K. and Lercher J.A., (2000) *Topics in Catal.*, **10**, 295
[43] Heifets E., Kotomin E.A. and Orlando R., (1996) *J. Phys. Condens. Matter*, **8**, 6577
[44] Trampert A., Ernst F., Flynn C.P., Fishmeister H.F. and Rühle M., (1992) *Proc. Int. Conf. On Metal-Oxide Interfaces; Acta Metall. Mater. Suppl.*, **40**, S227
[45] Schönberger U. , Andersen O.K. and Methfessel M., (1992) *Proc. Int. Conf. On Metal-Oxide Interfaces; Acta Metall. Mater. Suppl.*, **40**, S1
[46] Li C., Wu R., Freeman A.J and Fu C.L., (1993) *Phys. Rev.* **B48**, 8317
[47] Duffy D.M., Harding J.H. and Stoneham A.M., (1994) *J. Appl. Phys.*, **76**, 2791
[48] Boys S.F. and Bernardi F., (1970) *Mol. Phys.*, **19**, 553
[49] Bromley S. T., French S. A., Sokol A. A., Sushko P. V. and Catlow C. R. A., (2000) *Applications of density functional theory in solid state chemistry*, Recent advances in Density Functional Methods, Vol. 1, Part 3, V. Barone A. Bencini, P. Fantucci (eds.), World Scientific Publishing Company,
[50] Ferrari A. M., Neyman K. M., Mayer M., Staufer M., Gates B. C., Rösch N., (1999) *J. Phys. Chem. B*, **103**, 5311
[51] Bromley S. T. and Catlow C. R. A., *to be submitted to Phys. Rev. B.*
[52] Lopez N., Pacchioni G., Maseras F., Illas F., (1998) *Chem. Phys Lett.*, **294**, 611
[53] Bromley S. T., Sankar G., Maschmeyer T., Thomas J. M., Catlow C. R. A., (2001) *Micro. Meso. Mater.*, **44-45**, 395.
[54] Bromley S. T., Sankar G., Catlow C. R. A., Macshmeyer T., Johnson B. F. G., Thomas J. M., (2001) *Chem. Phys. Lett.*, **340**, 524
[55] Bromley S. T., Sankar G., Catlow C. R. A. *to be submitted to J. Phys. Chem B.*
[56] Stone P., Bennett R.A., Poulston S. and Bowker M., (1999) *Surf. Sci.*, **435**, 501
[57] Bowker M. and Bennett R.A.; *private communication*
[58] Yudanov I., Pacchioni G., Neyman K. and Rösch N., (1997) *J. Phys. Chem. B*, **101**, 2786
[59] Goniakowski J., (1998) *Phys. Rev. B*, **58**, 1189

[60] Goniakowski J. and Noguera C., *Phys. Rev. B.* (1999) **60**, 16120
[61] Narui K., Furuta K., Yata H., Nishida A., Kohtoku Y. and Matsuzaki T., (1998) *Catalysis Today*, **45**, 173
[62] Hoang D.L. and Lieske H., (1994) *Catalysis Lett.*, **27**, 33
[63] Alfredsson M., (1999) *Polar Molecules in Crystalline and Surface Environment*, thesis from Uppsala University, Uppsala, Sweden

ELEMENTARY STEPS OF CATALYTIC PROCESSES ON METALLIC AND BIMETALLIC SURFACES

F. ILLAS[1], C. SOUSA[1], J.R.B. GOMES[1], A. CLOTET[2], J.M. RICART[2]
1)Departament de Química Física i Centre Especial de Recerca en Química Teòrica, Universitat de Barcelona, C/Martí i Franquès 1, E-08028 Barcelona, Spain
2)Departament de Química Física i Inorgànica i Institut d'Estudis Avançats, Universitat Rovira i Virgili, Pl. Imperial Tàrraco 1, E-43005 Tarragona, Spain

1. INTRODUCTION

Heterogeneous catalysis is widely used in the chemical industry to produce thousands of different products in massive amounts. On a global scale, the technological and economic importance of heterogeneous catalysis is immense because it provides the backbone of the world's chemical and oil industries. This is the reason for intense research in both academic and industrial laboratories. Depending on the particular compounds to be synthesized, the chemical active constituents of a catalysts can be metals[1-4], supported metals[5-9], sulfides and oxides[10] or zeolites[11-13]. We must advert that the literature about heterogeneous catalysis is enormous and the previous list is restricted to references that are particularly relevant either to the microscopic aspects of the mechanisms of heterogeneous catalysis or to some of the applications presented as examples of the particular approach considered in this chapter.

The innate complexity of practical catalytic systems has lead to "trial and error" procedures as the common approach for the design of new and more proficient catalysts. Unfortunately, this approach is far from being efficient and does not permit to reach a deep insight into the chemical nature of the catalytic processes. The consequence of this difficulty is a rather limited knowledge about the molecular mechanisms of heterogeneous catalysis. To provide information about catalysis on a molecular scale, surface science experiments on extremely well controlled conditions have been designed and resulted in a new research field in its own. However, even under these extremely controlled conditions it is still very difficult, almost impossible, to obtain precise information about the molecular mechanisms that underlie catalytic processes without an unbiased theoretical guide. The development of new and sophisticated experimental techniques that enable resolution at

M.A. Chaer Nascimento (ed.), Theoretical Aspects of Heterogeneous Catalysis, 149–181.

the atomic scale with the assistance of ab initio theoretical support are aimed to produce a deep impact on the understanding of heterogeneous catalysis at a molecular level. This is also because the spectacular advances in computers and supercomputers together with new developments in electronic structure theory have broken the practical limitations that have constrained practical applications of ab initio theory to systems of academic interest. Nowadays, accurate first-principles calculations on realistic systems of industrial and technological interest are becoming possible. For good reason the 1998 Nobel prize in chemistry was awarded to Walter Kohn "for his development of the density-functional theory" and to John Pople "for his development of computational methods in quantum chemistry".

This chapter is devoted to review some of the theoretical methods and models that are currently being applied to the study of elementary steps in heterogeneous catalysis. We will limit ourselves to describe the cluster model approach and do it from a critical point of view, signaling its strength as well as its weakness. Likewise, we will focus on the description of some of the non-standard techniques of analysis of the interaction of an adsorbate with a surface, which are thought to be crucial to understand the mechanisms of heterogeneous catalysis from a microscopic point of view. This chapter is planned as follows: First, the cluster model approach is presented discussing its use and limitations, the methods of analysis and of the computation of the potential energy surface. Next, the concepts introduced previously will be used to describe a few representative examples of increasing complexity. These examples are chosen to show how the cluster model approach can be used to help to solve problems in surface science and in heterogeneous catalysis.

2. THE CLUSTER MODEL APPROACH

The precise features of real catalysts at a microscopic scale are rather unknown but in all cases the main interactions occur through a surface. Two different theoretical models are often used to describe the electronic and other microscopic features of a surface. On the one hand, there is the solid state physics approach in which a surface is considered as a slab of a given thickness, finite in the direction perpendicular to the surface and infinite in the two other dimensions with perfect two-dimensional periodical symmetry. On the other hand, one has the cluster model approach which represents the surface with a finite number of atoms and the surface-adsorbate interaction as a supermolecule; this is essentially a quantum chemical approach. It is important to realize that both approaches are crude representations of physical reality because real surfaces are far from being perfect, usually containing a rather large amount of dislocations, steps and kinks and other point defects so that the periodic approach can be questioned. Of course, under ultra-high-vacuum well-controlled conditions, as those currently employed in surface science experiments, the slab approach may be more adequate. However, the surface of real catalysts has little resemblance to a perfect surface and the cluster model approach can be used to capture the essential features of a given catalytic system. The physical basis of the cluster model approach relies on the hypothesis that some properties are of local character and hence are determined by a reduced number of atoms of the surface of the catalysts.[14-16] We will provide strong evidence that some relevant adsorbate-surface properties are indeed local. The finite number of atoms defining the local region to study the property of interest delimits the cluster model of the catalyst. The remaining part of this chapter is aimed to provide a detailed discussion of the cluster model approach applied to the study of elementary steps in metallic and bimetallic surfaces.

2.1 Designing and Testing Cluster Models

The cluster model approach assumes that a limited number of atoms can be used to represent the catalyst active site. Ideally, one would like to include a few thousands atoms in the model so that the cluster boundary is sufficiently far from the cluster active site thus ensuring that edge effects are of minor importance and can be neglected. Unfortunately, the computational effort of an ab initio calculation grows quite rapidly with the number of atoms treated quantum mechanically and cluster models used in practice contain 20 to 50 atoms only. It is well possible that with the advent of the N-scaling methods[17-25] this number can dramatically increase. Likewise, the use of hybrid methods able to decompose a very large system in two subsets that are then treated at different level of accuracy, or define a quantum mechanical and a classical part, are also very promising.[26] However, its practical implementation for metallic systems remains still indeterminate.

The weak point of the cluster model approach lies precisely in the control of the inopportune edge effects. For ionic and covalent substrates very simple yet efficient embedding schemes are possible.[14-16,27-30] However, for metallic surfaces the embedding problem is largely an unsolved problem although we must mention the pioneering work of Whitten[31,32] and the promising approach recently developed by Carter and coworkers.[33,34] This later approach combines slab and cluster approaches treated by density functional theory to extract embedding potentials which can be later used in more elaborate calculations including the description of adsorbate electronic excited states.

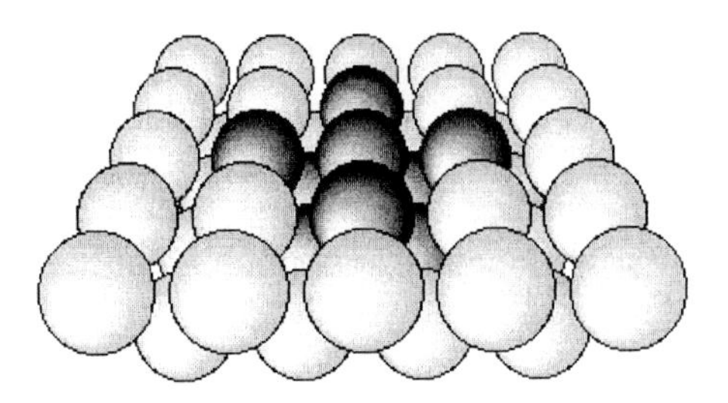

Figure 1. A two layer Pt_{41} (25,16) cluster model representation of an atop site of Pt(100). Dark spheres represent the nine (5,4) atoms in the local region whereas light spheres represent the metal atoms in the outer region.

To a large extent, the common approach consists in dividing the cluster into local and outer regions and then treat these regions differently. The local region should contain the atoms that are thought to play the key role in the particular phenomenon being studied, the rest of atoms in the model attempting to provide an adequate environment to the local region thus contributing to minimize the edge effects.

Usually the atoms on the local region include all electrons or a small core pseudopotential whereas the atoms in the outer region include a larger core and most often only the outermost *ns* electrons necessary to represent the metal conduction band are explicitly included in the quantum mechanical calculation. This is nowadays a standard procedure to model Ni, Cu, Ag and Pt surfaces.[35-44] In addition, a mixed basis scheme is usually employed to further reduce the computational cost while attempting to preserve the quality of the cluster model ab initio calculations.[45-47] In this scheme, the atoms in the outer region are treated with a rather limited, even minimal, basis set whereas atoms in the local region are treated with a more extended basis set.

Several strategies have been suggested to stablish the ability of a given cluster to provide a meaningful representation of the property of interest. A first trivial test consists on testing the influence of the cluster size and shape on the property being studied. Other aspects concern the cluster ionization potential and electron affinity. In a real metal surface both are equal and coincide with the surface work function. In general, the cluster ionization potential is close to the experimental surface work function even at the simplest Hartree-Fock level of theory but the electron affinity is usually too small and requires larger basis sets and the inclusion of electron correlation effects. Therefore, special care must be taken when studying processes where electron donation to the metal surface is the main physical mechanism.[48-50] A particular property that is characteristic of metal surfaces is its response to a test charge and it is customary to compare to the jellium model.[51] However, recent work has shown that the cluster model description of a metal surface goes well beyond the simple jellium model adding the atomistic character and showing that the definition of an image plane is not adequate.[52,53]

Another important aspect of the cluster model approach concerns the selection of the cluster electronic state. In principle, low-spin states should be chosen to represent the electronic structure of a non-magnetic metal surface and it is customary to use the lowest energy low-spin cluster electronic state for the naked cluster as well as for the cluster plus its corresponding adsorbate. This choice is respected even if in some cases the cluster plus adsorbate ground state does not directly correlate with the electronic ground state of the separated systems. The argument in favor of this choice is that in an extended surface the electronic states are close packed and consequently are accessible at almost no energy cost. Siegbahn et al. have shown that in a cluster this energy cost is not always negligible and proposed to compute the adsorption energy by referring to the cluster electronic state that is prepared to bond the adsorbate.[54-56] This bond-preparation idea is very attractive and has been used by other authors[57,58] showing that it leads to adsorption energies that are stable with the cluster size. The bond preparation rule arises naturally from the careful analysis of Bagus et al. of the adsorption energy of CO on cluster models representing the Cu(100) surface.[59] These authors have shown that the oscillation in the final adsorption energy is due to the different representation of the surface conduction band arising from the different clusters, causing a strong oscillation in the Pauli repulsion to the bond mechanism. Choosing a cluster electronic state different from the ground state permits to monitor this effect and to minimize cluster size effects on the calculated adsorption energy. However, there is an important consequence of using assumptions about the bond character. One of the most powerful features of the ab initio cluster model approach is lost. This lost feature is the ability to determine the leading physical interaction mechanism with an ab initio quantum mechanical formulation which requires as input only the cluster geometry and the level of cluster wave function or density. On the other hand, with the bond preparation, the overall analysis depends on choosing the "right" electronic state. Implicit in the idea of bond preparation is also the belief that bonding mechanisms, and hence the physical description of the chemisorption bond, depend on the electronic state chosen to represent the real system. Therefore, one would expect that the same adsorbate will exhibit different bonding mechanisms depending on the electronic state. It has been shown that this is not the case and that even if different electronic states may result in a qualitatively different behaviour ranging from attractive and bounded to repulsive and, clearly, unbounded they share a common bonding mechanism. For a fixed geometry, normally close to the experimental one if available, the different contributions to the bonding mechanisms arising from these different electronic states are remarkably similar except for the one arising from Pauli repulsion. Hence, the use of different electronic states does not necessarily introduce any change in the physical description of the chemisorption bond.[60] The fact that the chemisorption bond mechanism does not seem to strongly depend on the particular choice of the cluster model electronic state is in line with the common belief that chemisorption is a local phenomena and can be reasonably described with rather small cluster models regardless the cluster electronic state chosen to carry out the study, provided a low-lying state is used. Finally, we note that the bond-preparation rule faces some conceptual problems when dealing with species such as NO or F that are often adsorbed as anions and, hence, the chemisorption process necessarily implies an avoided cross between two adiabatic potential energy surfaces.[61,62]

Once a proper cluster has been built and tested as commented above, ab initio calculations at different levels of theory can be carried out to predict several properties of local character. One of the main advantages of the cluster model approach is that it permits the use of either wave function or density functional theory based computational methods to obtain the properties of interest.[14-16] The fact that accurate explicitly correlated wave functions can be obtained for the cluster model permits to assess the accuracy of current exchange-correlation functionals[63], to explore the effects of electronic correlation[64,65] and, in principle, to study adsorbate excited states as well.[66] Among the several properties of local character that can be studied by means of the cluster model we quote the determination of adsorbate geometries and vibrational frequencies, the interpretation of X-ray and UHV photoelectron spectra and the analysis of the bonding mechanisms.[14-16,67] An important particular feature of the cluster model approach is that it permits to explicitly include external electric field effects enabling the study of adsorbates in an electrochemical environment[68-70] and opening the way to the theoretical study of electrocatalysis. Most of the previous work studying the influence of an external electric field has been carried out on simple adsorbates such as CO or CN on different metal surfaces[68-76] but recent applications involve more complex adsorbates such as carbonate[77,78] and urea[79] on Pt surfaces.

2.2 Exploring adsorbate-substrate potential energy surfaces by means of the cluster model approach

The fact that in the cluster model approach the adsorbate plus the substrate are treated as a supermolecule permits to use the computational tools and strategies that are routinely used in molecular quantum chemistry. However, this system cannot be considered as a molecule and geometry optimization procedures have to be carried out with special caution. In most cases, it is convenient to fix the geometry of the substrate to prevent substrate geometries with little or no resemblance at all to the system being modeled. This geometrical constraint has important consequences because stable structures cannot always be univocally characterized as true stationary points of the potential energy surface. The common approach consists in performing a constrained geometry optimization with the adsorbate center of mass fixed above some special points of the surface and fully optimizing the adsorbate geometry and its distance above the surface. Occasionally, the final geometry can be compared with experimental data arising from surface science quantitative structure techniques such as Low Energy Electron Diffraction, LEED, or X-ray Photoelectron Diffraction, XPD. However, the use of these techniques requires the formation of an ordered adsorbed phase and this may not be the case for the system of interest. Catalytic reactions often involve many intermediates and isomers, the cluster model approach offers unique unbiased information about the structure and stability of these adsorbed species.

In a molecular system, the characterization of a stationary point of the potential energy surface is accomplished by the analysis of the eigenvalues of the hessian matrix that contains the second derivatives of the energy with respect to the cartesian coordinates of all atoms. In a cluster model of a given adsorbate-substrate system this approach is not always convenient because of the fixed geometry of the substrate. Nevertheless, the situation is changing because the landing of analytical second derivatives of the energy with respect to nuclear displacements — even when pseudopotentials are used to describe the inner cores of the metal atoms — permits to obtain the hessian matrix at a rather low computational cost. In this case, it is important to decouple the adsorbate and substrate modes with a two-fold purpose. On one hand, to facilitate the characterization of a given stationary point and, on the other hand, to assist the assignment of the normal modes of the adsorbate. A simple yet efficient way to decouple adsorbate and substrate modes consists in assuming a very large mass for the cluster metal atoms. Here, it is worth pointing out that this approach has been indirectly used in previous works dealing with the vibrational modes of diatomic molecules.[68-76] In this simple case the internal stretching and frustrated translation normal modes can be obtained by allowing nuclear displacements that maintain the molecule center of mass fixed. This procedure has been recently generalized to polyatomic adsorbates.[78]

2.3 Methods of analysis of the adsorbate-surface chemical bond

The exploration of the potential energy surface provides information based on total energy, it permits to find possible reaction intermediates and possible transition energy structures leading to a rather detailed picture of the catalytic reaction energy profile. While this information is no doubt of enormous importance it constitutes just the first step of cognition of the chemistry of a given process. The detailed understanding of the chemical mechanisms needs to go beyond the accurate description of the energy profile and analyze the electronic structure in detail. Different methods of analysis have been proposed to stablish the nature of a given chemical process. Most of these methods are indistinctly applied to gas-phase, organometallic chemistry or to surface catalyzed reactions.

The Mulliken population analysis[80] is unquestionably the most popular method. This population analysis is based in assigning the electron population of a given molecular orbital — either mono or multiconfigurational Hartree-Fock (HF and MCSCF, respectively) or Kohn-Sham — expressed in term of a Local Combination of Atomic Orbitals, LCAO, precisely to the atoms involved in the LCAO. In the Mulliken analysis the total population is hence divided into net and overlap populations and the overlap population arbitrarily equally distributed between the two overlapping atoms. In spite of this limitation and on the fact that the Mulliken analysis severely depends on the atomic basis sets[81] Mulliken charges are reported in almost every quantum chemical paper and most often presented as absolute measures. In the cluster model approach to chemisorption and catalysis the Mulliken analysis exhibits an additional shortcoming that is not usually mentioned. The use of a cluster model to represent a surface necessarily implies that some metal atoms have the proper bulk or surface coordination and other do not. The atoms with larger coordination have more contributions to the overlap population that the atoms at the cluster edge. Consequently, the atoms on the local region usually exhibit a very large net charge. Clearly, this is an artifact of the Mulliken population analysis and has no physical meaning. Charges on the adsorbate do not suffer from this artifact and can be used to unravel trends or tendencies.

Several alternatives to the Mulliken population have been presented that attempt to provide more rigorous estimates of the charges on atoms in molecules or clusters although not all have been applied in chemisorption and catalysis. We quote the Natural Bond Order analysis,[82] and the elegant topological analysis of the electron density[83] or of the electron localization function, ELF, introduced by Becke and Edgecombe.[84] ELF analysis has been extensively used to study the chemical bond in various situations[85-88] although so far there are no applications concerning the surface chemical bond. Based on an earlier idea of Davidson,[89] a projection operator technique has also been used to obtain an indication of the charge motion between a substrate and an adsorbate. In this procedure P_ϕ, the expectation value $P_\phi = \langle\Psi|\phi^+\phi|\Psi\rangle$, for an adsorbate orbital ϕ, occupied or virtual in the free molecule, is computed from the total Hartree-Fock wave function or Kohn-Sham determinant. It has been proposed that P_ϕ gives an indication of the extent to which the adsorbate orbital is occupied in the total wave function or density of the supermolecule.[90,91] This procedure is less basis set dependent and used in combination with other techniques of analysis permits to obtain a rather accurate picture of the charge transfer flow between adsorbate and substrate.[92,93]

The analysis of the dipole moment curves for the motion of the adsorbate perpendicular to the surface provides additional information about the degree of ionicity of a given surface chemical bond. Moreover, the analysis of the dipole moment curves is also related to the interpretation of variations of the surface work function induced by the presence of the adsorbate.[14,90,94] However, the response of the surface to the presence of the adsorbate does not permit to extract directly adsorbate charges from the dipole moment curve.[52,53] A procedure based in the use of frozen densities has been proposed[95] that permits to avoid the effect of the surface polarization on the dipole moment curve. Unfortunately, this method has not been yet extensively used. To close this short discussion about the different procedures commonly used to interpret the chemical bond between chemical species and the surface of a catalyst in terms of net charges we mention the valence bond reading of Hartree-Fock and Configuration Interaction wave functions. This procedure has been used to interpret the electronic correlation effect on the surface chemical bond.[96]

The methods of analysis of the surface chemical bond are not limited to the partition of the electronic structure into net charges and related quantities. Bagus et al. [97,98] have proposed to decompose the energy in different contributions by means of the use of constrained variations. The Constrained Space Orbital Variation, CSOV, method permits to decompose the Hartree-Fock interaction energy arising from two different units into Pauli repulsion (plus electrostatic interactions if units contain charges), intra-unit or polarization terms and inter-unit or charge donation contributions. At variance of other energy partition methods,[99,100] in the CSOV technique each step includes the effect of the previous one and, hence, avoids double counting. Initially designed for dative covalent bonds, the CSOV method has been later generalized and can be applied to decompose the interaction energy of almost any chemical bond.[101] The CSOV method was developed to decompose the Hartree-Fock interaction energy and, hence, its main limitation was the inability to decompose the correlation energy except in some simple MCSCF cases.[102] However, the advent of modern density functional theory permitted to extend the CSOV method to analyze either HF and DFT correlated energy routinely.[103] This is because of the single determinant nature of the Kohn-Sham, KS, procedure, almost universally used in practice to carry out DFT calculations. The CSOV technique is based in the partition of the orbital space and therefore it can also be applied to DFT calculations based on the Kohn-Sham approach in a rather straightforward way. It is worth to mention that a similar method to decompose the DFT energy was earlier developed by Ziegler and Rauk,[104] this analysis is implemented in the Amsterdam Density Functional code.[105]

A few applications have been reported where the CSOV technique has been employed within DFT methods. [106-110] However, these previous works were devoted to the study of different adsorbate-substrate interactions and not to compare the relative importance of the various bonding contributions as obtained from HF or DFT calculations. Recently, Márquez et al.[111] have reported a systematic comparison of the different contributions to the chemisorption bond that are obtained using HF and two different DFT based methods. Those are the gradient corrected, GC, exchange and correlation functionals obtained by combining the Becke gradient corrected, GC, exchange functional[112] with the GC correlation functional of Lee, Yang, and Parr[113], BLYP, and the widely used hybrid three-parameters functional, B3LYP, which mixes the HF single-determinant exchange with the Becke GC exchange.[114] The interaction of CO and of NH_3 on Cu(100) and on Pt(111) were chosen as representative examples and the various contributions to the interaction energy analyzed by means of the CSOV method.

The different contributions to the HF, BLYP and B3LYP interaction energy of CO on the $Pt_{13}(7,3,3)$ cluster model representation of the Pt(111) surface as obtained at the various steps of the CSOV method are reported on Figure 2. The most important result of this comparison is that the qualitative picture of the chemisorption bond arising from ab initio HF and DFT quantum chemical approaches is essentially the same; the relative importance of the different mechanisms remaining unchanged. This is an important conclusion because it validates many previous analysis of the chemisorption bond carried out in the framework of Hartree-Fock cluster model wave functions.[14,15]

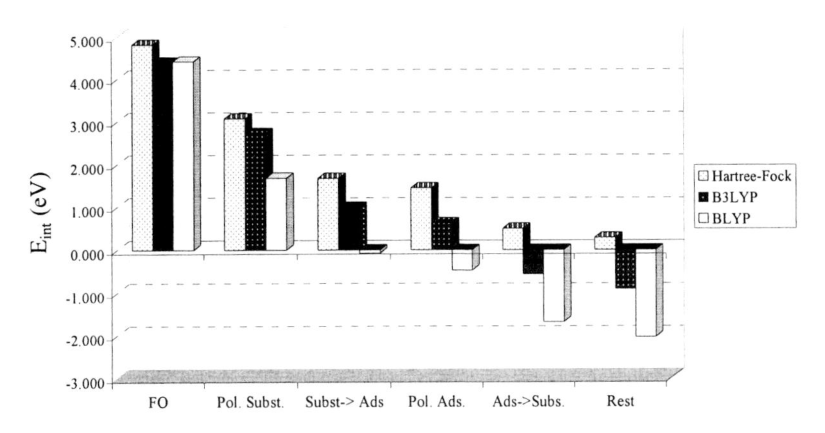

Figure 2. The interaction energy, E_{int} in eV, of CO on $Pt_{13}(7,3,3)$ at the various steps of the CSOV method at HF and DFT levels of theory.

The conclusion above is also in line with previous analysis of electronic correlation effects on the description of the chemisorption bond carried out by comparing correlated and uncorrelated cluster model wave functions.[64,96] Finally, we comment on the relationship between the CSOV method and the Charge Decomposition Analysis, CDA, proposed by Dapprich and Frenking[115] as a quantitative expression of the well-known donation-backdonation mechanism. Analyzing the mathematical expressions of CDA one can see that, in principle, the CDA method provides the equivalent of the CSOV method but in a partition of the total electron density instead of the total energy. Therefore, one would expect that CDA and CSOV complement each other so that a given interaction could be analyzed in terms of charge transfer and its energy contribution simultaneously. In a recent paper, García-Hernández et al.[116] have shown that unfortunately this is not the case. For the interaction of CO on Cu(100), Ag(100) and Pd(100), the CDA method predicts the σ-donation to be more important that the π-backdonation. This is contrary to chemical intuition and to the analysis of the chemisorption bond resulting from the CSOV energy decomposition method and from the use of projection operators. This failure of CDA to properly describe the interaction of CO on a transition metal surface has been related to the presence of near degeneracies in the basis set. Interestingly enough, the threshold for appearance of these artifacts is much higher than the one having undesirable effects on the total energy. Consequently, the application of CDA to chemisorption problems has to be carried out with extreme care.

3. FROM SURFACE SCIENCE TO HETEROGENEOUS CATALYSIS

This section reports a series of examples of application of the cluster model approach to problems in chemisorption and catalysis. The first examples concern rather simple surface science systems such as the interaction of CO on metallic and bimetallic surfaces. The mechanism of H_2 dissociation on bimetallic PdCu catalysts is discussed to illustrate the cluster model approach to a simple catalytic system. Next, we show how the cluster model can be used to gain insight into the understanding of promotion in catalysis using the activation of CO_2 promoted by alkali metals as a key example. The oxidation of methanol to formaldehyde and the catalytic coupling of propyne to benzene on copper surfaces constitute examples of more complex catalytic reactions.

3.1 The interaction of CO on metallic and bimetallic surfaces

The interaction of CO on metal surfaces is of importance because it provides a rather simple surface science system but also because catalysts containing noble metals in different forms and ratio have been used widely to lower the CO and hydrocarbon emissions in the automobiles exhaust.[117] These catalysts promote the oxidation of CO and C_xH_y species to CO_2 and H_2O. At low temperature, CO is chemisorbed on many single crystal well-defined metal surfaces as a stable species. The shift of the CO vibrational frequency, defined as the difference with respect to the gas phase value, is often used as a guide to the adsorption mode, site and to the adsorption mechanism. Many authors using a variety of theoretical and experimental approaches have studied the bonding mechanism of CO to a metal surface. Nearly forty years ago, Blyholder[118] extended to metal surfaces the well known donation-backdonation mechanism earlier proposed by Dewar-Chat-Duncanson, DCD, for metal-ligand interactions.[119,120] This model assumes that the bonding mechanism of CO to a metal surface is dominated by two different contributions; these are the σ-donation from the CO 5σ orbital to the metal conduction band and the π-backdonation from the metal d_π orbitals towards the CO $2\pi^*$ antibonding orbitals. Blyholder also assumed that the CO vibrational shift was a result of the competition of these two mechanisms, σ-donation favors CO^+ and tends to raise the CO vibrational frequency whereas the opposite holds for π-backdonation. Bagus et al.[59,93,121] studied in detail several aspects of the CO-surface interaction by means of the Constrained Space Orbital Variation, CSOV, method described in the previous section. For the interaction of CO on Cu(100) and Pd(100) surfaces their CSOV analysis revealed that the bonding interactions are indeed dominated by the π-backdonation contribution. They also found that the origin of the CO vibrational shift results from a cancellation between the Pauli repulsion effect that always tends to increase the CO vibrational frequency and the π-backdonation, they also devised the "*wall effect*" term to denote the Pauli repulsion effect on the CO vibrational stretch.[70,122,123] Recently, it has been shown that for CO on Pt surfaces the σ-donation does also play a significant role. This is because the electronic configuration of Pt contains an open d-shell and hence is able to receive electronic density. The different importance of the energy contributions to the HF or DFT bonding between CO and a Pt surface cluster model is clearly evidenced in Figure 2.

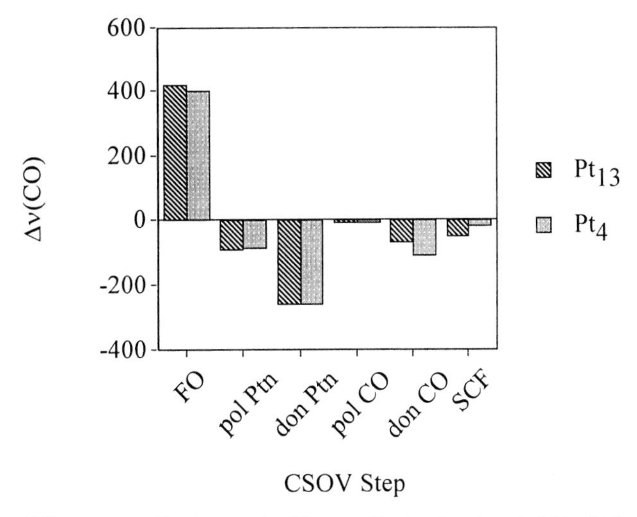

Figure 3. The different contribution to the Hartree-Fock vibrational shift of chemisorbed CO. FO stands for the Pauli repulsion or Frozen Orbital, FO, contribution, "pol Pt_n" and "don Pt_n" correspond to the substrate polarization and donation from the subtrate to the CO adsorbed molecule (π-backdonation), "pol CO" and "don CO" stand for to the CO polarization and donation from CO to the surface (σ-donation), SCF contains the rest of contributions due to coupling between the different mechanisms.

Two important points were missed in the Blyholder model, the first one is the role of the Pauli repulsion and the second is that, contrarily to what is assumed in this model, the σ-donation also contributes to the red shift of CO chemisorbed on low index platinum surfaces.[124,125]. In fact, all bonding mechanisms, other than Pauli repulsion, but specially π-backdonation, contribute to observed red-shift. Figure 3 reports the CSOV analysis of the CO vibrational shift, $\Delta\nu(CO)$ on top of two different cluster models representing the Pt(111) surface.

Moreover, the analysis of the interaction of CO on the different active sites of the low index Pt surfaces shows that the π-backdonation contribution to the red-shift is very similar for CO on different sites. Therefore, the π-backdonation cannot be the responsible for the observed difference between the CO vibrational frequency on on-top and bridge sites. The CSOV decomposition has permitted to reveal that the leading term contributing to this difference in vibrational frequency of chemisorbed CO is the initial Pauli repulsion or "wall effect"; this was a new, important and unexpected conclusion.

Next, we discuss the interaction of CO on bimetallic PdCu surfaces of different composition. This system has also been studied by means of the CSOV method complemented by the use of the projection operator technique.[126] The key point to extract from the CSOV analysis is that many important contributions are present in the CO-bimetallic surface bonding. This is in contrast to results presented by Hammer et al.[127] for a more extended series of alloys and bimetallic overlayers. In addition, it is observed that the interaction energy decreases with copper content in PdCu systems and this seems to be mainly related with variations in the Pauli repulsion and, more importantly, in electronic correlation contributions.[126] This has an important consequence, the experimentally observed linear dependence between the interaction energy and the core-level shifts of the isolated alloys is clearly originated by the Pauli repulsion and by the correlation effects. In fact, although the magnitude of the Pd 4s shift is dominated by the initial-state Pd polarization (hybridization) in response to the presence of copper, the correlation contribution to the shift has the larger variation between the bimetallic clusters studied; on going from Pd_4Cu_6 to Pd_1Cu_{12}, the differential polarization of the electronic clouds in response to the heterometallic bond formation has almost a null contribution due to a compensation effect between Pd and Cu.

Figure 4 shows that contrarily to what it has been proposed for metal overlayers,[127] the CO ($2\pi^*$) backdonation contribution does not permit to explain the experimental, and also calculated, linear relationship between the interaction energy and the Pd core level shift, pointing out to a different chemical interpretation of the phenomenon in alloys, at least PdCu alloys, and overlayers. Finally, the CSOV and projection analyses permit to explain the C-O stretch insensitivity to alloy composition. Since charge-transfer processes and Pauli repulsion do not vary significantly with copper content, only the correlation contribution is expected to influence this observable, but its variation with the internal C-O coordinate is usually of small to moderate magnitude.

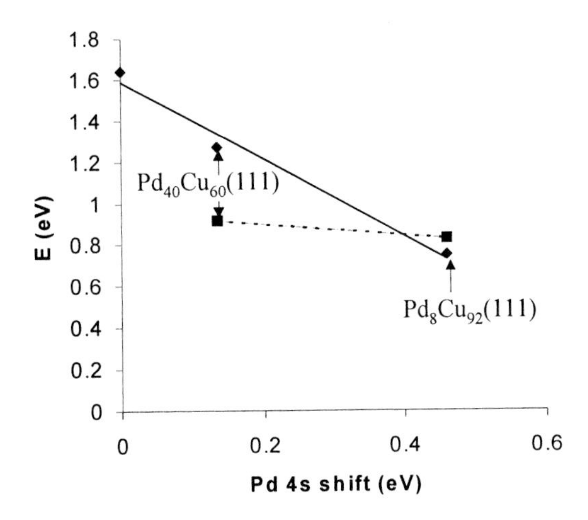

Figure 4. Correlation between Pd core-level shifts (eV) in PdCu alloys and CO adsorption energy (rhombus) or d(metal) to $2\pi^*$ CO interaction energy (squares) as obtained from the CSOV calculations, in eV. Results are BSSE corrected.

3.2 The dissociation of H_2 of PdCu(111)

Bimetallic catalysts have been long used in hydrocarbon industry due to the possibility of exploit and combine the different properties of various metals towards a given chemical reaction.[1,128,129] The bimetallic PdCu system is an example of catalyst which is employed efficiently in industry. Its main application is to catalyze the dissociation of molecular hydrogen, a very important reaction in heterogeneous catalysis since it is currently used in hydrogenation processes. PdCu catalysts are also of great interest in the automotive industry, since those are able to simultaneously oxidize CO and reduce NO[130], in the alkene oxidation[131], in the ethanol decomposition[132] and in several hydrogenation processes;[130-135] for instance of CO, benzene and toluene. These current applications make PdCu bimetallic catalysts the main target of many researchers. The nature of the heterometallic bond in these alloys have been investigated theoretically by means of the cluster model approach and first principles density functional calculations.[136] These authors applied the CSOV method to the bimetallic clusters in such a way that each component of the alloy was considered as a different unit. From this analysis it was concluded that there is charge transfer from Cu(4sp) to the Pd(5sp) orbitals, Pd(4d)→Pd(5sp) rehybridization and almost negligible changes in the Cu(3d) population. Likewise, it was found that surface Pd atoms carry on a small but noticeable negative charge which do not contradict the positive shift of the Pd core-levels. This shift was found to arise from the motion of electron density away from the core region due to the large Pd(4d)→Pd(5sp) rehybridization. The changes in the electronic structure described above have a marked influence in the chemical reactivity of these binary alloys. Additional information about the reactivity of PdCu bimetallic surfaces requires the use of a probe molecule such of CO or H_2. The bonding of CO with these bimetallic surfaces has been discussed in the previous section. Here, we will discuss the interaction of H_2 with the same cluster models of the PdCu(111) surface. The choice of H_2 permits to test the reactivity of a probe molecule which experiences some chemical reaction and is involved in breaking and forming bonds. It is important to stress again that H_2 dissociation on PdCu catalysts is an important step in industrial catalysis.

The dissociation of the hydrogen molecule was studied for three different adsorption sites on the catalyst surface. The active sites are defined as the final position occupied by the single hydrogen atoms. These are the bridge and two hollow sites depicted in Figure 5. The capacity of two different PdCu bimetallic catalysts to dissociate the hydrogen molecule has been analyzed and compared with that of Cu and Pd surfaces. The two different alloys are $Pd_{40}Pd_{60}$ and Pd_8Cu_{92} represented by Pd_4Cu_6 and $PdCu_{12}$ cluster models, respectively. Likewise, Pd_{10} and Cu_{10} cluster models are used to represent Pd(111) and Cu(111). The analysis of results is based on cluster model calculations carried out at the hybrid B3LYP density functional theory based method. Additional information about these approximations and also about the computational scheme followed to retrieve information concerning the several steps of hydrogen dissociation in the surfaces considered can be found elsewhere.[137]

The cluster approach within the B3LYP exchange-correlation functional properly predicts the dissociative character of the H_2 chemisorption on Pd(111) and the low reactivity of Cu(111) towards the same molecule. Moreover, the final position of atomic hydrogen on Pd(111) predicted by the present calculations is in agreement with previous theoretical studies using a completely different computational approach based on a plane wave basis pseudopotential slab model.[138] The reactivity of H_2 towards the different copper clusters, with atomic hydrogen located at the cluster edge, agrees with the calculations of Triguero et al. [139] for the interaction of atomic H with optimized Cu clusters. These results clearly show the adequacy of the cluster model approach to study the reactivity of molecular hydrogen with PdCu bimetallic surfaces.

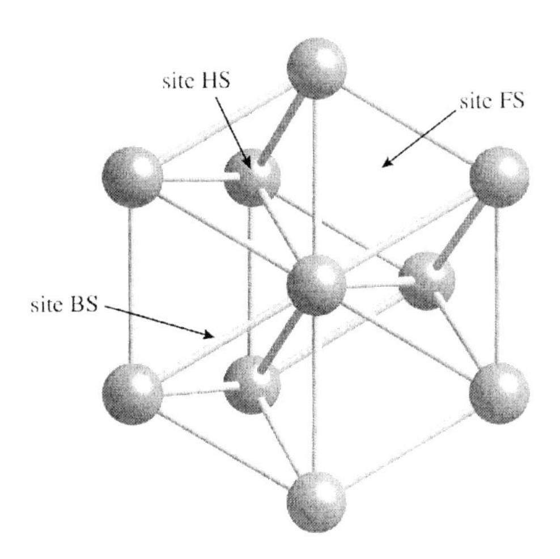

Figure 5. Schematic representation of the bridge, BS, tetrahedral hollow site, HS, and octahedral, FS, active sites considered for H_2 dissociative chemisorption. The M_{10} cluster model depicted in the figure can either represent Pd_{10}, Cu_{10}, Pd_4Cu_6 or the two first layers of $PdCu_{12}$. In the two former cases all atoms are equal. In Pd_4Cu_6 the first layer central atom and those on the second layer are Pd and the six remaining atoms on the first layer are Cu and, finally, in $PdCu_{12}$ the first layer central atom is Pd and all remaining atoms are Cu.

Another important result is that in spite of the similar surface morphology of the Pd_4Cu_6 and $PdCu_{12}$ bimetallic clusters both exhibit a rather different reactivity towards molecular hydrogen. The cluster model calculation show that a single surface Pd atom coordinated to other Pd atoms in the second layer is able to trap and dissociate molecular hydrogen with a very low energy cost thus being a potential active site for catalysis. It is worth pointing out that the existence of isolated Pd atoms in the surface that are active towards molecular hydrogen has been invoked to explain experimental data.[140] However, the cluster model study does also show that not all isolated Pd surface atoms of the PdCu(111) surface are equally active. The reactivity of a single Pd atom coordinated to Cu atoms, both in the surface and second layer, is very different from that described above for Pd_4Cu_6. Therefore, an important conclusion reached from the theoretical study is that a necessary condition for single Pd atoms at the PdCu surface is to be coordinated to other Pd atoms below the surface, i.e., forming Pd microclusters that are surrounded by the other component of the binary alloy. This is an important result that points out that electronic, or ligand, effects do also play an important role in the activity of the PdCu(111) surface sites towards molecular hydrogen.

3.3 The activation of CO_2 by metal alkali-promoted surfaces

The interaction of carbon dioxide with metal surfaces and its possible activation is of enormous potential interest because atmospheric CO_2 provides the most abundant and cheapest source of C-containing compounds. In addition, CO_2 activation would also help to reduce the greenhouse effect. Unfortunately, CO_2 is very stable and unreactive, its natural activation occurs in the different steps of photosynthesis only and therefore the industrial applications are very limited. In spite of about 40 years of interest in the CO_2 surface chemistry,[141,142] the microscopic information about the chemistry of CO_2 at metal surfaces related to heterogeneous catalysis is very limited. The existence of only a few theoretical works devoted to this issue is probably due to the difficulty to describe the interaction energy.[143,144] Again, the key step for heterogeneously catalysed reactions involving CO_2 is its activation. A possible activation mechanism consists in a partial charge transfer from the metal to the CO_2 molecule, leading to a basic change in the bonding and structure of adsorbed CO_2. In the particular case of platinum group metals, this process has a very high activation barrier due, in part, to the large work function of these metal surfaces. Thus, on these surfaces CO_2 adsorption is weak and non-dissociative [144,145]. By means of high resolution electron energy loss spectroscopy, HREELS, Liu et al.[145] established that wile CO_2 is not adsorbed on clean Pt(111), both physisorbed and chemisorbed CO_2 species exist on the surface when preadsorbed potassium is present. In these conditions, the presence of negatively charged CO_2 was suggested by the vibrational analysis. Similar results were obtained by Wohlrab et al.[146] for CO_2 on Pd(111) with coadsorbed Na. The chemisorption of CO_2 on Pd(111) and on clean and K-covered Pt(111) has been studied by using cluster models. The coordination sites, adsorption modes, vibrational frequencies and the effect of coadsorption were analyzed in detail.[143,144]

To study CO_2 on clean Pd(111), two different clusters $Pd_{10}(7,3)$ and $Pd_{15}(10,5)$ were selected to represent mono-coordinated and bi-coordinated adsorption modes respectively. The local/outer separation described above was employed, pseudopotentials and basis sets chosen according to this partition. The hybrid B3LYP density functional method was used to explore the potential energy surface. The different optimizations converged to three unique species corresponding to two coordination models only. For the η^1-C coordination two different species were found, one being a physisorbed and almost undistorted CO_2 with a concomitant adsorption energy of about 3 kcal/mol and the other being a highly bent CO_2 molecule, with a chemisorption energy of 6 kcal/mol. Since about 27 kcal/mol are necessary to bend the molecule, the interaction energy with respect to the bent molecule is noticeable (32 kcal/mol). The second coordination involves a η^2-CO with the C-O bond parallel to the surface and with the C and one O atom interacting directly with different surface metal atoms. The adsorption energy of this η^2-CO species is –3 kcal/mol, thus suggesting an endothermic process. However, as the energy cost to bend CO_2 in this configuration is ~ 47 kcal/mol, even larger than in the previous case, the interaction energy with the activated (bent) molecule will be ~40 kcal/mol.

The same strategy has been used to study the interaction of CO_2 with clean and K covered Pt(111).[143] The $Pt_{10}(7,3)$ and $Pt_{16}(10,6)$ cluster models were used and partitioned in local and outer regions. Using these two cluster models and the same starting points used in the study of the interaction of CO_2 with Pd surface cluster models it is predicted that CO_2 does not chemisorb on the clean ideal Pt(111) surface. Two adsorbed stable structures were found; those are η^1-C (parallel to the surface) and η^1-O (perpendicular to the surface) but in both cases CO_2 is undistorted, far away from the surface, and with an interaction energy of ~ 3 kcal/mol only. To model the effect of the presence of co-adsorbed potassium, at different coverage, one and three potassium atoms were placed in a threefold, hcp, position around a CO_2 molecule in η^1-C coordination on top of the central Pt surface atom. Both CO_2 and K were permitted to move on the surface. The preferred situation is shown in Figure 6, the CO_2 molecule is highly bent and the K atom moves to a bridge position, approaching the molecule. The interaction energy of CO_2 with respect Pt_{16}+K is only 2 kcal/mol, and becomes negative if the basis set superposition error is properly corrected.

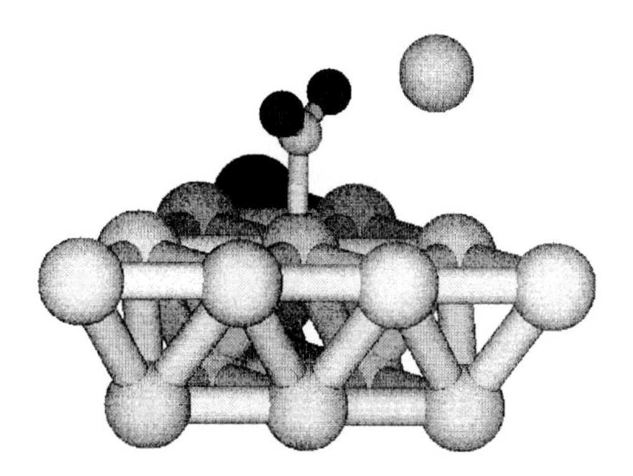

Figure 6. Absolute minimum for CO_2 chemisorbed on $Pt_{16}K$.

In the basis of the small interaction energy it may be argued that this structure is biased by the use of a cluster model. However, it is very unlikely that the highly distorted structure of CO_2 could be just the result of a computational artifact arising either from the cluster model approach or from BSSE. The same line of reasoning used in the case of Pd shows that the binding energy arises from two different effects, the energy necessary to activate the molecule (which is now ~ 46 kcal/mol) and the interaction energy of the distorted molecule, and this is a large value. The final energy balance may be affected by the use of a cluster but the general trend is not. This balance is important because it shows the large influence of the co-adsorbed potassium as an activator. Indeed, CO_2 desorbs or leads to the physisorbed species if K is not present on the surface. For a higher K coverage, represented by a cluster model having three K atoms on the surface, the same picture is obtained, Figure 7.

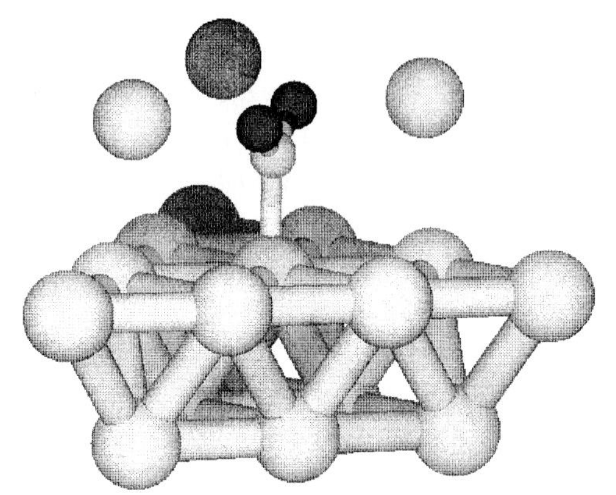

Figure 7. Absolute minimum for CO_2 chemisorbed on $Pt_{16}K_3$.

The main difference with respect to the previous case is that the binding energy is now of 27 kcal/mol. This fact is attributed to a decrease in the surface work function induced by K atoms and shows the importance of K coverage on the chemisorption of CO_2.[145]

Once the structure of the stable chemisorbed species has been established, it is important to analyze the nature of the chemisorption bond, with special emphasis on the CO_2-K interaction. Using the projection operator technique, the net charge on CO_2, K and Pt_{16} are about -0.9 e, 0.7 e, and 0.2 e, respectively. This is a strong indication that K is the main donor agent. However, a point that remains unclear is whether the donation from K to CO_2 is direct or indirect and mediated by the metal surface. It may be claimed that the effect of K is simply to reduce the work function of the substrate, thus enhancing the charge donation to CO_2 which is activated without a direct K-CO_2 interaction. A simple theoretical experiment shows that this is not the case. A K atom was placed at the opposite face of the cluster, thus reducing the work function but not interacting with CO_2. The re-optimization of the system leads to CO_2 desorption. This is a rather strong indication that in order to activate CO_2, the K atom needs to be directly involved in the interaction with the incoming molecule. This interaction is evidenced by means of the use of the Bader's theory of atoms in molecules[83]. The most salient feature is the appearance of (3,-1) saddle critical points between C and the Pt atom directly below and between K and oxygen atoms. This demonstrates the existence of chemical bonds between CO_2 and K and therefore reinforces the hypothesis of a direct participation of K in the activation of CO_2.

3.4 The oxidation of methanol to formaldehyde

Industrially, methanol is produced in huge quantities from syn-gas — a mixture of CO, CO_2, and H_2 — over oxide supported copper catalysts; ZnO or/and Al_2O_3 are typically used as support.[147-150] However, methanol production over the clean copper surface is also posible[151-153]. This alcohol is used in the chemical industry to obtain formaldehyde by its oxidation over silver catalysts.[154,155] Furthermore, methanol oxidation over transition metal catalysts is also of interest because this is a promising reactant to be employed in fuel cells.[156]

Due to the importance of this heterogeneously catalyzed reaction, several studies devoted to the understanding of the interaction and reaction of methanol on transition metal surfaces have been reported.[157-159] Unfortunately, results from experimental studies are difficult to interpret because several reaction intermediates are formed and, in addition, a number of different competitive reaction pathways exist. Obviosuly, obtaining a clear picture of the total reaction mechanisms in surface science problems is a formidable and difficult task. Therefore, a theoretical approach may be of invaluable interest to get insight into these cases, since it permits to model each specific reaction without having to worry about the competitive reactions as in the experimental situation. The mechanism of catalytic oxidation of methanol over a Cu(111) surface has been studied by means of a DFT cluster model approach. The energetics of the different elemental steps indicated and the preferred adsorption modes for the $CH_3OH \rightarrow H_2CO$ reaction taking place above the Cu(111) surface are shown in Figure 8.

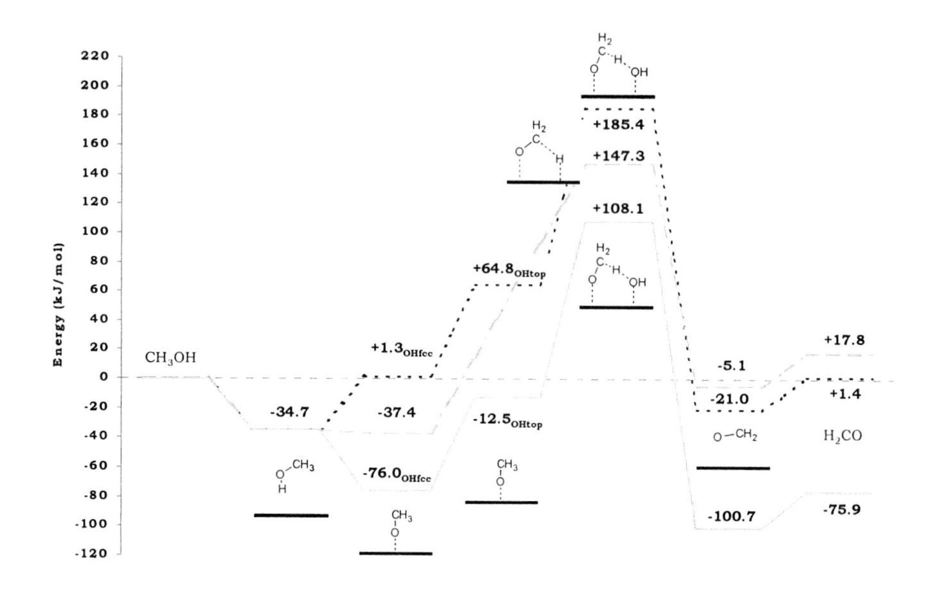

Figure 8. Different possible reaction routes for the oxidation of methanol to formaldehyde.

The first reaction mechanism shown schematically below (scheme 1) concerns the co-adsorption and reaction of methanol with hydroxyl species present above the Cu(111) catalyst. The hydroxyl species are responsible for the dehydrogenation, in a first step, of the methanol molecule and, in a second step, of the methoxy radical. A second reaction route comes from the methanol reaction with a Cu(111) surface covered with atomic oxygen, producing water and molecular hydrogen (reaction scheme 2). Water is produced from the interaction of two methanol molecules with an oxygen atom present on the surface while molecular hydrogen is formed after combination of two hydrogen atoms released in the dehydrogenation of methoxy to formaldehyde. A third possible reaction mechanism starts with the interaction of methanol with atomic oxygen present on the metallic catalyst, yielding a surface hydroxyl species. Then, this hydroxyl species reacts with the methoxy radical formed yielding one water molecule per each formaldehyde molecule formed (reaction scheme 3).

Reaction scheme 2 is proposed based on experimental results coming from the application of different experimental methods.[160,161] The energy data was obtained by consideration of the total energy of each gas-phase species and also from each adsorbate interacting with a $Cu_{22}(14,8)$ cluster substrate used to model the perfect Cu(111)' surface. The geometric parameters of the transition state structures were taken from published data[162] while details for the other structures considered, as well as the computational approach used to obtain the energetic data, can be found in a previous work.[163]

$$CH_3OH_{(g)} \longrightarrow CH_3OH_{(ads)} \xrightarrow[-H_2O_{(g)}]{+2OH_{(ads)}} CH_3O_{(ads)} \xrightarrow[-H_2O_{(g)}]{+OH_{(ads)}} H_2CO_{(ads)} \longrightarrow H_2CO_{(g)} \quad (1)$$

$$2\,CH_3OH_{(g)} \longrightarrow 2\,CH_3OH_{(ads)} \xrightarrow[-H_2O_{(g)}]{+O_{(ads)}} 2\,CH_3O_{(ads)} \xrightarrow[-H_{2\,(g)}]{} 2\,H_2CO_{(ads)} \longrightarrow 2\,H_2CO_{(g)} \quad (2)$$

$$CH_3OH_{(g)} \longrightarrow CH_3OH_{(ads)} \xrightarrow{+O_{(ads)}} CH_3O_{(ads)} + OH_{(ads)} \xrightarrow[-H_2O_{(g)}]{} H_2CO_{(ads)} \longrightarrow H_2CO_{(g)} \quad (3)$$

The comparison of the last and initial stages in the energetic variation depicted in Figure 8 shows that according to scheme 1, i.e. if the reaction proceeds over an hydroxylated Cu(111) surface, the energy required for the oxidation of methanol to formaldehyde is almost zero. In fact, only 1.4 kJ are consumed in the oxidation of one molecule of methanol originating one formaldehyde molecule. Through reaction scheme 2, the energy needed for the oxidation of one molecule of methanol is higher, ~18 kJ. The oxidation of methanol to formaldehyde is highly favorable (exothermic) for the reaction following the third mechanism. From a thermodynamic point of view, scheme 3 seems to be the preferred mechanism for the production of formaldehyde. Looking at the transition state energies, the reaction barrier from methoxy to the TS is similar, both for CH_3O decomposition on the OH covered surface or on the clean copper surface, the values being 184.1 kJ/mol or 184.7 kJ/mol, respectively. Thus, the three reaction mechanisms might be expected to be competitive. The energetic barrier for reaction schemes 1 and 3 are calculated from the energy required to promote one OH species from a fcc hollow site to a top site. The fcc hollow was calculated to be more favorable for OH adsorption on the Cu(111) surface and the energetic cost to promote it to the most unfavorable top site is of about 63.5 kJ/mol. In a previous work,[163] the energy barriers calculated for the reaction of CH_3O with OH adsorbed on a contiguous fcc or hcp-long sites are of up to 18 kJ/mol smaller than the calculated energy for the same reaction with OH adsorbed on a top site. However, the latter mechanism proceeds through two small reaction barriers and is thought to be more favorable for reaction completeness. Also, one should keep in mind that for reaction schemes 1 and 3, the methoxy islands formed above the surface should be close to hydroxyl species. Scanning Tunneling Microscopy, STM, results for CH_3OH adsorption on the Cu(110) surface[164-166] show that methoxy species formed at the catalytic surface do not react with nearby oxygen islands. This observation is consistent with Temperature Programmed Desorption (TPD) data[167] which show the release of small quantities of water during the decomposition of methoxy. Unfortunately, the theoretical results cannot be directly compared with the available STM and TPD data because the experimental and theoretical surfaces exhibit different Miller indexes. Bowker and coworkers do also report the presence of added Cu atoms in the middle of oxygen and methoxy islands for the Cu(110) surface which will be able to capture the hydrogen atom formed in the decomposition of methoxy to formaldehyde.[164,165] Additionally, isolated oxygen atoms are proposed to activate the added Cu atoms that participate in the dehydrogenation of methoxy. This explains why the direct dehydrogenation mechanism, where one of the hydrogen atoms of methoxy is adsorbed directly over the copper surface, is preferred. Similarly, STM and TPD studies would be important to complete the study of the Cu(111) surface where hollow sites are much closer and the space available for added Cu atoms is smaller. In summary, in contrast with data from the Cu(110) surface, it is expected that methanol oxidation catalyzed by the Cu(111) surface can occur by three competitive reaction mechanisms. Moreover, the cluster calculations predict that on the Cu(111) surface larger quantities of water than on the Cu(110) surface will be formed during the $CH_3O \rightarrow H_2CO$ reaction step.

3.5 The catalytic coupling of propyne to benzene on Cu(111)

Small hydrocarbon species chemisorbed on metal surfaces are involved in a variety of important catalytic processes. A detailed knowledge of the elementary steps involved in the reactivity is essential in the understanding of the microscopic mechanisms of these processes. In this final section some aspects of the reactivity of propyne chemisorbed on Cu(111) will be discussed. Propyne shows a rather more complex reactivity than that of acetylene which is the simplest alkyne. In particular it has been shown that while ethyne chemisorbed on Cu, Pd, Pd/Au trimerizes leading to benzene through a mechanism involving a C_4H_4 metallocycle intermediate,[168-170] propyne undergoes less efficient coupling reactions and yields either benzene or C_6 dienes of 82 atomic mass units (amu).[171] These two processes occur with approximately equal probability. A proposed mechanism for benzene[171] formation involves head(C_1)-to-tail(C_3) interaction of two propyne molecules. It is likely that 1,4 cyclohexadiene is the intermediate in this process, its formation being rate determining. Head-to-head and tail-to-tail coupling of two propyne molecules is also possible. The resulting metallocycle is unable to add a third propyne (or acetylene) molecule.

Instead, it undergoes either hydrogenation to C_6 dienes or disproportionation to butadiene and ethyne (a minor pathway).

The coupling mechanism of two propyne molecules on the Cu(111) surface has also been studied using the B3LYP cluster model approach within the usual partition of the cluster regions. These calculations are aimed to explore the different reaction paths and the thermodynamic feasibility of the reactivity in both gas phase and on the surface.[172] To couple two propyne molecules a strong distortion of the isolated molecules is necessary. In the gas phase this can be achieved by promotion to the triplet state. In addition, to form 1,4 cyclohexadiene an intramolecular hydrogen transfer is required leading to vinylcarbene, a biradical species. Next, vinylcarbene can be head-to-tail or head-to-head coupled to yield either 1,3- or 1,4-cyclohexadiene, which, after H_2 elimination lead to benzene. In the gas phase, the energy necessary to promote propyne to the triplet state is very high (~3.75 eV) (See Figure 9). From this promoted species the formation of vinylcarbene is exothermic. Once the two vinylcarbenes are formed they can couple to give the six ring intermediates with a further energy lowering and, finally, molecular hydrogen can be easily eliminated.

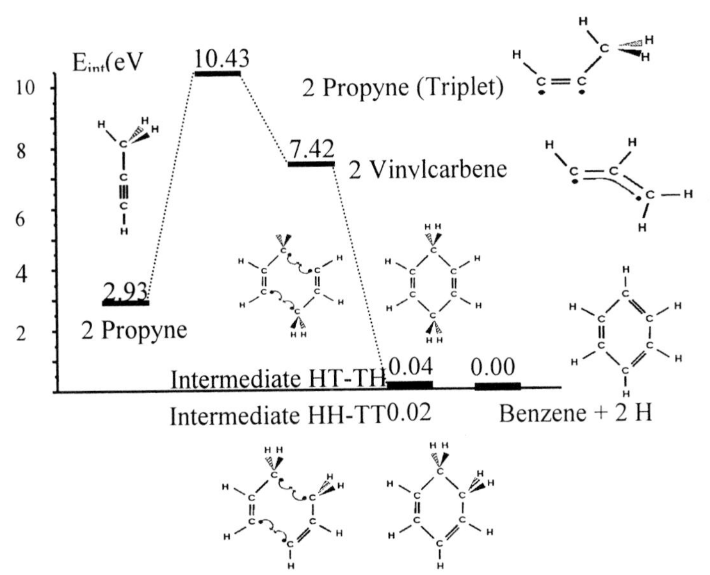

Figure 9. Schematic energy profile of the gas phase propyne dimerization reaction.

However, it is seen that while the overall reaction is thermodynamically favorable, the formation of the *activated* propyne requires a very high energy and, hence the reaction is hindered. The fact that the reaction spontaneously takes place on Cu(111) must be due to a direct activating role played by the surface, i.e. lowering the energy necessary to give the intermediates. The dimerization reaction on the surface has been studied with the help of a Cu_{22}(12,7,3) cluster model with the three innermost surface metal atoms defining the local region. Both propyne and vinylcarbene species were found to be chemisorbed and stable on the surface cluster model. Moreover, chemisorbed propyne appears to be not linear, but strongly bent and structurally close to the gas phase propyne in its triplet state.

The chemisorption energy is 0.22 eV and can be decomposed in the energy necessary to activate the molecule (3.24 eV) and the interaction energy of this activated species with the surface (3.46 eV) thus leading to a exothermic process. For the chemisorbed vinylcarbene radical the changes in structure with respect the gas phase are small. Vinylcarbene is already distorted and well suited to interact with the surface, the binding energy being now of 2.70 eV, or 0.45 eV with respect to the gas-phase propyne. Therefore, it is reasonable that adsorbed propyne spontaneously isomerizes to adsorbed vinylcarbene.

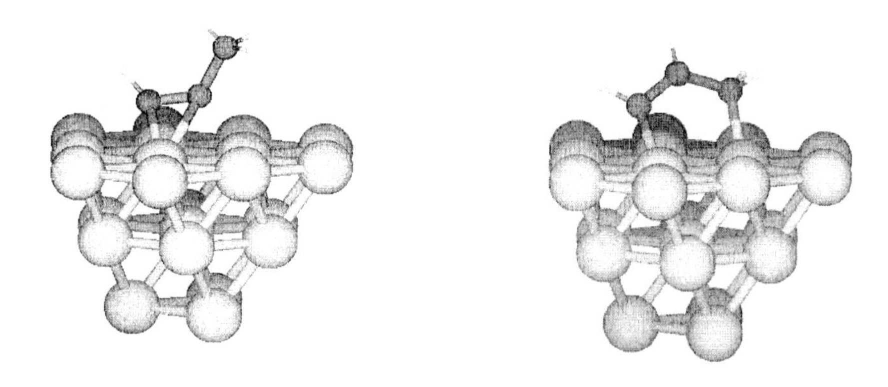

Figure 10. Schematic representation of the structure of propyne and vinylcarbene chemisorbed on the Cu_{22} cluster model.

Once the structures of the two species were obtained, calculations were carried out for a pair of vinylcarbene molecules on the model surface starting from both HH-TT and HT-TH orientations (H and T refer to head and tail, respectively). The geometry optimizations led to 1,4- and 1,3-cyclohexadiene, which are indeed the intermediates suggested by Middleton and Lambert.[171] These intermediates are stable with respect to dissociation to gas-phase benzene plus hydrogen by about 0.5 eV. Thus the energy cost is relatively low and the reaction may be expected to occur but, in agreement with experimental evidence, the conversion should be low. These intermediates could also be hydrogenated to cyclohexene, (82 amu). However, this possibility must be rejected because the desorption profile and mass spectral fragmentation pattern of cyclohexene are very different from that of the 82 amu product observed in this process. Further exploration of the potential energy surface considering the interaction of two propynes or two vinylcarbenes starting from a geometry in which coupling involved only one end of each molecule in HH, HT or TT possibilities. The calculations predict that only the corresponding HH coupling intermediates led to a stable non cyclic C_6H_8 structure, which after hydrogenation produces 2,4-hexadiene, independently if the starting fragments were propyne or vinylcarbene. Thus, a rather clear explanation of the coupling reaction emerges from the cluster model theoretical study.

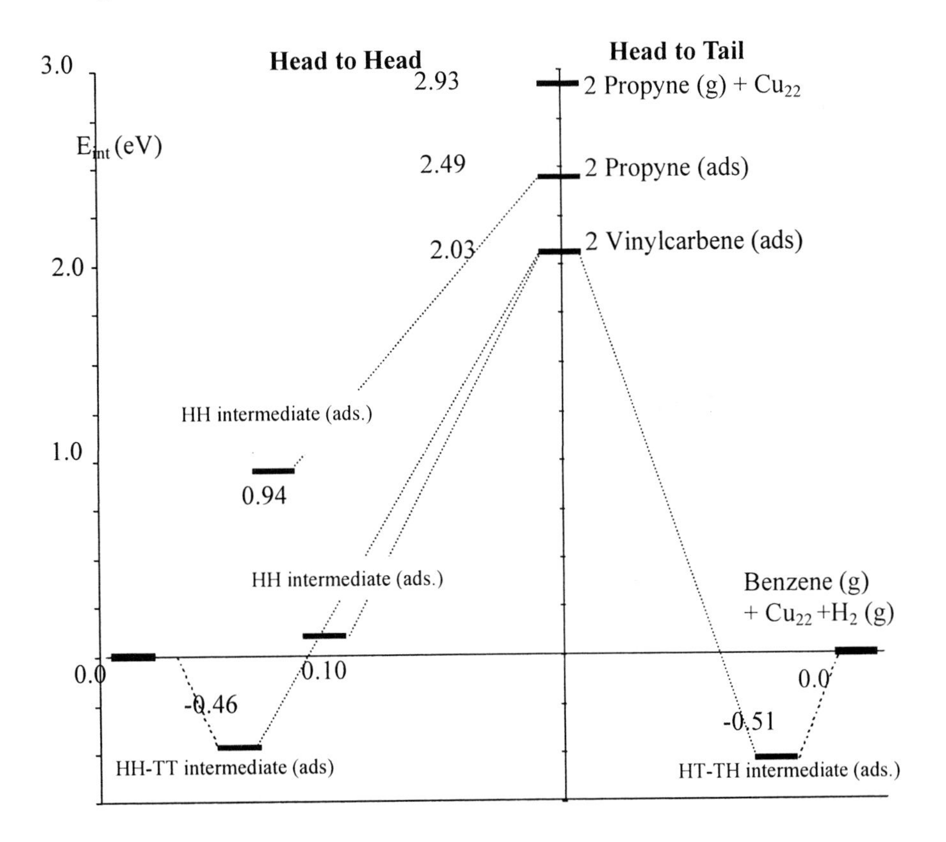

Figure 11. Schematic energy profile of the Cu(111) catalyzed propyne dimerization.

From the energy profile of the possible reaction paths commented above, cf. Figure 11, it is seen that starting with two isolated propyne molecules on top of the cluster model, the adsorbate-surface interaction is strong enough to lower the energy by 0.5 eV. The transformation to two non-interacting vinylcarbenes represents an additional gain of 0.5 eV. Next, the formation of cyclic HH-TT or HT-TH intermediates decreases the energy again by almost 2.5 eV resulting in rather stable adsorbed cyclohexadiene, precursor of benzene. Moreover, HH linear intermediates arising from both propyne or vinylcarbene are the responsible of the 82 amu product that should be 2,4-hexadiene.

The present study focused on the energy profiles that are obtained by means of a B3LYP cluster model approach and hence is not free from computational and physical limitations. Whilst improving the level of calculation may lead to some variations in the energy profile, it is most unlikely that qualitative changes would result. A complete description of the reaction requires determination of the various isomers and transition state structures, as well as the energy barriers between the different steps. Adsorbed structures other than propyne and vinylcarbene are not totally excluded. In this respect, propyne exhibits a much richer surface chemistry than ethyne and other mechanisms are possible, at least in principle. The present study enables to illuminate additional key features of the mechanism thus providing a satisfactory and plausible account of a complex and interesting heterogeneously catalysed reaction that has many analogies in homogeneous catalysis.[173]

4. CONCLUDING REMARKS

The cluster model approach and the methods of analysis of the surface chemical bond have been presented and complemented with a series of examples that cover a wide variety of problems both in surface science and heterogeneous catalysis. In has been show that the cluster model approach permits to obtain qualitative trends and quantitative structural parameters and energetics of problems related to surface chemistry and more important, provide useful, unbiased information that is necessary to interpret experiments. In this way, the methods and models discussed in the present chapter are thought to be an ideal complement to experiment leading to a complete and detailed description of the mechanism of heterogeneous catalysis.

ACKNOWLEDGEMENTS.

Financial support from Spanish CICyT PB98-1216-C02 (01 and 02), and Catalan 1999SGR00182, 1999SGR00040, projects is acknowledged. Part of the computer time was provided by the "Centre de Supercomputació de Catalunya",CESCA-CEPBA.

REFERENCES

1 Somorjai, G.; *Introduction to Surface Chemistry and Catalysis,* JohnWiley & Sons, Inc., New York, **1994**.

2 Ponec, V.; Bond, G.C.; *Catalysis by Metals and Alloys*, Elsevier, Amsterdam, **1995**.

3 Rodriguez, J.A.; *Surf. Sci. Rep.*, **1996**, *24*, 223.

4 Ormerod, R.M.; Lambert, R.M. in *Surface Reactions*, Springer Series in Surf. Sci., Vol. 34, ed. R.J. Madix (Springer, Berlin, **1994**) ch. 4.

5 *Chemisorption and Reactivity on Supported Clusters and Thin Films*, R. M. Lambert, G. Pacchioni, Eds., NATO ASI Series E, Vol. 331, Kluwer: Dordrecht, **1997**.

6 Gates, B.C.; *Chem. Rev.* **1995**, *95*, 511.

7 Goodman, D.W.; *Chem. Rev.* **1995**, *95*, 523.

8 Gates, B. C.; *Catalytic Chemistry*; Wiley: New York, **1992**.

9 Henry, C. R.; *Surf. Sci. Rep.* **1998**, *31*, 231.

10 Henrich, V.E.; Cox, P.A.; *The Surface Science of Metal Oxides*, Cambridge University Press, Cambridge, **1994**.

11 Rabo, J.A.; *Zeolite Chemistry and Catalysis*, ACS Monograph 171, **1976.**

12 Hartmann, M.; Kevan, L.; *Chem. Rev.*, **1999**, *99*, 635.

13 Shelef, M.; *Chem. Rev.*, **1995**, *95*, 209.

14 Bagus, P.S.; Illas, F.; *The Surface Chemical Bond* in *Encyclopedia of Computational Chemistry*, P.V. Schleyer, N.L., Allinger, T. Clark, J. Gasteiger, P.A. Kollman, H.F. Schaefer III and P.R. Schreiner, (Eds.), Vol. 4, p. 2870, John Willey & Sons, Chichester, **1998**, UK.

15 Pacchioni, G., *Heterogenous Chemistry Reviews,* **1996**, *2*, 213.

16 Sauer, J.; *Chem. Rev.*, **1989**, *89* ,1999.

17 Strain, M.C.; Scuseria, G.E.; Frisch, M.J., *Science*, **1996**, *271*, 51.

18 Guerra, C.F.; Snijders, J.G.; Tevelde, G; Baerends, E.J.; *Theoret. Chem. Acc.*, **1998**, *99*, 391.

19 Goh, S.K.; Gallant, R.T.; St.Amant A.; *Int. J. Quantum Chem.*, **1998**, *69*, 401.

20 Helgaker, T.; Larsen, H.; Olsen, J.; Jorgensen, P.; *Chem. Phys. Lett.*, **2000**, *327*, 397.

21 Bowler, D.R.; Gillan, M.J.; *Mol. Simul.*, **2000**, *25*, 239.

22 Haynes, P.D.; Payne, M.C.; *Mol. Simul.*, **2000**, *25*, 257.

23 Ayala, P.Y.; Scuseria, G.E.; *J. Chem. Phys.*, **1999**, *110*, 3660.

24 Millam, J.M.; Scuseria, G.E.; *J. Chem. Phys.*, **1997**, *106*, 5569.

25 White, C.A.; Johnson, B.G.; Gill, P.M.W.; Head-Gordon, M.; *Chem. Phys. Lett.*, **1996**, *253*, 268.
26 Maseras, F., Morokuma, K.; *J. Comput. Chem.*, **1995**, *16*, 1170.
27 Pacchioni, G.; Ferrari, A.M.; Marquez, A.M.; Illas, F.; *J.Comput.Chem.*, **1996**, *18*, 617.
28 Illas, F.; Roset, L.; Ricart, J.M.; Rubio, J.; *J.Comput.Chem.*, **1993**, *14*, 1534.
29 Sauer, J.; Ugliengo, P.; Garrone, E.; Saunders, V.R.; *Chem. Rev.* **1994**, *94*, 2095.
30 López, N.; Illas, F. *J. Phys. Chem. B* **1998**, *102*, 1430.
31 Whitten, J.L.; Pakkanen, T.A.; *Phys. Rev. B*, **1980**, *21*, 4357.
32 Whitten, J.L.; Yang, H.; *Surf. Sci. Rep.*, **1996**, *24*, 59.
33 Govind, N.; Wang, Y.A.; Carter, E.A.; *J. Chem. Phys.*, **1999**, *110*, 7677.
34 Govind, N.; da Silva, A.J.R.;Wang, Y.A.; Carter, E.A.; *Chem. Phys. Lett.*, **1998**, *295*, 129.
35 Stoll, H.; Fuentealba, P.; Dolg, M.; Flag, J.; von Scentpaly, L.; Preuss, H.; *J. Chem. Phys.*, **1983**, *79*, 5532.
36 Igel G.; Wedig U.; Dolg M.; Fuentealba P.; Preuss H.; Stoll H., Frey R.; *J. Chem. Phys.*, **1984**, *81*, 2737.
37 Jeung, G.H.; Malrieu, J.P.; Daudey, J.P.; *J. Chem. Phys.*, **1982,** *77*, 3571.
38 Illas, F.; Rubio, J.; Barthelat, J.C.; *Chem. Phys. Lett.*, **1985**, *119*, 397.
39 Illas, F.; Rubio, J.; Ricart, J.M., Daudey, J.P.; *J. Chem. Phys.*, **1990**, *93*, 2521.
40 Wahlgren, U.; Pettersson, L.G.M.; Siegbahn, P.; *J. Chem. Phys.*, **1989**, *90*, 4613.
41 Pettersson, L.G.M.; Åkeby, H; *J. Chem. Phys.*, **1991**, *94*, 2968.
42 Rubio, J.; Zurita, S.; Barthelat, J.C.; Illas, F.; *Chem. Phys. Lett.*, **1994**, *217*, 283.
43 Zurita, S.; Rubio, J.; Illas F.; Barthelat J. C.; *J. Chem. Phys.*, **1996**,*104*, 8500.
44 Panas, I.; Schüle, J.; Siegbahn, P. E. M. ; Wahlgren, U.; *Chem. Phys. Lett.*, **1988**, *149*, 265.
45 Illas, F.; Pouplana, R.; Rubio, J.; Ricart, J. M.; *J. Electranal. Chem.*, **1987**, *216*, 29.
46 Illas, F.; Bagus, P. S.; *J. Chem. Phys.*, **1991**, *94*, 1236.
47 Bagus, P. S.; Bauschlicher, Jr C. W.; Nelin, C. J.; Laskowski, B.C.; Seel, M.; *J. Chem. Phys.*, **1984**, *81*, 3594.
48 Pacchioni, G.; Bagus, P.S.; *Surf. Sci.*, **1992**, *269/270*, 669.
49 Pacchioni, G.; Bagus, P.S.; *Surf. Sci.*, **1993**, *286*, 317.
50 Bagus, P.S.; Pacchioni, G.; *J. Chem. Phys.*, **1995**, *102*, 879.
51 Lang, N.D.; Kohn, W.; *Phys. Rev. B*, **1973**, *7*, 3541.
52 García-Hernández, M.; Bagus, P.S.; Illas, F.; *Surf. Sci.*, **1998**, *409*, 69.
53 Bagus, P.S.; García-Hernández, M.; Illas, F.; *Surf. Sci.*, **1999**, *429*, 345.
54 Panas, I.; Schule, I.; Siegbahn, P.E.M.; Wahlgren, U.; *Chem. Phys. Lett.*, **1988**, *149*, 265.
55 Siegbahn, P.E.M.; Pettersson, L.G.M.; Wahlgren, U.; *J. Chem. Phys.*, **1991**, *94*, 4024.
56 Siegbahn, P.E.M.; Wahlgren, U.; *Int. J. Quantum Chem.*, **1992**, *42*, 1149.
57 Sellers, H.; *Chem. Phys. Lett.*, **1990**, *170*, 6.
58 Sellers, H.; *Chem. Phys. Lett.*, **1991**, *178*, 351.
59 Hermann, K.; Bagus, P. S.; Nelin, C. J.; *Phys. Rev. B*, **1987**, *35*, 9467.
60 Ricart, J.M.; Rubio, J.; Illas, F.; Bagus, P.S.; *Surf. Sci.*, **1994**, *304*, 335.
61 Illas, F.; Ricart, J. M.; Fernández-García, M., *J. Chem. Phys.*, **1996**, *104*, 5647.
62 Illas, F.; Rubio, J.; Sousa, C.; Povill, A.; Zurita, S.; Fernández-García, M.; Ricart, J.M.; Clotet, A.; Casanovas, J.; *J. Mol. Struct. (THEOCHEM)*, **1996**, *371*, 257.
63 Illas, F.; Rubio, J.; Ricart, J. M.; Pacchioni, G., *J. Chem. Phys.*, **1996**, *105*, 7192.
64 Ricart, J. M.; Clotet, A.; Illas, F.; Rubio, J.; *J. Chem. Phys.*, **1994**, *100*, 1988.
65 Illas, F.; Rubio, J.; Ricart, J.M.; *J. Chem. Phys.*, **1988**, *88*, 260.
66 Klüner, T.; Govind, N.; Wang, Y.A., Carter, E.A.; preprint.
67 Bagus, P.S.; Illas, F.; Pacchioni, G.; Parmigiani, F.; *J. Electr. Spectrosc. Relat. Phenom.*, **1999**, *100*, 215.

68 Bagus, P.S.; Nelin, C. J.; Müller, W.; Philpot, M.R.; Seki, H., *Phys. Rev. Lett.*, **1987**, *58*, 559.
69 Bagus, P.S.; Nelin, C. J.; Hermann, K; Philpot, M.R.; *Phys. Rev. B*, **1987**, *36*, 8169.
70 Bagus, P.S.; Pacchioni, G.; *Surf. Sci.*, **1990**, *236*, 233.
71 Illas, F.; Mele, F.; Curulla, D.; Clotet, A; Ricart, J.M.; *Electrochim. Acta*, **1988**, *44*, 1213.
72 Curulla, D.; Clotet, A.; Ricart, J. M.; Illas, F.; *Electrochim. Acta*, **1999**, *45*, 639.
73 García-Hernández, M.; Curulla, D.; Clotet, A.; Illas, F.; *J. Chem. Phys.*, **2000**, *113*, 364.
74 Koper, M.T.M.; van Santen, R.A.; Wasileski, S.A.; Weaver, M.J.; *J.Chem. Phys.*, **2000**, *113*, 4392.
75 Koper, M.T.M.; van Santen, R.A.; *J. Electroanal. Chem.*, **1999**, *476*, 64.
76 Wasileski, S.A.; Weaver, M.J.; Koper, M.T.M.; *J. Electroanal. Chem.*, **2001**, *500*, 344.
77 Markovits, A.; García-Hernández, M.; Ricart, J.M.; Illas, F.; *J. Phys. Chem.*, **1999**, *103*, 509.
78 García-Hernández, M.; Markovits, A.; Clotet, A.; Ricart, J.M.; Illas, F.; Kluwer Series on Theoretical Chemistry and Physics, Kluwer, *Progr. Theor. Chem. Phys. B*, **2001**, *7*, 211.
79 García-Hernández, M.; Birkenheuer, U.; Hu, A.; Illas, F.; Rösch, N.; *Surf. Sci.*, **2001**, *471*, 151.
80 Mulliken, R. S.; *J. Chem. Phys.*, **1955**, *23*, 1833.
81 Bagus, P. S.; Illas, F.; Sousa, C.; Pacchioni, G.; *The ground and excited state of oxides* in *Electronic properties of solids using cluster models*, p. 93, Edited by Kaplan T. A. and Mahanti S D., Plenum, New York, **1995**.
82 Reed, K A.E.; Curtiss, L. A.; Weinhold, F.; *Chem. Rev.*, **1988,** *88*, 899.
83 Bader, R. F.W.; *Atoms in Molecules: A Quantum Theory*, Oxford Univ. Press: Oxford 1990.
84 Becke, A. D.; Edgecombe, K. E. *J. Chem. Phys.* **1990**, *92*, 5397.
85 Silvi, B.; Savin, A. *Nature* **1994**, *371*, 683.
86 Savin, A.; Nesper, R.; Wengert, S.; Fässler, T. P. *Angew. Chem. Int. Ed. Engl.* **1992**, *36*, 1809.
87 Krokidis, X.; Noury, S.; Silvi, B. *J. Phys. Chem. A* **1997**, *101*, 7277.
88 Noury, S.; Colonna, F.; Savin, A.; Silvi, B. *J. Mol. Struct.* **1998**, *450*, 59.
89 Davidson, E. R.; *J. Chem. Phys.*, **1967**, *46*, 3320.
90 Nelin, C. J.; Bagus, P. S.; Philpott, M. R.; *J. Chem. Phys.*, **1987**, *87*, 2170.
91 Bauschlicher, C. W.; Bagus, P. S.; *J. Chem. Phys.*, **1984**, *81*, 5889.
92 Bagus, P.S.; Illas, F., *Phys. Rev. B*, **1990,** *42*, 10852.
93 Bagus, P. S.; Pacchioni, G.; Philpott, M. R.; *J. Chem. Phys.*, **1989**, *90*, 4287.
94 Pettersson, L.G.M.; Bagus, P.S., *Phys. Rev. Lett.*, **1986,** *56*, 500.
95 Bagus, P.S.; Clotet, A.; Curulla, D.; Illas, F.; Ricart, J.M.; *J Mol. Catal.*, **1997**, *119*, 3.
96 Clotet, A.; Ricart, J.M.; Rubio, J.; Illas, F.; *Chem. Phys.*, **1993**, *177*, 61.
97 Bagus, P. S.; Hermann, K.; Bauschlicher, Jr C.W.; *J.Chem.Phys.*, **1984**, *81*, 1966.
98 Bagus, P. S.; Hermann, K; *Phys. Rev. B*, **1986**, *33*, 2987.
99 Morokuma, K.J.; *J. Chem. Phys.*, **1971**, *55*, 1236.
100Morokuma, K.J.; *Acc. Chem. Res.*, **1977**, *10*, 249.
101Bagus, P.S.; Illas, F.; *J. Chem. Phys.*, **1992**, *96*, 8962.
102Bauschlicher, C.W.; Bagus, P.S.; Nelin, C.J.; Ross, B.O.; *J.Chem.Phys.*, **1986**, *85*, 354.
103 Dupuis, M.; Johston, F.; Márquez, A.; *HONDO 8.5 for CHEM-station.* IBM corporation, Neighborhood Road, Kingston, NY 12401. Hartree-Fock CSOV adaptation by Illas, F.; Rubio, J.; Márquez, A.. DFT implementation by Márquez, A.
104 Ziegler, T.; Rauk, A.; *Theor. Chim. Acta*, **1977**, *46*, 1.

105 Amsterdam Density Functional,User's Guide Examples Utilities, Theoretical Chemistry Vrije Universiteit, Amsterdam, Netherlands.
106 Neyman, K.M.; Nasluzov, V.A.; Zhidomirov, G.M.; *Catal.Lett.*, **1996**, *40*, 183.
107 Albert, K., Neyman, K.M.; Nasluzov, V.A.; Ruzankin, S. Ph.; Yeretzian, C.; Rösch, N.; *Chem. Phys. Lett.*, **1995**, *245*, 671.
108 Neyman, K.M.; Ruzankin, S. Ph.; Rösch, N.; *Chem. Phys. Lett.*, **1995**, *246*, 546.
109 Neyman, K.M.; Strodel, P.; Ruzankin, S.Ph.; Schlensog, N.; Knoezinger, H.; Rösch, N.; *Catal. Lett.*, **1995**, *31*, 273.
110 Chung, S.C.; Kruger, S.; Ruzankin, S. Ph.; Pacchioni, G.; Rösch, N.; *Chem. Phys. Lett.*, **1996**, *248*, 109.
111 Márquez ,A.M.; López, N.; García-Hernández, M.; Illas, F.; *Surf. Sci.*, **1999**, *442*, 463.
112 Becke, A.D.; *Phys. Rev. A*, **1988**, *38*, 3098.
113 Lee, C.; Yang ,W.; Parr, R.G.; *Phys. Rev. B*, **1988**, *37*, 785.
114 Becke, A.D.; *J. Chem. Phys.*, **1993**, *98*, 5648.
115 Dapprich, S.; Frenking, G.; *J. Phys. Chem.*, **1995**, *99*, 9352.
116 García-Hernández, M.; Beste, A.; Frenking, G.; Illas, F., *Chem. Phys. Lett.*, **2000**, *320*, 222.
117 Taylor, K. C. In *Catalysis Science and Technology*; Anderson, J. R., Boudart, M., Eds.; Springer-Verlag: **1984**; Vol. 5.
118 Blyholder, G.; *J. Phys. Chem.*, **1964**, *68*, 2772.
119 Dewar, M.J.S.; *Bull. Soc. Chim. Fr.*, **1951**, *18*, c79.
120 Chatt, J.; Duncanson, L.A.; *J. Chem. Soc.*, **1953**, 2939.
121 Bagus, P.S.; Pacchioni, G., *Surf. Sci.*, **1992**, *278*, 427.
122 Bagus, P.S.; Müller, W.; *Chem. Phys. Lett.*, **1985**, *115* 540.
123 Müller, W.; Bagus, P.S.; *J. Vac. Sci. & Technol. A*, **1985**, *3*, 1623.
124 Illas, F.; Zurita, S.; Márquez, A.M.; Rubio J.; *Surf. Sci.*, **1997**, *376*, 27.
125 Curulla, D.; Clotet, A.; Ricart, J. M.; Illas, F.; *J. Phys. Chem. B*, **1999**, *103*, 5246.
126 Illas, F.; López, N.; Ricart, J.M., Clotet, A.; Conesa J.C.; Fernández-García M.; *J. Phys. Chem.*, **1998**, *102*, 8017.
127 Hammer, B.; Morikawa, Y.; Norskov, J.K. *Phys. Rev. Lett.* **1996**, *76*, 2141.
128 Rodriguez, J.A. *Surf. Sci. Rep.* **1996**, *24*, 223.
129 Ponec, V.; Bond, G.C. *Catalysis by Metals and Alloys*, Elsevier, Amsterdam, 1995.
130 Debauge, Y.; Abon, M.; Bertolini, J.C.; Massardier, J.; Rochefort, A. *Appl. Surf. Sci.* **1995**, *90*, 15.
131 Espeel, P.H.; De Peuter, G.; Trelen, M.C.; Jacobs, P.A. *J. Chem. Phys.* **1994**, *98*, 11588.
132 Skoda, F.; Astier, M.P.; Pajonk, G.M.; Primet, M. *Catal. Lett.* **1994**, *29*, 159.
133 Anderson, J.A.; Fernández-García, M.; Haller, G.L. *J. Catal.* **1996**, *164*, 477.
134 Meléndrez, R.A.; Del Angel, G.; Bertin, V.; Valenzuela, M.A.; Barbier, J. *J. Mol. Catal. A* **2000**, *154*, 143.
135 Leon y Leon, C.A.; Vannice, M.A. *Appl. Catal.* **1991**, *69*, 305.
136 Fernández-García, M.; Conesa, J.C.; Clotet, A.; Ricart, J.M.; López, N.; Illas, F. *J. Phys. Chem. B* **1998**, *102*, 141.
137 Sousa, C.; Bertin, V.; Illas, F.; *J. Phys. Chem. B*, **2001**, *105*, 1817
138 Dong, W.; Hafner, J. *Phys. Rev. B* **1997**, *56*, 15396.
139 Triguero, L.; Wahlgren, U.; Bousard, P., Siegbahn, P. *Chem. Phys. Lett.* **1995**, *237*, 550.
140 Noordermeer, A.; Kok, G.A.; Nieuwenhuys, B.E. *Surf. Sci.* **1986**, *172*, 349.
141 Solymosi, F.; *J. Mol. Catal.*, **1991**, *65*, 337.
142 Freund, H.J.; Roberts, M.W.; *Surf. Sci. Rep.*, **1996**, *25*, 225.
143 Ricart, J.M.; Habas, M.P.; Clotet, A.; Curulla, D.; Illas, F.; *Surf. Sci.*, **2000**, *460*, 170.

144 Habas, M.P.; Mele, F.; Sodupe, M.; Illas, F.; *Surf. Sci.*, **1999**, *431*, 208.
145 Liu, Z.M.; Zhou, Y.; Solymosi, F.; White, J.M.; *Surf. Sci.*, **1991**, *245*, 289.
146 Wohlrab, S.; Ehrlich, D.; Wambach, J.; Kuhlenbeck, H.; Neumman, M.; *Surf. Sci.*, **1990**, *198*, 243.
147 Klier, K., *Adv. Catal.*, **1981**, *31*, 243.
148 Nakamura, I.; Nakao, H.; Fujitani, T.; Uchijima, T.; Nakamura, J.; *J. Vac. Sci. Technol. A*, **1999**, *17*, 1592.
149 Taylor, P.A.; Rasmussen, P.B.; Ovesen, C.V.; Stoltze, P.; Chorkendorff, I.; *Surf. Sci.*, **1992**, *261*, 191.
150 Taylor, P.A.; Rasmussen, P.B.; Chorkendorff, I.; *J. Chem. Soc. Faraday Trans.*, **1995**, *91*, 1267.
151 Yoshihara, J.; Parker, S. C.; Schafer, A.; Campbell, C. T.; *Catal. Lett.*, **1995**, *31*, 313.
152.Yoshihara , J.; Campbell, C. T.; *J. Catal.*, **1996**, *161*, 776.
153.Rasmussen, P. B.; Holmblad, P. M.; Askgaard, T. ; Ovesen, C. V. ; Stoltze, P.; Norskov, J. K.; Chorkendorff, I.; *Catal. Lett.*, **1994**, *26*, 373.
154 Walker, J. F.; *Formaldehyde*, 3rd edn., Reinhold, New York, **1964**.
155 Barteau, M.A.; Madix, R. J.; *The Surface Reactivity of Silver Oxidation Reactions, The Chemical Physics of Solid and Heterogeneous Catalysis,* ESPC, vol. 4, **1982**, pp. 96-142.
156 Hogarth, M.; Christensen, P.; Hamnet, A.; Shukla, A.; *J. Power Sources*, **1997**, *69*, 113.
157 Canning, N.D. S.; Madix, R. J.; *J. Phys. Chem.*, **1984**, *88*, 2437.
158 Friend, C.M.; Xu, X.; *Annu. Rev. Phys. Chem.*, **1991**, *41*, 251.
159 Gomes, J.R.B.; Gomes, J.A.N.F.; *J. Molec. Struct. (Theochem)*, **2000**, *503*, 189.
160 Canning, N.D.S., Madix, R.J.; *J. Phys. Chem.*, **1984**, *88*, 2437.
161 Jones, A.H.; Poulston, S.; Bennet, R.A.; Bowker, M.; *Surf. Sci.*, **1997**, *380*, 31.
162 Gomes, J.R.B.; Gomes, J.A.N.F.; Illas, F.; *Surf. Sci.*, **1999**, *443*, 165.
163 Gomes, J.R.B.; Gomes, J.A.N.F.; *Surf. Sci.*, **2001**, *471*, 59.
164 Leibsle, F.M.; Francis, S.M. ; Haq, S.; Bowker, M.; *Surf. Sci.*, **1994**, *318*, 46.
165 Jones, A.H.; Poulston, S.; Bennett, R.A.; Bowker, M.; *Surf. Sci.*, **1997**, *380*, 31.
166 Silva, S.L.; Lemor, R.M. ; Leibsle, F.M.; *Surf. Sci.*, **1999**, *421*, 135.
167 Francis, S.M.; Leibsle, F.M.; Haq, S.; Xiang, N.; Bowker, M.; *Surf. Sci.*, **1994**, *315*, 284.
168 Lomas, J.R.; Baddley, C.J.; Tichov, M.S.; Lambert, R. M.; *Langmuir*, **1995**, *11*, 3048.
169 Ormerod, R. M.; Lambert, R.M.; in *Surface Reactions, Springer Series in Surf. Sci.*, Vol. *34*; Madix, R.J.; Ed.; Springer, Berlin, **1994**, Chapter 4.
170 Pacchioni, G.; Lambert, R.M.; *Surf. Sci.*, **1993**, *98*, 5648.
171 Middleton, R.L.; Lambert, R.M.; *Catal. Lett.*, **1999**, *59*,15.
172 Clotet, A. ; Ricart, J.M.; Illas, F.; Pacchioni, G.; Lambert, R.M. ; *J. Am.Chem.Soc.*, **2000**, *122*,7573.
173 Gevorgyan, V.; Takeda, A.; Homma, M.; Sadayori, N.; Radhakrishnan, U.; Yamamoto, Y. *J. Am. Chem. Soc.* **1999,** *121*, 6391.

ROLE OF POINT DEFECTS IN THE CATALYTIC ACTIVATION OF Pd ATOMS SUPPORTED ON THE MgO SURFACE

G. PACCHIONI, L. GIORDANO
Dipartimento di Scienza dei Materiali, Università di Milano-Bicocca, and Istituto Nazionale per la Fisica della Materia, via R. Cozzi, 53 - I-20125 Milano, Italy

A. M. FERRARI
Dipartimento di Chimica IFM, Università di Torino, via P. Giuria 5, I-10125 Torino, Italy

S. ABBET, U. HEIZ
Institute of Surface Chemistry and Catalysis, University of Ulm, D-89069 Ulm, Germany

1. Introduction

The chemical properties of oxide surfaces are dominated to large extent by the presence of extended or point defects. Depending on the oxide electronic structure, more ionic or more covalent, on its crystalline structure and surface reconstruction, and on the preparation of the material (polycrystalline, thin film, etc.), a great variety of defect centers can form at the surface. For the case of a stoichiometric oxide, like MgO, nine different point defects have been identified and described in the literature [1]. These defect centers can have rather different properties and characteristics and exist in various concentrations. Low-coordinated O anions at steps and kinks are known to exhibit a pronounced basic character and to be more reactive than terrace sites with incoming molecules via electron transfer from the surface to the adsorbate [2]. Oxygen vacancies, also called F^+ and F-centers, contain one or two electrons trapped in the cavity left by the missing O ion [3,4]. These centers can act as good electron sources for charge transfer reactions [5,6] or can facilitate the heterolytic dissociation of H_2 or H_2O [7]. Cation vacancies result in localized holes on surface O anions, which then exhibit enhanced reactivity because of their radical character [8]. Hydroxyl groups, OH, can favor the nucleation of metal clusters [9] and introduce inhomogeneous electric fields at the surface [10]. The complexity of the problem is increased by the fact that the point defects can be located at various sites, terraces, edges, steps, and kinks [11] and that they can be isolated, occur in pairs, or even in "clusters". Furthermore, the concentration of the defects is usually low, making their detection by integral surface sensitive spectroscopies very difficult. A microscopic view of the metal/oxide interface and a detailed analysis of the sites where the deposited metal atoms or clusters are bound become essential in order to rationalize the observed phenomena and to design new materials with known concentrations of a given type of defects.

Each surface defect has a direct and characteristic effect on the properties of adsorbed species. This becomes particularly important in the analysis of the chemical reactivity of supported metal atoms and clusters [12,13]. The defects not only act as nucleation centers in the growth of metal islands or clusters [14,15], but they can also modify the catalytic activity of the deposited metal by affecting the bonding at the

M.A. Chaer Nascimento (ed.), Theoretical Aspects of Heterogeneous Catalysis, 183–198.

interface [16,17,18]. In the last two years we have addressed the general question of the role of the point defects at the surface of MgO in promoting or modifying the catalytic activity of isolated metal atoms or clusters deposited on this substrate. As a test case, we have considered a particular reaction, the cyclization of acetylene to form benzene, $3C_2H_2 \rightarrow C_6H_6$, a process which has been widely studied on single crystal surfaces from UHV conditions (10^{-12}-10^{-8} atm) to atmospheric pressure (10^{-1}-1 atm) [19,20,21,22]. A Pd(111) surface is the most reactive one [23,24] and it has been shown unambiguously that the reaction proceeds through the formation of a stable C_4H_4 intermediate, resulting from addition of two acetylene molecules [25]. This intermediate has been characterized experimentally [26,27,28] and theoretically [29] and its structure is now well established. Once the C_4H_4 intermediate is formed, it can add a third acetylene molecule to form benzene that then desorbs from the surface at a temperature of about 230 K [19,20]. The same reaction has been studied on size selected Pd clusters deposited on a MgO surface.

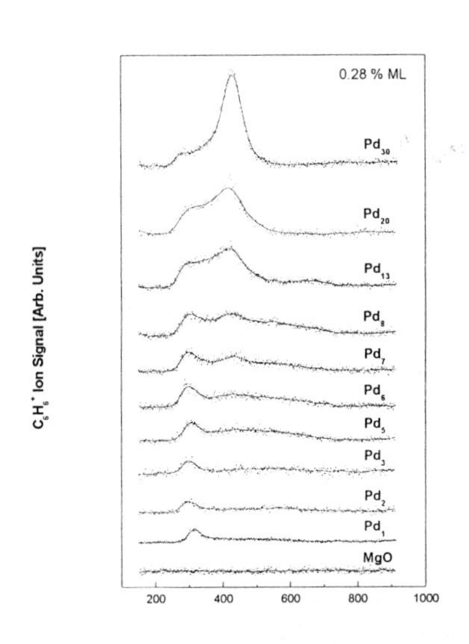

Figure 1. Catalytic C_6H_6 formation for different Pd cluster sizes obtained from temperature programmed reaction experiments. The bottom spectrum shows that for clean MgO(100) films no benzene is formed. Dots: data; full line: data smoothing with adjacent averaging (25 points). Cluster coverage is 0.28 % of a monolayer for all cluster sizes, where one monolayer corresponds to 2.25×10^{15} atoms/cm^2.

The Pd clusters have been produced by a recently developed high-frequency laser evaporation source, ionized, then guided by ion optics through differentially pumped vacuum chambers and size-selected by a quadrupole mass spectrometer [16-18]. The monodispersed clusters have been deposited with low kinetic energy (0.1-2 eV) onto a MgO thin film surface. The clusters-assembled materials obtained in this way exhibit peculiar activity and selectivity in the polymerization of acetylene to form benzene and aliphatic hydrocarbons [30].

Figure 1 shows the Temperature Programmed Reaction (TPR) spectra for the cyclotrimerization of acetylene on supported Pd_n ($1 \leq n \leq 30$) clusters. Pd_1, Pd_2, and Pd_3, form exclusively benzene at temperatures around 300 K, while for cluster sizes up to Pd_8 a broad feature between 400 K and 700 K is observed in the TPR, Figure 1 [17]. For Pd_7, an additional desorption peak of benzene is clearly observed at about 430 K. For Pd_8 this feature becomes as important as the peak at 300 K while for Pd_{30} benzene mainly desorbs around 430 K. We note that on a clean MgO(100) surface no benzene is produced at the same experimental conditions. This strongly size-dependent behavior is related to the distinct electronic and geometric properties of the metal clusters, making this new class of materials extremely interesting for the understanding of the structure-property relationship. Interestingly, according to these results even a single Pd atom can produce benzene. This is an important result which allows one to investigate the activity of the Pd atoms as function of the support where it is deposited or of the sites where it is bound.

In the following we review the most important aspects of the cyclization reaction over supported Pd atoms, looking in particular at the role of the various types of morphological defects present on the MgO surface, and performing an accurate analysis of the electronic effects involved in the metal-support interaction. The review is largely based on density functional (DF) cluster model calculations to interpret the experimental data, either thermal desorption spectra (TDS) or infra-red spectra (IR). The results provide an indication of the importance of defects like low-coordinated oxygen anions, neutral and charged oxygen vacancies, etc. in changing the reactivity of a supported Pd atom.

2. Theoretical approach

The theoretical calculations have been carried out with the help of density functional theory using either the gradient-corrected BP functional (Becke's exchange functional [31] in combination with Perdew's correlation functional [32]) or the B3LYP functional (Becke3 functional for exchange [33] and the Lee-Yang-Parr functional for correlation [34]). For further details about the computational aspects see for instance ref. [17,18]. The MgO (001) surface has been represented by cluster models [35], an approach which has been found to reproduce in a sufficiently accurate way the electronic structure and the binding properties of surface complexes. Due to the highly ionic nature of MgO, the truncation of the lattice in cluster calculations implies the use of an external field to represent the long-range Coulomb potential. The model clusters considered have been embedded in arrays of point charges (PC = ±2 e) in order to reproduce the correct Madelung potential at the adsorption site under study [36]. The complete models, ions and PCs taken together, are electrically neutral. The positions of most substrate atoms and PCs were kept fixed at bulk-terminated values of MgO, with a measured bulk Mg-O distance of 2.104 Å [37]. Previous studies [38,39] have demonstrated that the use of PCs for embedding significantly affects the calculated adsorption properties when the positive PCs are nearest neighbors to the highly polarizable oxygen anions of the clusters. In that case, an artificial polarization of the oxygen anions at the cluster borders results; however, this artifact can be essentially eliminated by surrounding the anions with total ion model potentials (TIMPs) of Mg^{2+} instead of positive PCs.

The positions of the supported Pd atom and of the surface ions closest to it and the structure of the adsorbed hydrocarbon molecules have been fully optimized using analytical energy gradients. The calculations have been performed with the GAUSSIAN98 [40] program package.

3. Results and discussion

3.1. Isolated Pd atom

The possibility to catalyze the acetylene trimerization depends critically on the ability of the metal center to coordinate and activate two C_2H_2 molecules and then to bind the C_4H_4 intermediate according to reaction:

$$Pd + 2C_2H_2 \rightarrow Pd(C_2H_2)_2 \rightarrow Pd(C_4H_4) \quad (1)$$

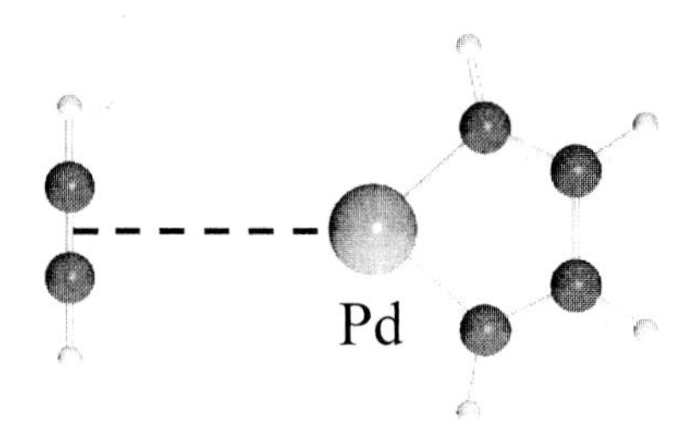

Figure 2. Structure of a $Pd(C_4H_4)(C_2H_2)$ complex. The third acetylene molecule is weakly bound and is not activated (linear shape as in the gas-phase).

The activation of the acetylene molecules is easily monitored, for instance, by the deviation from linearity of the HCC angle due a change of hybridization of the C atom from sp to sp^2, or by the elongation of the C-C distance, d(C-C), as a consequence of the charge transfer from the metal 4d orbitals to the empty π^* orbital of acetylene. This process has been considered by optimizing the geometry of one and two acetylene molecules coordinated to an isolated Pd atom, and of the corresponding $Pd(C_4H_4)$ complex, Figure 2. One acetylene is strongly bound to Pd, by 1.84 (the results refer to the BP functional); the second is bound by 1.09. Also the C_4H_4 intermediate is quite strongly bound to the Pd atom and is 0.93 more stable than two adsorbed acetylene molecules. A similar result has been obtained for the Pd(111) surface [29]. The level of activation of acetylene is clearly larger when a single C_2H_2 is adsorbed, consistent with the idea that the electron density on the metal is essential for promoting the molecular activation. However, it should be kept in mind that a stronger bonding may imply a larger barrier for conversion of two C_2H_2 units into C_4H_4.

The following step, $C_4H_4 + C_2H_2 \rightarrow C_6H_6$, requires the capability of the metal center to coordinate and activate a third acetylene molecule according to reaction:

$$Pd(C_4H_4) + C_2H_2 \rightarrow Pd(C_4H_4)(C_2H_2) \rightarrow Pd(C_6H_6) \quad (2)$$

However, the third C_2H_2 molecule interacts very weakly with $Pd(C_4H_4)$ with a binding of 0.3 eV (actually an overestimate because of the occurrence of the basis set superposition error, BSSE) [41]. The structure of the third molecule is not deformed compared to the gas-phase, with d(C-C) = 1.216 Å and a HCC angle of 177°. The molecule is bound to the $Pd(C_4H_4)$ complex by dispersion forces only and its distance from the Pd atom is therefore very long, $\approx$ 2.6 Å. This result indicates that *an isolated Pd atom is not* a catalyst for the cyclization process, at variance with the experimental observation for Pd_1/MgO, Figure 1.

The role played by the support becomes therefore a critical aspect of the interaction. Several experimental and theoretical studies have been dedicated to the interaction of metal atoms and clusters with oxide surfaces and in particular with MgO [42-45]. It is now well established that the preferred sites for metal binding are the oxygen anions of the surface [46]. The Pd-MgO bonding, however, is not characterized by a pronounced charge transfer and is better described as covalent polar. However, only regular adsorption sites have been considered and the bonding mode and character at defect or low-coordinated sites has attracted less attention so far [47,48]. In this respect one can formulate two hypotheses. One of them states that the surface oxygen anions of the MgO surface still have some oxidizing power to deplete charge from the metal atom. The other possibility is that the O anions are fully reduced, O^{2-}, and donate charge to the metal atom, thus leading to electron enriched species (the surface acts as a Lewis base). In the two cases one would end up with positively or negatively charged supported Pd atoms, respectively. In principle, both situations can lead to an activation of the supported metal. In fact, the bonding of unsaturated hydrocarbons to metal complexes and metal surfaces is classically described in terms of σ donation from a filled bonding level on acetylene to empty states on the metal and π back-donation from the occupied d orbitals on the metal to empty antibonding π^* orbitals of the ligand [49,50]. Both mechanisms result in a weakening of the C-C bond and a distortion of the molecule. A simple model study where the charge on Pd has been artificially augmented has shown that the presence of a charge on Pd, either positive or negative, reinforces the bonding of acetylene to the $Pd(C_4H_4)$ complex; however, only for negatively charged Pd atoms (electron-rich) a substantial activation of the third acetylene molecule is observed [17]. This result shows that the increase of the electron density on Pd is a key mechanism to augment the catalytic properties of the metal atom. The role of the substrate is just that of increasing the "basic" character of the Pd atoms (or clusters).

3.2. Supported Pd atom

To establish the role of the regular and low-coordinated surface anions of MgO in modifying the properties of a supported Pd atom we have considered three simple models. They represent five-, four-, and three-coordinated O anions at terrace, edge and corner sites, respectively; a $Pd(C_2H_2)$ complex has been added on-top of these oxygen atoms. The results have been compared with those for an isolated $Pd(C_2H_2)$ complex. A Constrained Space Orbital Variation, CSOV, [51,52] analysis, performed at the Hartree-Fock level, has permitted to decompose the interaction energy of the system into a sum of various bonding contributions, like intra-unit polarization and charge transfer. When Pd is deposited on MgO the activation of C_2H_2 is much more efficient than for an isolated Pd atom. The structural distortion of adsorbed C_2H_2 follows the trend corner>edge>terrace>free atom. The change in interaction energy and in dipole moment for each CSOV step provides a measure of the role of the MgO substrate in activating the acetylene molecule. The contribution of the charge transfer from acetylene to Pd or Pd_1/MgO, the σ donation in the Dewar-Chatt-Duncanson model [49-50], is similar for the four cases, and is much smaller than the charge transfer in the other direction, i.e. from Pd or Pd_1/MgO to acetylene. This confirms that the interaction with the MgO substrate increases the basic character of the Pd atom. The donor capability of Pd increases as a function of the adsorption site on MgO in the order terrace<edge<corner, the same order found for the deformation of the acetylene molecule. This qualitative analysis shows unambiguously that (a) the substrate plays a direct role in the modification of the properties of the supported Pd atom, and (b) low-coordinated sites are more active than the regular terraces. This is consistent with the idea that the degree of surface basicity in MgO is larger for the low-coordinated anions because of the lower Madelung potential at these sites [2].

TABLE 1. Properties of free and MgO supported $Pd(C_2H_2)$ complexes.[a]

	free Pd	Pd/O_{5c}	Pd/O_{4c}	Pd/O_{3c}
d(C-C), Å	1.201	1.212	1.219	1.237
d(C-Pd), Å	2.452	2.204	2.158	2.075
∠(HCC)°	173.8	167.8	164.2	157.4
$\Delta E(Pd \rightarrow C_2H_2)$, eV[(b)]	0.34	0.97	1.20	2.03
$\Delta\mu(Pd \rightarrow C_2H_2)$, D[(b)]	-0.72	-1.78	-2.16	-3.14

(a) HF results.

(b) Energy contribution, ΔE, and dipole moment change, Δμ, associated to Pd → C_2H_2 charge transfer mechanism.

In principle there are several defect sites which can be active in promoting the catalytic activity of a deposited Pd atom. The surface of polycrystalline MgO presents in fact a great variety of irregularities [53] like morphological defects, anion and cation vacancies [10,54] divacancies [55] impurity atoms, etc. [1]. Each of these defects can then have a different coordination number and be located at terraces, steps, kinks, corners, etc. Figure 3 gives a schematic representation of oxygen vacancies at terrace, F_{5c}, edge, F_{4c}, and corner, F_{3c}, sites. The electron localization in these sites is shown, for instance, by electron density plots. Recent studies have shown that even MgO thin films deposited on a conducting substrate are not completely defect free, as shown by the different adsorption properties of CO on MgO thin films and single crystal surfaces [56].

It is therefore difficult to answer the question which defect sites of the MgO surface are more likely involved in the acetylene trimerization. To contribute to a clarification of the role of defects, we have considered the structure and stability of Pd atoms, $Pd(C_2H_2)$, $Pd(C_2H_2)_2$, $Pd(C_4H_4)$ and $Pd(C_2H_2)(C_4H_4)$ complexes formed on various regular and defect sites at the MgO surface, in particular O sites, neutral oxygen vacancies (the F centers), and charged oxygen vacancies, the F^+ centers. Recent theoretical [11,57] and experimental [54] studies suggest that F centers form preferentially at low coordinated sites like kinks and corners. F_s centers, where the subscript s indicates the location of the defect at the surface, are characterized by the presence of trapped electrons in the cavity [3].

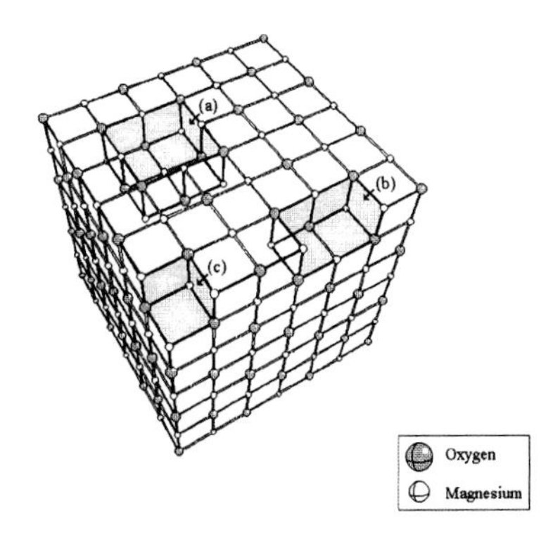

Figure 3. Schematic representation of oxygen vacancies (F centers) formed at the terrace (a), edge (b), and corner (c) sites of MgO.

Let us consider the adsorption energy of an isolated Pd atom on the various MgO adsorption sites (we give here DFT-B3LYP results, the same level of theory used for CO/Pd/MgO, see § 5). On a terrace, O_{5c}, the binding energy is ≈1 eV; it is slightly higher on edge and corners, ≈1.3 eV for O_{4c}, ≈1.5 eV for O_{3c}. Much stronger Pd_1/MgO interactions are found on F centers: 3.4 eV on F_{5C}, 3.6-3.7 eV on F_{4C} and F_{3C}. On charged F^+ centers the adsorption of Pd is intermediate between that of the surface O anions and of the neutral F centers: 2.1 eV on F_{5C}^+, 2.4 eV on F_{4C}^+ F_{3C}^+. The strong bonding in case of F centers arises from the delocalization of the trapped electrons into the 5s orbital of Pd which is empty in the atomic ground state ($4d^{10}5s^0$) [58]. These results show that Pd atoms deposited on MgO have a greater chance to be stabilized at defect sites where the bonding is stronger. Since the barrier for diffusion is estimated to about 0.5-0.6 eV, it is likely that the Pd atoms deposited at the terrace sites diffuse until they become trapped at some defect center; the oxygen vacancies are very good candidates but preliminary results show that also the presence of an hydroxyl group, OH, considerably increases the adsorption energy of Pd [59].

We consider now the formation of the surface complexes at O anions, starting with the five-coordinated oxygens at the (001) terraces. A Pd atom bound at these sites is able to add and activate one or two acetylene molecules and to form a stable $O_{5c}/Pd(C_4H_4)$ complex. However, the addition of the third acetylene molecule does not occur. The situation is reminiscent of that of an isolated Pd atom where the third acetylene is weakly bound by dispersion forces, Figure 4. Therefore, the complex is unstable and acetylene does not bind to $O_{5c}/Pd_1(C_4H_4)$. Thus, even if we assume that Pd atoms can be stabilized at the MgO terraces, they are not active in promoting the cyclization reaction. The situation is only slightly better on the low-coordinated O sites. The formation of $Pd(C_4H_4)$ on four- and three-coordinated oxygens and the consequent addition of the third acetylene molecule, results in a surface complex, $O_{4c}/Pd(C_2H_2)(C_4H_4)$ or $O_{3c}/Pd(C_2H_2)(C_4H_4)$ which is very weakly bound or even unbound, depending on the details of the calculation (exchange-correlation functional used, size of the MgO cluster, etc.). Thus, also the O anions at low-coordinated defects are not good candidates for increasing the catalytic activity of supported Pd atoms.

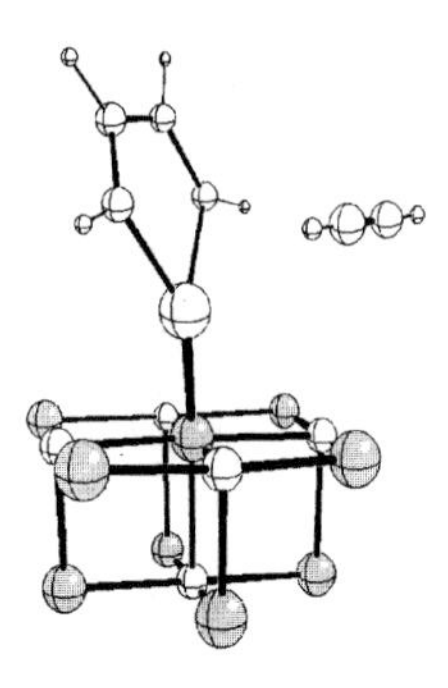

Figure 4. Optimal structure of a $Pd(C_4H_4)(C_2H_2)$ complex formed on a five-coordinated O anion at the MgO terrace. As for a gas-phase Pd atom the third acetylene molecule is weakly bound and not activated.

We have then considered the F and F^+ centers located on various sites. They all show a similar activity, with formation of a stable $Pd(C_4H_4)$ complex and substantial stabilization of the third acetylene molecule; the strength of the bonding of C_2H_2 to the supported $Pd(C_4H_4)$ complex is larger than for the morphological defects and becomes quite large, 2.43 eV (DFT BP), for a neutral F_{3C} center. We know that the formation of strongly bound species can then result in high barriers for the conversion process, and can therefore block the catalytic process or considerably affect the kinetics (see below). The degree of activation is comparable in the three cases, with C-C distances of about 1.31 Å and HCC angles of about 140 degrees.

As a last point in the analysis of the cyclization reaction we consider the reaction barriers. These have been obtained for all the steps of the process and all the morphological sites considered above. In general the results can be summarized as follows. On the morphological sites, O_{5c}, O_{4c}, or O_{3c}, we found that the binding of one acetylene is very strong, the addition of two acetylene molecules results in a very stable complex, and the formation of the C_4H_4 intermediate is very favorable. However, as we mentioned above, the $MgO/Pd/(C_4H_4)$ complex is unable to bind another C_2H_2 molecule in a stable way. Attempts to overcome the barrier for reaction (2) assuming a direct interaction of the $MgO/Pd/(C_4H_4)$ complex with a gas-phase acetylene molecule failed. Thus, Pd atoms adsorbed on the O sites of the MgO surface do not seem to be good candidates for the reaction. The situation is quite different on the F centers. Here, the first acetylene molecule is only weakly bound to Pd. This can be explained with the excess of electron density present on Pd which prevents the first acetylene to overlap efficiently with the localized 4d orbitals of Pd (Pauli repulsion). Once one acetylene is bound, the reduced electron density on Pd favors the addition of the second molecule and the consequent stabilization. On a F_{5c} center the barrier for reaction (1), $2C_2H_2 \rightarrow C_4H_4$, is of 0.48 eV only, Figure 5. The complex can then add a third acetylene molecule with a gain of 1.2 eV and form a stable $Pd(C_4H_4)(C_2H_2)$ intermediate. From this structure, the system evolves into $Pd + C_6H_6$ with a barrier of 0.98 eV, which roughly corresponds to a reaction temperature of 300 K, in qualitative agreement with the experimental observation [17]. Once formed, the benzene molecule easily desorbs from Pd. In fact, while the binding of benzene to a Pd(111) surface or even a Pd atom is quite strong, of the order of 1 eV [29], it is very weak on a Pd atom supported on a F center for the same reasons described above (excess of electron density on Pd). Thus, the barrier for the process (2), $C_4H_4 + C_2H_2 \rightarrow C_6H_6$, is the rate determining step, at variance with the mechanism for the same reaction on Pd(111) where the rate determining step is the benzene desorption from the surface.

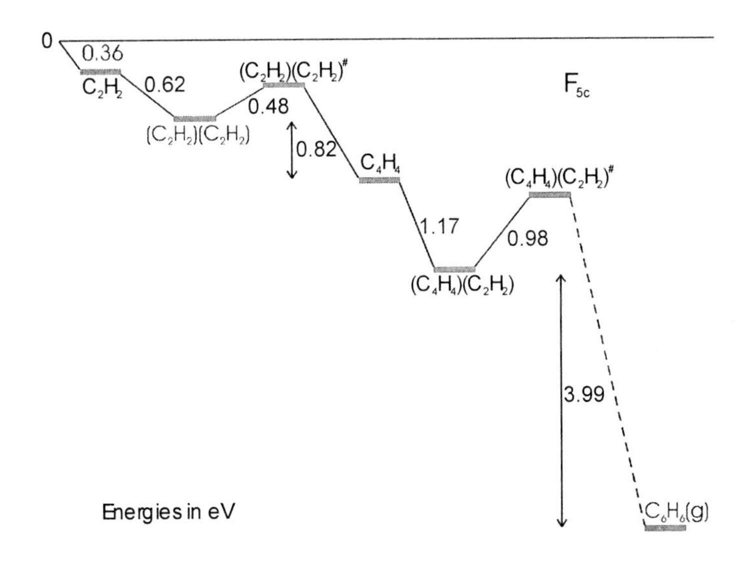

Figure 5. Computed reaction path for the formation of benzene starting from acetylene promoted by a Pd atom supported on a neutral oxygen vacancy at a MgO terrace, F_{5c} (DFT BP results).

On F^+ centers the mechanism is similar. The two acetylene molecules bind easily to the Pd/F_{5c}^+ complex, they form C_4H_4 with a barrier of about 1 eV, then add a third acetylene which reacts with C_4H_4 to form benzene with a barrier of 0.9 eV. Thus, the barriers are at most of 1 eV, consistent with a reaction temperature of 300 K. Also in this case, once formed, benzene desorbs easily. It should be noted that the barriers for the reaction occurring on Pd atoms adsorbed on low-coordinated F or F^+ centers are always significantly higher, hence inconsistent with the TPR experiment. This suggests that the most likely Pd adsorption sites, at least from the point of view of the reaction barriers, are the F and F^+ centers located at the terraces of the MgO surface.

4. Activity of MgO supported Pd atoms: a summary

The study of the cyclyzation reaction of acetylene to benzene on Pd atoms supported on MgO allows one to clarify some of the experimental aspects and to draw some general conclusions about the role of defects on oxide surfaces. The results show that *only in the presence of surface defects a single Pd atom becomes an active catalyst for the reaction*; in fact, an isolated Pd atom is not capable to add and activate three acetylene molecules, an essential step for the process. The change in the electronic structure of supported Pd is connected to the electron donor ability of the substrate, which does not simply act as an inert substrate. The oxide surface acts in pretty much the same way as a ligand in coordination chemistry and provides an additional source of electron density, which increases the capability of the Pd atom to back donate charge to the adsorbed hydrocarbon. In this respect it is remarkable that on transition metal complexes the acetylene trimerization reaction follows a very similar mechanism as on heterogeneous supported catalysts [60]. The oxide anions of the MgO (001) surface, located either on terraces, steps, or corners, do not change significantly the Pd catalytic activity. The basicity of these sites in fact is too low, and in any case not sufficient to significantly increase the density on the metal. Things are completely different on oxygen vacancies, the F or F^+ centers. Due to the electron(s) trapped in the cavity left by the missing oxygen, these centers are good basic sites and promote the activity of an adsorbed Pd atom. Indeed, it has been shown recently that F centers on the surface of polycrystalline MgO are able to reduce very inert molecules like N_2 leading to metastable N_2^- radicals [6].

The great variety of defect sites at the MgO surface, low-coordinated sites, cation and anion vacancies, divacancies, impurity atoms, etc. makes it difficult to clearly identify the centers responsible for the activation of the metal atom. This opens a new problem which is connected to the proper identification of the MgO sites where the Pd atom are most likely bound.

5. Identification of active MgO defect sites

In order to characterize the MgO sites where the Pd atoms are stabilized after deposition by soft-landing techniques, we used CO as a probe molecule [61]. The adsorption energy, E_b, of CO has been computed and compared with results form thermal desorption spectroscopy (TDS). The vibrational modes, ω, of the adsorbed CO molecules have been determined and compared with Fourier transform infrared (FTIR) spectra. From this comparison one can propose a more realistic hypothesis on the MgO defect sites where the Pd atoms are adsorbed.

The binding energies and the vibrational frequencies of CO molecules adsorbed on Pd atoms have been obtained at the gradient corrected DFT level using the hybrid B3LYP method. The MgO sites, O_{nc}, F_{nc}, F_{nc}^+ (n = 3, 4, 5), have been modeled by various clusters. A vibrational analysis has been performed to compute the harmonic frequency of CO. With the 6-311G* basis set used to describe the CO molecule, the C-O ω_0 is 2221 cm^{-1}, while the experimental harmonic frequency is 2170 cm^{-1} [62]. Therefore all the frequencies have been scaled by a factor 2170/2221=0.977 in order to take into account the overestimate of the free CO ω_0. All the computed vibrational shifts are given with respect to the harmonic ω_0 for free CO, 2170 cm^{-1}; all the experimental shifts are given with respect to the anharmonic ω_e, 2143 cm^{-1} [62].

5.1. MgO/Pd/CO: Experimental results

The vibrational spectrum of ^{13}CO adsorbed on Pd_1/MgO and measured at 100 K reveals a broad band at 2045 cm^{-1} with a shoulder at 2010 cm^{-1} (not shown). The existence of the two bands indicates the presence of two different CO adsorption sites. In view of the very low Pd coverage on the surface, the two different adsorption sites are therefore likely to exist on a single Pd atom and can be tentatively identified with Pd(CO) and $Pd(CO)_2$ surface complexes. The two observed frequencies are typical for on-top bound CO molecules. A second band at 1830 cm^{-1}, typical for CO adsorbed on bridge sites, appears at 300 K. This suggests that at 300 K the Pd atoms become mobile on the surface and form small Pd clusters. CO can leave the on-top sites and bind at the bridge or open sites of the cluster where it is more strongly bound. At 350 K most of the on-top bound CO has desorbed from the sample, whereas the bridge-bound CO desorption is completed only at 500 K.

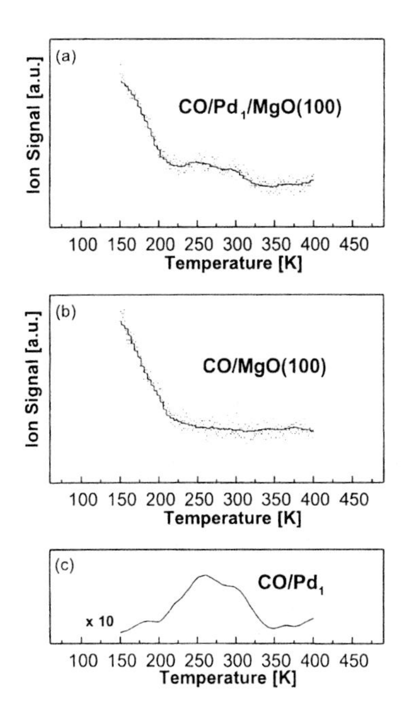

Figure 6. Thermal desorption spectra of CO adsorbed on (a) a sample of MgO(100) thin film with deposited 2% of monolayer of Pd atoms (b) a clean MgO(100) film. In (c) the difference spectrum is shown.

Desorption of CO at these two temperatures is independently observed by TDS. Figure 6a shows the CO-desorption spectrum up to 400 K of ~2 % ML Pd-atoms deposited on the MgO(100) film (1 ML = 2.25×10^{15} clusters/cm^2). For comparison Figure 6b shows the CO-desorption spectrum from a clean MgO(100) thin film. The intense peak around 150 K is due to CO bound at defect sites, most likely Mg cations at steps. The difference spectrum (Figure 6c) clearly indicates that CO is desorbing from the deposited Pd-atoms between 200 K and 340 K. A Redhead analysis of the TDS spectrum using a temperature of maximal desorption rate of 260 K, a heating rate of 2 K/s, and a pre-exponential factor of 10^{-13} s^{-1} results in an energy of desorption of 0.70 eV.

By adsorbing ^{12}CO instead of ^{13}CO onto the same surface held at 95 K, one observes a peak at 2090 cm^{-1} and a shoulder around 2055 cm^{-1}. These are the same vibrational fingerprints as for ^{13}CO, but shifted by the well-known isotopic shift. As for ^{13}CO, the peak at 2090 cm^{-1} disappears and the feature around 2055 cm^{-1} gains intensity after annealing the sample to 180 K. This phenomenon is reversible indicating that the system is thermally stable, i.e., Pd monomers remain trapped at the defect sites, up to at least 180 K. The change between the spectra taken before and after annealing to 180 K can be explained by the existence of a vibrational coupling. The two bands at around 2090 cm^{-1} and 2055 cm^{-1} couple, resulting in high intensity of the 2090 cm^{-1} absorption band at 95 K. In summary, these results revealed two CO molecules bound near or on the supported Pd atoms with observed vibrational frequencies of 2090 cm^{-1} (2045 cm^{-1} for ^{13}CO) and 2055 cm^{-1} (2010 cm^{-1} for ^{13}CO). *The frequency at 2055 cm^{-1} (2010 cm^{-1} for ^{13}CO) is attributed to a single CO molecule adsorbed on a Pd atom deposited on MgO(100).* This frequency is very close to the frequency (2052 cm^{-1}) attributed to single CO molecules on Pd atoms deposited on thin Al_2O_3 films [63]. In addition, the vibrational frequency of CO of a PdCO complex in Ne, Ar, and Kr are 2056.4 [64], 2044.2 [65]/2050 [66], and 2045 [67] cm^{-1}, respectively.

5.2. MgO/Pd/CO: DFT results

On a free, unsupported Pd atom CO is bound by 1.71 eV with a frequency shift with respect to the gas-phase molecule of –97 cm^{-1} due to the occurrence of a substantial back donation of charge from the metal 4d levels to the $2\pi^*$ antibonding orbital of CO. The computed shift is in excellent agreement with that measured from matrix isolation, from –87 cm^{-1} in Ne [64] to –98 cm^{-1} in Kr [67]. These properties change when Pd is deposited on the MgO substrate. Let us consider first the case of Pd interacting with anion sites, O_{5c}, O_{4c}, or O_{3c}. We have seen before that the binding of a Pd atom with these sites increases monotonically from ≈1 eV (O_{5c}) to ≈1.5 eV (O_{3c}), and consequently the distance of the Pd atoms from the surface decreases. On these sites the Pd-CO bond is quite strong and goes from 2.2 eV on Pd/O_{5c} to 2.6 eV on Pd/O_{3c}, Table 2. Therefore, the bond strength of CO on the three sites is rather similar. The CO frequency is even more red shifted than in the case of free Pd; $\Delta\omega$ is –140 cm^{-1} for Pd on O_{5c} and O_{4c} sites, and –175 cm^{-1} for Pd/O_{3c}. These values are sufficiently different from the experimental data to allow the conclusion that the Pd atoms *do not* sit on these MgO sites. This conclusion is fully consistent with the results of the acetylene cyclyzation reaction described above. The computed Pd-CO adsorption energies, 2.2-2.6 eV, are 3-4 times larger than the measured ones and the vibrational frequency shifts are twice as large. It is apparent that when Pd is bound on the O anions its electron density is higher than in the gas-phase. This results in a reinforced bonding and in an increased back donation.

TABLE 2 - Adsorption properties of Pd and CO/Pd complexes adsorbed on O anions, F and F^+-centers located at terrace (5c), edge (4c), and corner (3c) sites of the MgO surface (DFT BP results).

	Terrace			Edge			Corner		
	O_{5c}	F_{5c}	F_{5c}^+	O_{4c}	F_{4c}	F_{4c}^+	O_{3c}	F_{3c}	F_{3c}^+
E_b(Pd), eV	0.96	3.42	2.10	1.35	3.64	2.41	1.51	3.66	2.35
E_b(CO), eV	2.23	0.52	0.71	2.24	0.37	0.64	2.56	0.31	0.61
$\omega_0(^{12}CO)$, cm^{-1}	2030	2014	2092	2030	2026	2124	1995	2001	2105
$\Delta\omega(^{12}CO)$, cm^{-1}	-140	-156	-78	-140	-144	-46	-175	-169	-65

As a second class of defects we consider the neutral F-centers. As we mentioned above on these sites the bonding of Pd is about three times stronger than on the O anions in agreement with other studies [68,69]. There is no large difference in Pd atom adsorption energy when the F-center is located at a terrace, at a step, or at a corner site ($E_b \approx 3.5\pm0.1$ eV). Thus the Pd atoms are likely to diffuse on the surface until they become trapped at defect sites like the F-centers where the bonding is so strong that only annealing to high temperatures will induce further mobility. The formation of a strong bond at the MgO/Pd interface has a dramatic consequence on the Pd-CO binding energy, which is between 0.3 and 0.5 eV, depending whether the F-center is at a corner or at a terrace site, Table 2. This binding energy is 6-8 times smaller than for Pd adsorbed on O_{nc}. The strong Pd/F_{nc} bond results in a weak Pd-CO interaction. The two electrons in the vacancy are largely delocalized over the empty 5s orbital of Pd, thus increasing its density [58]. This leads to a stronger Pauli repulsion with the CO molecule; the Pd-C bond distance is in fact 0.10-0.15 Å longer than for low-coordinated O sites. The augmented electron density on the metal, however, reinforces also the back donation mechanism so that the CO ω_0 for CO/Pd/F_{nc} is similar as for the low-coordinated O sites despite the longer Pd-C distance, Table 2. Thus, while the computed CO adsorption energies, 0.3-0.5 eV, are consistent, although somewhat underestimated, with the experimental TDS value, 0.7 eV, the vibrational frequencies are somewhat too low compared to the experimental data: the computed shifts are of $-150/-170$ cm^{-1} while a shift of about -90 cm^{-1} is deduced from FTIR. However, some care is necessary in interpreting these data since vibrational shifts can depend on various details of the calculations.

The last group of defect sites considered in this work is that of the paramagnetic F^+-centers. On these sites the binding of Pd is between that of the O_{nc} sites and of the F-centers, Table 2. For instance, on F_{5c}^+, Pd is bound by 2.10 eV, while E_b for Pd on O_{5c} is 0.96 eV and that for F_{5c} is 3.42 eV. CO is bound on Pd/F_{nc}^+ by 0.6-0.7 eV in close agreement with the experimental TDS spectrum. Also the vibrational frequencies are closer to those measured by FTIR than for the case of Pd on an F-center. For CO normal to the surface plane the frequency shifts go from ≈-50 cm^{-1} for CO/Pd/F_{4c}^+ to ≈-80 cm^{-1} for CO/Pd/F_{5c}^+. The scaled harmonic frequency in this case is 2072 cm^{-1}; taking into account the anharmonic correction, ≈15 cm^{-1}, on finds a value very close to the broad band centered around 2055 cm^{-1} in the experimental spectrum.

6. Active defect sites at the MgO surface: a summary

On a single crystal MgO(100) surface with a low concentration of defects CO exhibits only one desorption peak at 57 K which corresponds to CO weakly bound to terrace sites (E_b = 0.14 eV) [56]. On a defective MgO(100) surface CO interacts preferentially with defect sites, the low-coordinated Mg^{2+} cations at steps and kinks [70,53] and gives rise to a feature in the TDS spectrum at 150 K. In the presence of deposited Pd atoms, a weak shoulder appears at around 260 K due to the desorption of CO from the deposited Pd atoms. The corresponding binding energy is about 0.7 eV. The supported Pd-CO complexes exhibit a ^{12}CO ω_e of 2055 cm^{-1}, red-shifted with respect to gas-phase CO by 88 cm^{-1}.

These data provide important information for the identification of the surface sites where the Pd atoms are likely to be trapped. The Pd-CO complexes bound to surface O_{nc} anions exhibit a strong Pd-CO bond, > 2 eV, and a large CO frequency shift. This rules out the O anions as the sites where Pd is bound. This is also consistent with the fact that the binding energy of Pd atoms on-top of the O anions is not very high, 1-1.5 eV. The diffusion of the Pd atoms through the O-O channels on the surface can imply barriers of 0.5 eV or less [71] suggesting that the Pd atoms are likely to diffuse on the surface until they become trapped by some active defect. The neutral F-centers are very good candidates from this point of view since the binding of Pd to these centers is of the order of 3.5 eV. This means that to detrap a Pd atom from one of these sites a temperature of about 900 K is required. The Pd-CO units bound to the F-centers exhibit a much weaker Pd-CO bond, of the order of 0.3-0.5 eV. This is close to the experimental estimate, 0.7 eV, and suggests that the F-centers could be possible adsorption sites for Pd atoms. On the F^+ centers Pd is bound more strongly than on the O sites but not as strongly as on the F-centers; the binding energy of Pd on F^+ defect centers is consistent with a temperature-induced diffusion of 400 K. The Pd-CO complexes at F_{nc}^+ sites have bonding characteristics very close to the measured ones. The Pd-CO dissociation energy is in fact of 0.6-0.7 eV and a vibrational shift very close to the experimental value.

To summarize, the adsorption properties of Pd-CO complexes formed on MgO thin films cannot be reconciled with the picture of Pd atoms bound to the surface O anions, located either on terraces or on low-coordinated sites. Much more consistent with the observation is the hypothesis that the Pd atoms are bound to the oxygen vacancies.

7. Conclusions

The results presented in the previous sections demonstrate the importance of point defects at the surface of oxide materials in determining the chemical activity of deposited metal atoms or clusters. A single Pd atom in fact is not a good catalyst of the cyclization reaction of acetylene to benzene except when it is deposited on a defect site of the MgO(100) surface. A detailed analysis of the reaction mechanism, based on the calculation of the activation barriers for the various steps of the reaction, and of a study of the preferred site for Pd binding, based on the MgO/Pd/CO adsorption properties, has shown that the defects which are most likely involved in the chemical activation of Pd are the oxygen vacancies, or F centers, located at the terraces of the MgO surface and populated by two (neutral F centers) or one (charged paramagnetic F^+ centers) electrons.

This combined theoretical-experimental approach shows the potentiality of generating defective surfaces to modify the chemical activity of supported nano-aggregates for catalytic applications. It also shows the importance to obtain an atomistic control on the species deposited on the surface for the proper characterization of their activity. Through a specifically designed experimental apparatus which makes it possible to deposit size-selected metal clusters it has become possible to study the properties of a supported cluster as a function of its size. This opens new, unprecedented perspectives for the understanding of size-dependent activity and selectivity of a supported metal catalyst. To reach this ambitious goal theory plays a role which cannot be overlooked.

Acknowledgment
This work has been supported by the Italian Istituto Nazionale per la Fisica della Materia through the PRA-ISADORA project and by the Ministry of University and Research through a PRIN project. The experimental work was supported by the Swiss National Science Foundation. We also would like to thank W.-D. Schneider for his support.

References

[1] Pacchioni, G.; *Theory of point defects on ionic oxides*, in: The Chemical Physics of Solid Surfaces, (Woodruff, P.; Ed.) Vol. 9, Elsevier: Amsterdam 2001.

[2] Pacchioni, G.; Ricart, J. M.; Illas, F. *J. Am. Chem. Soc.*, **1994**, *116*, 10152.

[3] Ferrari, A. M.; Pacchioni, G. *J. Phys. Chem.* **1995**, *99*, 17010.

[4] Scorza, E.; Birkenheuer, U.; Pisani, C. *J. Chem. Phys.* **1997**, *107*, 9645.

[5] Pacchioni, G.; Ferrari, A. M.; Giamello, E. *Chem. Phys. Lett.* **1996**, *255*, 58.

[6] Giamello, E.; Paganini, M. C.; Chiesa, M.; Murphy, D. M.; Pacchioni, G.; Soave, R.; Rockenbauer, A. *J. Phys. Chem. B* **2000**, *104*, 1887.

[7] D'Ercole, A.; Pisani, C.; *J. Chem. Phys.* **1999**, *111*, 9743.

[8] Sterrer, M.; Diwald, O.; Knözinger, E. *J. Phys. Chem. B* **2000**, *104*, 3601.

[9] Bogicevic, A.; Jennison, D. R. *Surf. Sci.* **1999**, *437*, L741.

[10] Giamello, E.; Paganini, M. C.; Murphy, D. M.; Ferrari, A. M.; Pacchioni, G. *J. Phys. Chem. B* **1997**, *101*, 971.

[11] Pacchioni, G.; Pescarmona, P. *Surf. Sci.* **1998**, *412/413*, 657.

[12] H.-J. Freund, *Angew. Chem. Int. Ed. Engl.* **1997**, *36*, 452.

[13] C. T. Campbell, *Surf. Sci. Rep.* **1997**, *27*, 1.

[14] Henry, C. *Surf. Sci. Rep.* **1998**, *31*, 231.

[15] Haas, G.; Menck, A.; Brune, H.; Barth, J. V.; Venables, J. A.; Kern, K. *Phys. Rev. B* **2000**, *61*, 11105.

[16] Sanchez, A.; Abbet, S.; Heiz, U.; Schneider, W.-D.; Häkkinen, H.; Barnett, R. N.; Landmann, U. *J. Phys. Chem. A* **1999**, *103*, 9573.

[17] Abbet, S.; Sanchez, A.; Heiz, U.; Schneider, W.-D.; Ferrari, A. M.; Pacchioni, G.; Rösch, N. *J. Am. Chem. Soc.* **2000**, *122*, 3453.

[18] Abbet, S.; Sanchez, A.; Heiz, U.; Schneider, W.-D.; Ferrari, A. M.; Pacchioni, G.; Rösch, N., *Surf. Sci.* **2000**, *454/456*, 984.

[19] Tysoe, W. T.; Nyberg, G. L.; Lambert, R. M. *J. Chem. Soc., Chem. Commun.* **1983**, 623-625.

[20] Sesselmann, W. S.; Woratschek, B.; Ertl, G.; Kuppers, J.; Haberland, H. *Surf. Sci.* **1983**, *130*, 245.

[21] Holmblad, P. M.; Rainer, D. R.; Goodman, D. W. *J. Phys. Chem. B* **1997**, *101*, 8883-8886.

[22] Abdelrehim, I. M.; Pelhos, K.; Madey, T. E.; Eng, J.; Chen, J. G. *J. Mol. Catal. A* **1998**, *131*, 107.

[23] Gentle, T. M.; Muetterties, E. L. *J. Phys. Chem.* **1983**, *87*, 2469-2472.

[24] Rucker, T. G.; Logan, M. A.; Gentle, T. M.; Muetterties, E. L.; Somorjai, G. A. *J. Phys. Chem.* **1986**, *90*, 2703.

[25] Hoffmann, H.; Zaera, F.; Ormerod, R. M.; Lambert, R. M.; Yao, J. M.; Saldin, D. K.; Wang, L. P.; Bennett, D. W.; Tysoe, W. T. *Surf. Sci.* **1992**, *268*, 1.

[26] Patterson, C. H.; Lambert, R. M. *J. Am. Chem. Soc.* **1988**, *110*, 6871.

[27] Ormerod, R. M.; Lambert, R. M. *J. Phys. Chem.* **1992**, *96*, 8111-8116.

[28] Ormerod, R. M.; Lambert, R. M.; Hoffmann, H.; Zaera, F.; Yao, J. M.; Saldin, D. K.; Wang, L. P.; Bennet, D. W.; Tysoe, W. T. *Surf. Sci.* **1993**, *295*, 277.

[29] Pacchioni, G.; Lambert, R. M. *Surf. Sci.* **1994**, *304*, 208-222.

[30] Abbet, S.; Sanchez, A.; Heiz, U.; Schneider, W.-D., *J. Catal.* **2001**, *198*,122.

[31] Becke, A.D. *Phys. Rev. A* **1988**, *38*, 3098.

[32] Perdew, J.P. *Phys. Rev. B* **1986**, *33*, 8622; ibidem **1986**, *34*, 7406.

[33] Becke, A. D. *J. Chem. Phys.* **1993**, *98*, 5648.

[34] Lee, C.; Yang, W.; Parr, R. G. *Phys. Rev. B* **1988**, *37*, 785.

[35] Pacchioni, G.; Bagus, P.S.; Parmigiani, F. Eds., *Cluster Models for Surface and Bulk Phenomena*, NATO ASI Series B, Vol. 283, Plenum Press: New York, **1992**.

[36] Pacchioni, G.; Ferrari, A. M.; Marquez, A. M.; Illas, F. *J. Comp. Chem.* **1997**, *18*, 617.

[37] Wyckoff, R. W. G. *Crystal Structures*, 2nd ed., Interscience Publisher: New York, **1965**.

[38] Yudanov, I. V.; Nasluzov, V. A.; Neyman, K. M.; Rösch, N. *Int. J. Quant. Chem.* **1997**, *65*, 975.

[39] Nygren, M. A.; Pettersson, L. G. M.; Barandiaran, Z.; Seijo, L. *J. Chem. Phys.* **1994**, *100*, 2010.

[40] *Gaussian98*, Frisch, M.J. *et al.*, Gaussian Inc.: Pittsburgh, PA, 1998.

[41] Boys, S. F.; Bernardi, F. *Mol. Phys.* **1970,** *19*, 553.

[42] Lopez, N.; Illas, F.; Rösch, N.; Pacchioni, G. *J. Chem. Phys.* **1999**, *110*, 4873.

[43] Musolino, V.; Selloni, A.; Car, R. *J. Chem. Phys.* **1998**, *108,* 5044.

[44] Lopez, N.; Illas, F. *J. Phys. Chem.* **1998**, *102*, 1430.

[45] Ferrari, A. M.; Xiao, C.; Neyman, K.; Pacchioni, G.; Rösch, N. *Phys. Chem. Chem. Phys.* **1999**, *1*, 4655.
[46] Flank, A. M.; Delanay, R.; Legarde, P.; Pompa, M.; Jupille, J. *Phys. Rev. B* **1996**, *53*, R1737.
[47] Matveev, A. V.; Neyman, K. M.; Yudanov, I. V.; Rösch, N. *Surf. Sci.* **1999**, *426*, 123.
[48] Goellner, J. F.; Neyman, K. M.; Mayer, M.; Nörtemann, F.; Gates, B. C.; Rösch, N. *Langmuir* **2000**, *16*, 2736.
[49] Dewar, M. J. S. *Bull. Soc. Chem. Fr.* **1951**, *18*, C71.
[50] Chatt, J.; Duncanson, L. A. *J. Chem. Soc.* **1953,** 2939.
[51] Bagus, P.S.; Nelin, C.J.; Bauschlicher, C. W. *J. Chem. Phys.* **1984**, *80*, 4378.
[52] Bagus, P. S.; Illas, F. *J. Chem. Phys.* **1992**, *96*, 8962.
[53] Pacchioni, G. *Surf. Rev. & Letters* **2000**, *7*, 277.
[54] Paganini, M. C.; Chiesa, M.; Giamello, E.; Coluccia, S.; Martra, G.; Murphy, D. M.; Pacchioni, G. *Surf. Sci.* **1999**, *421*, 246.
[55] Ojamäe, L.; Pisani, C. *J. Chem. Phys.* **1998**, *109*, 10984.
[56] Wichtendahl, R.; Rodriguez-Rodrigo, M.; Härtel, U.; Kuhlenbeck, H.; Freund, H. J. *Surf. Sci.* **1999**, *423*, 90.
[57] Kantorovich, L.N.; Holender, J. M.; Gillan, M. J. *Surf. Sci.* **1995**, *343*, 221.
[58] Ferrari, A. M.; Pacchioni, G. *J. Phys. Chem.* **1996**, *100*, 9032.
[59] Di Valentin C.; Pacchioni, G. to be published.
[60] Hardesty, J. H.; Koerner, J. B.; Albright, T. A.; Le, G. Y. *J. Am. Chem. Soc.* **1999**, *121*, 6055.
[61] Abbet, S.; Riedo, E.; Brune, H.; Heiz, U.; Ferrari, A.M.; Giordano, L.; Pacchioni, G.; to be published.
[62] Herzberg, G. *Molecular Spectra and Molecular Structure*, Vol. 1, Van Nostrand: Princeton, 1950.
[63] Frank, M., Bäumer, M., Kühnemuth, R., Freund, H.-J., *J Phys. Chem B*, submitted.
[64] Liang, B., Zhou, M., Andrews, L. *J. Phys. Chem. A* **2000**, *104*, 3905.
[65] Tremblay, B., Manceron, L., *Chem. Phys.* **1999**, *250*, 187.
[66] Kündig E.P., McIntosh, D., Moskovits, M., Ozin, G.A. *J. Am. Chem. Soc.* **1973**, *95*, 7234.
[67] Darling, J.H., Ogden, J.S., *J. Chem. Soc., Dalton Trans* **1973**, 1079.
[68] Yudanov, I. V.; Vent, S.; Neyman, K.; Pacchioni, G.; Rösch, N. *Chem. Phys. Lett.* **1997**, *275*, 245.
[69] Goniakovski, J. *Phys. Rev. B* **1998**, *58*, 1189.
[70] Nygren, M. A.; Pettersson, L. G. M. *J. Chem. Phys.* **1996**, *105*, 9339.
[71] Neyman, K. M.; Vent, S.; Pacchioni, G.; Rösch, *Nuovo Cimento* **1997**, *19D*, 1743.

THE VALENCY EFFECT ON REACTION PATHWAYS IN HETEROGENEOUS CATALYSIS: INSIGHT FROM DENSITY FUNCTIONAL THEORY CALCULATIONS

A. MICHAELIDES and P. HU
School of Chemistry,
The Queen's University of Belfast,
Belfast BT9 5AG, U. K.

Abstract: Density functional theory (DFT) has been used to determine reaction pathways for several reactions taking place on Pt and Cu surfaces. On Pt(111) all the elementary steps in the hydrogenation of carbon to methane, nitrogen to ammonia, and oxygen to water were investigated. Also on Pt(111) the reactions of C with O and N were studied, and on Cu(111) the reaction of C with H was examined. It is found that reaction pathways are related to the valency of the reactants involved. As the valency of a reactant decreases so too does the coordination number of the site at which the transition state is located. In the hydrogenation reactions on Pt(111) it was found that monovalent adsorbates (for example, CH_3, NH_2 and OH) tend to make transition states close to top sites; divalent adsorbates (CH_2, NH and O) near bridge sites; and tri- and tetra-valent adsorbates (CH, N and C) near three-fold hollow sites. This relationship between the transition state and the valency applies equally to the C+O and C+N reactions on Pt(111), the C+H reaction on Cu(111), and many others from the literature. On this basis, we have developed rules to facilitate the prediction of catalytic reaction pathways without recourse to experimental or theoretical investigations.

1. Introduction

Few fields of scientific endeavour can have contributed so dramatically to human existence over the last forty years as heterogeneous catalysis. Over 80 % of industrial chemical production proceeds with the aid of a catalyst and heterogeneous catalysis represents 20-30 % of global GNP annually. The key to future advances in heterogeneous catalysis must be to understand surface processes at an atomic level, and the surface catalysed reaction is of paramount importance. Acquiring this knowledge is not, however, an easy task. To determine microscopic reaction pathways high temporal (picosecond) as well as high spatial (Angstrom) resolution is simultaneously required, making experimental investigations in this area problematic. Computer simulations - chemistry in a "virtual laboratory" - are, therefore, an attractive alternative. Thanks to recent methodological and computational developments it is now possible to apply highly accurate quantum mechanical techniques to the study of catalytic reactions.[1] The

M.A. Chaer Nascimento (ed.), Theoretical Aspects of Heterogeneous Catalysis, 199–215.

accuracy of one particular technique, density functional theory (DFT), has improved dramatically over recent years with the effect that DFT calculations now yield precise structural and energetic information. This makes DFT particularly amenable to the study of reaction pathways and barriers.

By far the most common type of reaction in heterogeneous catalysis is the Langmuir-Hinshelwood reaction.[2] In this type of reaction the reactants adsorb on the face of the catalyst from where they diffuse and react. The practical desire to understand such reactions stems from the long held belief that a microscopic understanding may facilitate the design of more efficient, more selective, less expensive catalytic substrates. DFT has been helpful in this regard and the pathways of certain catalytic reactions have recently been determined, providing a clear indication as to how these reactions are likely to take place.[3-5] However, a general explanation as to how chemical reactions proceed on catalyst surfaces is lacking.

In this chapter we present the results of extensive DFT calculations for catalytic reactions taking place on Pt(111) and Cu(111) surfaces. On Pt(111) we have performed a systematic study of the pathways of nine different hydrogenation reactions. Microscopic reaction pathways have been determined for all the elementary steps in the hydrogenation of carbon to methane, nitrogen to ammonia, and oxygen to water. In addition we have examined the reaction of: (i) chemisorbed carbon and oxygen yielding chemisorbed CO; (ii) chemisorbed carbon with chemisorbed nitrogen to form chemisorbed CN; and (iii) on Cu(111) the reaction of chemisorbed C and H. Clearly, these are all very important reactions as C, N, O, and H are the principal building blocks of most of the species involved in the fundamental processes of heterogeneous catalysis. Through a determination of the microscopic reaction pathways of these prototype reactions we aim at a deeper understanding of the factors that govern catalytic reaction pathways in general.

Below we briefly describe details of our total energy calculations. Following this, in section 3.1, we present results for the chemisorption of the various atomic and molecular fragments involved. In sections 3.2 and 3.3 reaction pathways for the various reactions are presented and discussed, with particular emphasis on transition states. In section 4 we discuss our results in a wider context and in section 5 we draw our conclusions and introduce rules derived to facilitate the prediction of catalytic reaction pathways.

2. Calculation details

First-principle total energy calculations within the DFT framework were performed.[1] Ultrasoft pseudopotentials[6] were expanded within a plane wave basis set up to a cut-off of 300 eV. Electron exchange and correlation effects were described by the generalised gradient approximation (GGA) of Perdew and Wang (PW91).[7] A Fermi smearing of 0.1 eV was utilised and the corrected energy extrapolated to zero Kelvin. The Pt(111) and Cu(111) surfaces were modelled by a periodic array of slabs. Each slab consisted of three layers of metal atoms fixed at bulk truncated positions. The vacuum region between adjacent slabs was in excess of 12 Å. Calculated Pt and Cu lattice constants of 3.9711 Å (experimental = 3.9239 Å) and 3.6310 Å (experimental = 3.6147 Å) were used throughout. A p(3x2) surface unit cell was used. This cell is large enough to avoid bonding competition in the initial states of the reactions by allowing each reactant to chemisorb in three-fold hollow sites which do not 'share' metal atoms with each other.[8] A Monkhorst-Pack[9] mesh with 2x2x1 **k**-point sampling within the surface Brillouin zone, was found to offer sufficient accuracy.

Reaction pathways and transition states were searched with a well-tested constrained minimisation technique.[4] The approach employed in each reaction was to fix the distance between the two reactants and minimise the total energy with respect to all remaining degrees of freedom. Through a series of such constrained structure optimisations, with a different reactant separation in each case we determine energy profiles and reaction pathways. Since the only constraint is the distance between the reactants, the reactants are free to rotate and translate subject to the above constraint. The transition state is identified when (i) the forces on the atoms vanish and (ii) the energy is a maximum along the reaction coordinate but a minimum with respect to all remaining degrees of freedom. For each reaction several possible reaction channels were searched, and only the results of the lowest energy pathway, i.e., that which accesses the lowest energy transition state, are presented here.

Extensive testing calculations were conducted before deciding upon our chosen surface model. Calculations were performed with different **k**-point meshes, GGA functionals, cut-off energies and numbers of layers of metal atoms. In addition the effect of surface relaxation on the energetics and pathways of certain reactions was investigated. Some of the results of these testing calculations are shown in Tables 1-3. From Tables 2 and 3 it can be seen that reaction barriers (E_a) between our chosen model

TABLE 1. Testing calculations for a bridge site OH at 1/4 monolayer on Pt(111).

K points	Layers	E_{ads} (eV)[a]	O-Pt (Å)
2x2x1	2	2.30	2.16
2x2x1	3	2.27	2.17
2x2x1	4	2.25	2.17
2x2x1	4[b]	2.45	2.14
4x4x1	2	2.24	2.17
4x4x1	3	2.22	2.17
4x4x1	4	2.20	2.17
4x4x1	4[a]	2.23	2.14

[a]E_{ads} is the chemisorption energy. See section 3.1 for definition.
[b]refers to a surface model in which the top layer of Pt atoms were allowed to relax.

and more accurate models differs by *ca.* 0.1 eV (1 eV = ~23.06 Kcal/mol) and pathways and transition states were qualitatively similar. With specific reference to surface relaxation the testing calculations revealed that the surface degrees of freedom play only a minor role in the overall reactivity. Our particular surface model was selected because the testing calculations revealed that it offered the best compromise between accuracy and computational load. In addition, previous studies using this[10-13] and similar models[14,15] have shown that this approach provides a satisfactory description of the adsorbate-substrate interface.

Table 2. Testing calculations for the O+H reaction on Pt(111).

Cell	K points	Layers	Functional	E_a[a] (eV)	ΔH[b] (eV)
p(2x2)	2x2x1	3	PW91	0.99	+0.14
p(2x2)	2x2x1	4	PW91	0.94	+0.14
p(3x2)	2x2x1	4[c]	PW91	0.95	+0.07
p(2x2)	2x2x1	3	PW91	0.87	-0.07
p(2x2)	4x4x1	2	PW91	0.96	+0.10
p(2x2)	4x4x1	3	PW91	0.96	+0.06
p(2x2)	4x4x1	4[c]	PW91	0.94	-0.02
p(2x2)	4x4x1	4	PBE[d]	0.96	+0.03
p(2x2)	4x4x1	2	RPBE[e]	0.96	-0.10

[a]Reaction barriers (E_a) are determined by calculating the energy differences between the most stable initial and the lowest energy transition states.
[b]Enthalpy changes (ΔH) are determined by calculating the energy differences between the most stable initial and final states.
[c]refers to surface models in which the top layer of Pt atoms were allowed to relax.
[d]Reference 32
[e]Reference 33

Table 3. Testing calculations for the C+H reaction on Cu(111) in a p(3x2) unit cell.

K points	Layers	Cut-off (eV)	E_a (eV)
2x2x1	3	300	0.73
2x2x1	3	400	0.75
2x2x1	4[a]	300	0.80
2x2x1	4	300	0.76
4x3x1	3	300	0.72

[a]refers to surface models in which the top layer of Cu atoms were allowed to relax.

3. Results

3.1. CHEMISORPTION

Prior to the calculation of reaction pathways the chemisorption of each adsorbate was investigated. Chemisorption energies (E_{ads}) are calculated from

$$E_{ads} = E_A + E_S - E_{A/S}$$

where E_A, E_S and $E_{A/S}$ are the total energies of the isolated adsorbate, the clean surface and the chemisorption system, respectively. In this definition positive adsorption energies correspond to an exothermic process. The energy of the gas phase adsorbate is calculated by placing it in a large empty unit cell. A cell of at least 10-12 Å^3 guarantees a negligible interaction between species in adjacent cells. When required the energy of the gas phase adsorbate is obtained from a spin polarised calculation. The Pt(111) surface has four high symmetry sites where adsorption may occur: top, bridge, and two three-fold hollow sites, shown in Figure 1. The three-fold hollow sites differ only in their relationship with the second layer. The hcp three-fold hollow site has a metal atom

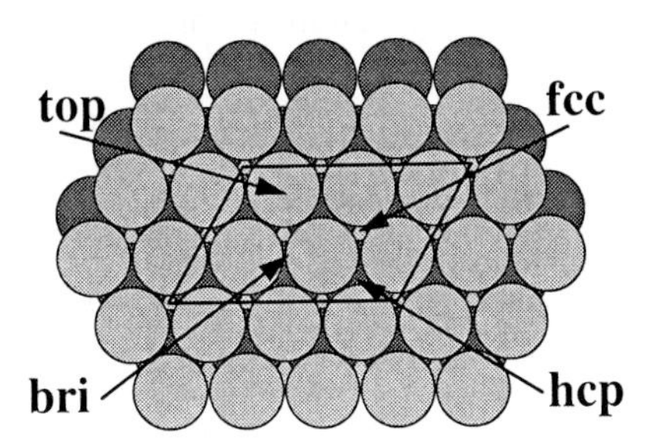

Figure 1. Plan view of the Pt(111) surface. The arrows indicate the location of the four high symmetry sites and the straight lines a p(3x2) unit cell. Only two layers of metal atoms are shown for clarity.

TABLE 4. Chemisorption energies in eV for 1/6 monolayer C, N, O, and H adsorption on the four high symmetry sites of a fixed three layer Pt slab.

Site	C	N	O	H
hcp	**6.86**[a]	4.27	3.59	2.79 (2.97)[b]
fcc	6.85	**4.43**	**3.97**	**2.85 (3.01)**
bridge	6.08	3.56	3.41	2.83 (2.94)
top	4.49	1.92	2.51	2.83 (2.87)

[a]The values shown in bold are chemisorption energies at the most stable site for each adsorbate.
[b]Values in parenthesis are obtained using a four layer Pt slab in which the top layer is allowed to relax.

TABLE 5. Chemisorption energies in eV for various molecules at 1/6 monolayer coverage on the four high symmetry sites of a fixed three layer Pt slab.

Site	CH	CH_2	CH_3	NH	NH_2	NH_3	OH	H_2O
hcp	6.76		1.25	3.76	1.52		1.82	
fcc	**6.84**[a]		1.38	**4.05**	1.72		2.00	
bridge	6.10	**4.06**	1.39	3.15	**2.58**		**2.27**	0.21
top	3.92	3.19	**2.05**	1.26	1.93	**1.05**	2.23	**0.34**

[a]The values shown in bold are chemisorption energies at the most stable site for each adsorbate.

directly below it, while the fcc three-fold hollow site has a metal atom only in the third layer. Structure optimisations for each adsorbate at the four high symmetry sites of the

Pt(111) surface were performed. Tables 4 and 5 list chemisorption energies for C, N, O, and H and $CH_{X(X=1\to3)}$, $NH_{X(X=1\to3)}$, $OH_{X(X=1\to2)}$ on the four high symmetry sites of Pt(111). From Table 4 we see that the atomic adsorbates have a strong preference for adsorption at three-fold hollow sites, with the exception of H which has a very smooth potential energy surface (PES). Table 5 reveals that the molecular adsorbates do not exhibit quite the same preference for adsorption at three-fold hollow sites, binding preferentially at a variety of sites: CH and NH favour adsorption at fcc three-fold hollow sites; CH_2, NH_2 and OH at bridge sites; and CH_3, NH_3 and H_2O at top sites.

3.2. HYDROGENATION REACTIONS ON Pt(111)

After calculating stable adsorption sites for each adsorbate low energy reaction pathways were determined. First we consider the hydrogenation reactions on Pt(111), following this the C+O and C+N reactions on Pt(111) and the C+H reaction on Cu(111).

3.2.1. *Reaction Pathways*

All the elementary steps in the hydrogenation of carbon to methane, nitrogen to ammonia, and oxygen to water on Pt(111) were investigated, i.e., reaction pathways for the following reactions were determined:

$$C + H \rightarrow CH;\ CH + H \rightarrow CH_2;\ CH_2 + H \rightarrow CH_3;\ CH_3 + H \rightarrow CH_4$$

$$N + H \rightarrow NH;\ NH + H \rightarrow NH_2;\ NH_2 + H \rightarrow NH_3$$

$$O + H \rightarrow OH;\ OH + H \rightarrow H_2O$$

Reaction barriers and enthalpy changes have been calculated for each reaction; these are listed in Table 6.

TABLE 6. Reaction barriers (E_a) and enthalpy changes (ΔH) for the nine hydrogenation reactions on Pt(111).

Reaction	E_a (eV)[a]	ΔH (eV)[b]
$C + H \rightarrow CH$	0.78	-0.75
$CH+H \rightarrow CH_2$	0.72	0.58
$CH_2+H \rightarrow CH_3$	0.63	-0.19
$CH_3+H \rightarrow CH_4$	0.74	0.08
$N+H \rightarrow NH$	0.94	-0.42
$NH+H \rightarrow NH_2$	1.31	0.01
$NH_2+H \rightarrow NH_3$	0.74	-0.62
$O+H \rightarrow OH$	0.87	-0.07
$OH+H \rightarrow H_2O$	0.22	-0.56

[a]Reaction barriers (E_a) are determined by calculating the energy differences between the most stable initial and the lowest energy transition states.

[b]Enthalpy changes (ΔH) are determined by calculating the energy differences between the most stable initial and final states. The final state for each of the reactions, except CH_3+H, is a species chemisorbed upon the surface.

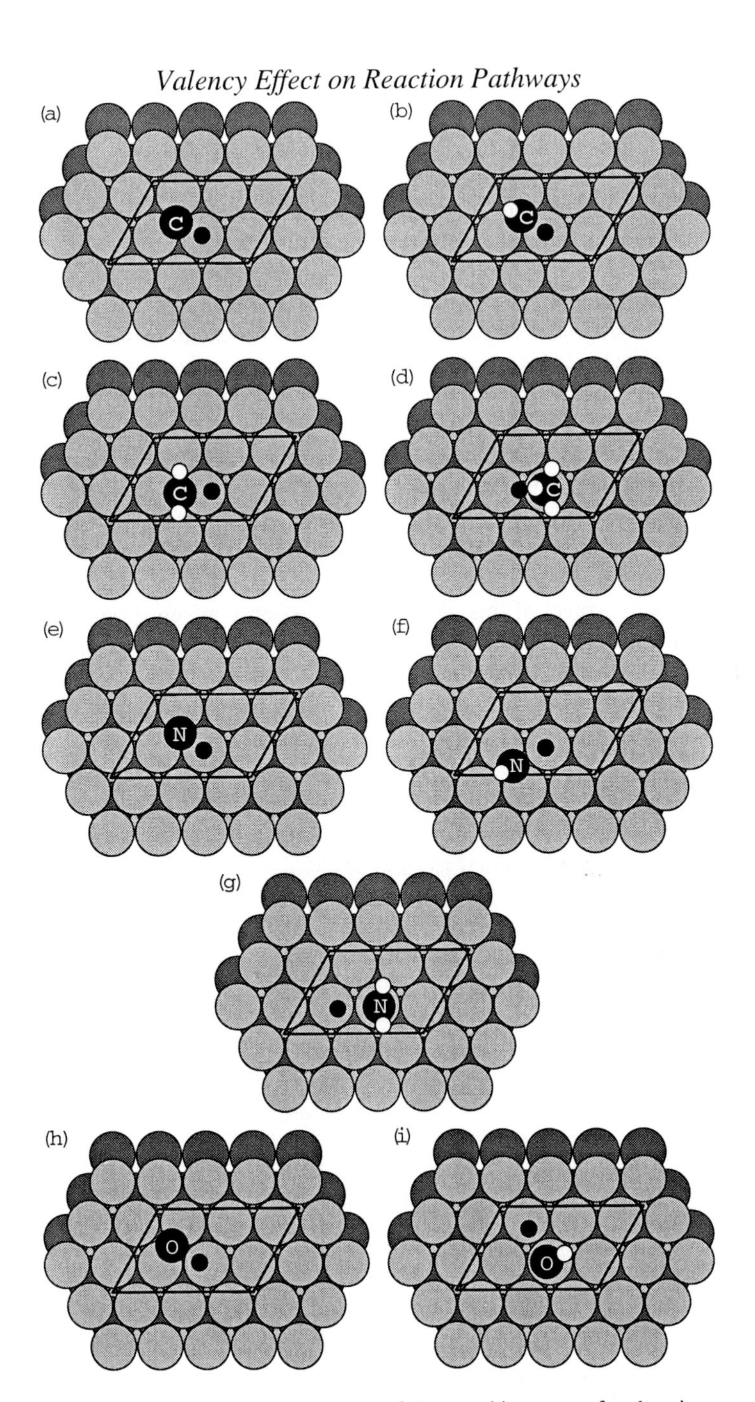

Figure 2. Basic structure of each of the transition states for the nine hydrogenation reactions on Pt(111): (a) C + H; (b) CH + H; (c) CH_2 + H; (d) CH_3 + H; (e) N + H; (f) NH + H; (g) NH_2 + H; (h) O + H; and (i) OH + H. The large grey circles represent Pt atoms. The small white circles are hydrogen atoms that are bonded to an adsorbate and the small black circles are the hydrogenating hydrogens.

Our principal finding is the identification of clear periodic trends in the microscopic pathways of each reaction. These are best observed by an examination of the transition state of each reaction. A simple representation of these is shown in Figure 2, a consideration of which yields the following interesting features.

1. There are only a limited number of classes of transition state.
Class I: The "heavier fragment", i.e., CH_X,NH_X,OH_X, is located close to a three-fold hollow site [Figure 2(a),(b),(e)].
Class II: The heavier fragment is located close to a bridge site [Figure 2(c),(f),(h)].
Class III: The heavier fragment is located close to a top site [Figure 2(d),(g),(i)]. The hydrogenating hydrogen atoms have a large degree of freedom and appear at several positions due to a flat potential energy surface on Pt(111).

2. Heavier fragments that are isoelectronic access similar transition states. Both CH and N, for example, are located at a hollow site in the transition state of their reaction with H. The most noteworthy aspect of this phenomenon probably arises upon consideration of the reverse of these reactions, i.e., dehydrogenation: The transition states for the removal of the first hydrogen from CH_4 (Figure 2(d)), NH_3 (Figure 2(g)), and H_2O (Figure 2(i)), respectively, are alike. The transition states for subsequent H removal are also similar.

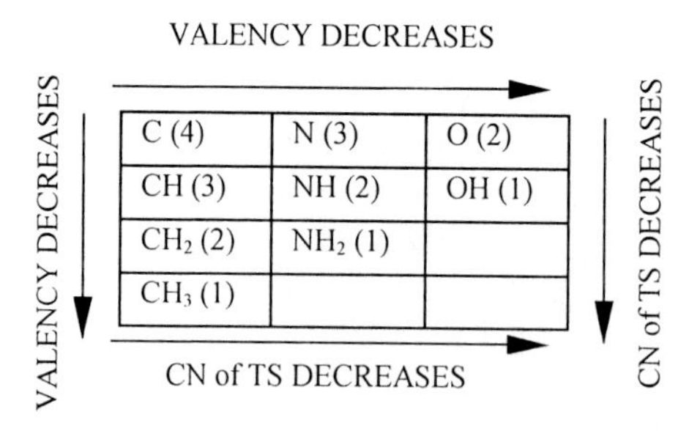

Figure 3. Relationship between the valency of an adsorbate and the location of the transition state when it reacts with a chemisorbed H on Pt(111). As the valency decreases, the coordination number (CN) of the site upon which the transition state (TS) is located decreases. The values in brackets refer to both the valency of the adsorbate *and* the coordination number of the site upon which the transition state is located, except for C, where this value refers to valency alone.

3. There is a clear relationship between the valency of the heavier adsorbates and the location of the transition state. The higher the valency of the adsorbate, the higher the coordination number (the coordination number of a surface site equalling the number of

surface atoms at that site) of the site upon which the transition state is located. This is illustrated in Figure 3. Monovalent adsorbates favour transition states at top sites, divalent adsorbates at bridge sites, and trivalent and tetravalent adsorbates at hollow sites. This relationship is rigorously followed, except for the NH + H reaction in which a hollow site transition state of similar energy to that on the bridge site has also been identified. A consequence of this relationship is that if the coordination number of the site at which an adsorbate is initially chemisorbed is greater than the adsorbate's valency, then it must diffuse to a site of lower coordination to form a transition state. For example, the O atom must move from a hollow site (initial chemisorption site) to a bridge site in order to react.

4. The directions in which the hydrogenating hydrogen atoms approach the heavier fragments are determined by the symmetries of the molecular orbitals of the heavier fragments (compare for example the location of H in the CH_2 + H (Figure 2(c)) and NH + H (Figure 2(f)) transition states). If the hydrogenating H atoms approach the heavier fragments from completely different directions to those shown then the reaction barriers will be much higher and the reactions may not even occur.

3.2.2. *Discussion of Hydrogenation Reactions*

What is the physical origin of the transition-state valency relationship? An analysis of the electronic structures of the initial, transition, and final states reveals that an answer to this question lies in the p orbitals of the adsorbates (except chemisorbed H). We find

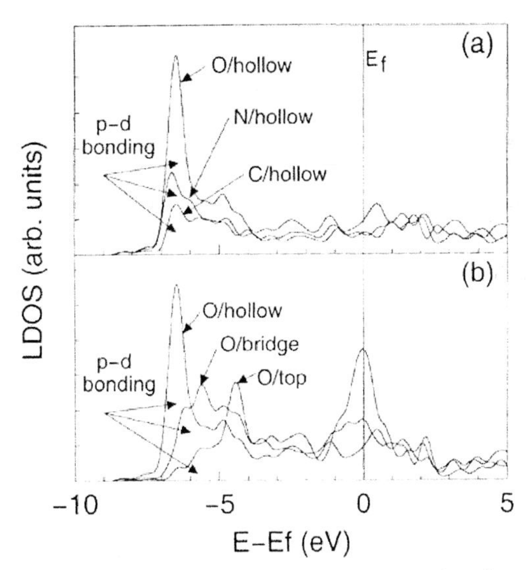

Figure 4. Local densities of states (LDOS) projected on to (a) C, N, and O on a hollow (fcc) site of Pt(111) and (b) O at hollow, bridge, and top sites. The almost purely localised 2s derived states which occur at >10 eV below the Fermi level (E_f) are not shown.

that the 2s orbitals of the adsorbates are localised and not involved in adsorbate-surface bonding. The 2p orbitals, however, mix significantly with metal states before reaction and also in the transition states the 2p orbitals mix with metal states and H. In other words, the 2p orbitals are involved in the bond formation and bond breakage processes and are thus strongly related to the reactivity of an adsorbate. It is found that an adsorbate is not reactive at a particular site, i.e., it cannot form a transition state at this site and must be activated to a site of lower coordination to achieve a transition state, if a large proportion of the p electrons around the adsorbate are at low energies. In contrast, an adsorbate is reactive at a site if a small proportion of its p electrons are at low energies. An example is shown in Figure 4(a) which displays the local densities of states (LDOS) for the valence electrons projected onto C, N, and O for these species chemisorbed on hollow sites. An examination of the real space distribution of the individual quantum states reveals that states between -8 eV and the Fermi level are mostly mixing states between adsorbate p and Pt d orbitals. It is clear that a greater proportion of the p electrons around O (not reactive at hollow sites) are at lower energies (about -7 eV) than those around C or N (reactive at hollow sites). This result can be rationalised as follows. If an adsorbate is to be reactive at a certain site, i.e., capable of forming a transition state, it is essential that it has available p orbitals with which another species can form a new bond. If all the p orbitals of the adsorbate at a surface site are at low energies, i.e., bonding with surface atoms, and none are available for new bond formation, then the adsorbate is unreactive and this is not a suitable site for a transition state. The large proportion of p electrons at low energies around the hollow site O (Figure 4(a)) indicates that at this site oxygen p orbitals are fully or nearly fully used up in bonding with metal d states. There are insufficient p orbitals for new bond formation thus rendering O unreactive. Oxygen, at the hollow site, can be considered to be "saturated".[16] Consequently, it was not possible to identify an O transition state on or near a hollow site. In contrast, the smaller proportion of p electrons at low energies (centred at -7 eV) around C and N, as a result of their increased valency, suggest that these adsorbates are not "saturated". Therefore, unlike O, C and N are reactive at hollow sites and transition states could be found on hollow sites (Figure 2(a),(e)). Another example is shown in Figure 4(b), which displays LDOS plots for O chemisorbed at various sites. When O is on the bridge site (reactive), the peak of p-d bonding states is much smaller than that which is observed when O is on the hollow site. Similar findings have been found for all other adsorbates studied.

Consistently throughout we see that two factors determine the proportion of low-energy p electrons around an adsorbate and ultimately the reactivity of the adsorbates. The first is the valency of the adsorbate; as the valency increases, the proportion of low-energy p electrons decreases (Figure 4(a)). The second is the adsorption site; the proportion of low-energy p electrons increases as the coordination of an adsorption site increases (Figure 4(b)). It is the combination of these two trends that yields the important relationship between valency and the location of the transition state.

3.3. SOME MORE REACTIONS

3.3.1. *C+O, C+N, and C+H Reaction Pathways*

The reactions reported, above, were all hydrogenation reactions and chemisorbed H was, therefore, a reactant in each case. As discussed chemisorbed H atoms had a large degree of freedom with their location at transition states determined by their reaction counterpart. This is due, in part, to an exceptionally smooth potential energy surface for H diffusion on Pt(111) (Table 4 and references 17 and 18). Reactions in which *both* reactants exhibit a corrugated potential energy surface are more common and encompass a wider variety of chemical processes. Obvious questions are therefore: (i) What principles govern the pathways of these common types of reaction? (ii) How does the transition state-valency relationship apply to these reactions? To assist with answering these questions we use the following prototype systems: (i) chemisorbed carbon and oxygen yielding chemisorbed CO; (ii) chemisorbed carbon with chemisorbed nitrogen to form chemisorbed CN; and (iii) on Cu(111), the reaction of chemisorbed carbon and chemisorbed hydrogen yielding chemisorbed CH.

The pathways and energy profiles for each of the three reactions investigated are displayed in Figures 5 and 6. Figure 5(a) displays the initial states of each reaction, each of which are the most stable coadsorption of the reactants in p(3x2) unit cells. In the coadsorption states shown in Figure 5(a), and when chemisorbed independently, all the reactants chemisorb preferentially at three-fold hollow sites. In the initial states of the C+O and C+N reactions, C atoms are at hcp three-fold hollow sites and O or N atoms are at fcc sites. In the C+H reaction it is most stable to have C atoms at fcc sites and H atoms at hcp sites.

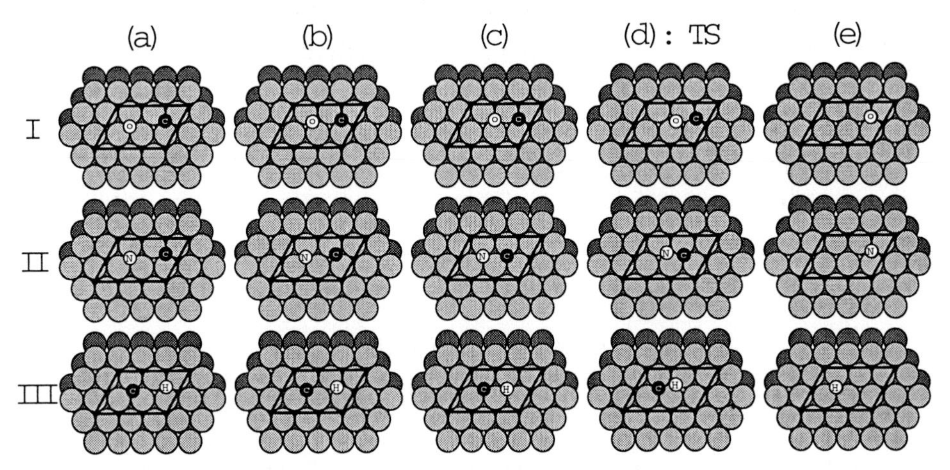

Figure 5. Reaction pathways for: (I) C+O to CO on Pt(111); (II) C+N to CN on Pt(111); and (III) C+H to CH on Cu(111). The large grey circles are surface atoms, black circles are C. The white circles are O, N and H atoms in I, II and III, respectively. The solid lines indicate the surface unit cell. For clarity, the periodic nature of the systems is not shown and only two layers of surface atoms are displayed. Panel (d) represents the transition states (TSs) of each reaction.

From Figure 5 it can be seen that the transition states of each reaction are *similar:* The separation of the reactants is about 1.8-2.0 Å and one adsorbate is close to a three-fold hollow site while the other has been activated from their favoured (hollow) site to a bridge site. We have here then a fourth class of transition state:

Class IV:. The first reactant is at a three-fold hollow site and the second is at a next nearest neighbour bridge site. Indeed, this represents a very common structure for a transition state on a (111) metal surface. Transition states with this structure are often observed in the dissociation of diatomic molecules - reactions which are the reverse of those currently under discussion. DFT calculations have shown that the dissociation of N_2,[19] CO,[3,20] and NO[21,22] on a variety of transition metal surfaces all proceed via this

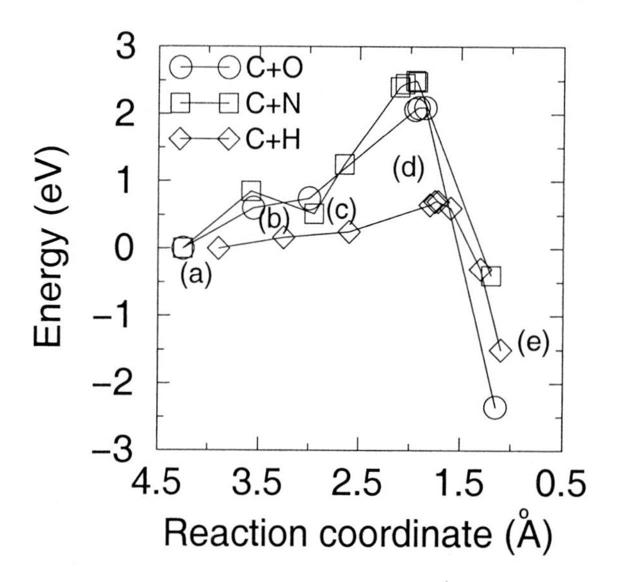

Figure 6. Energy profiles of the C+O, C+N reactions on Pt(111) and the C+H reaction on Cu(111). The points (a)-(e) of each reaction correspond to the points (a)-(e) of Figure 5.

type of transition state. Further, we report that the transition state for the N+N recombination reaction on Pt(111) is also of this type.

3.3.2. *Discussion of C+O, C+N, and C+H reactions*

In order to understand the structure of this important class of transition state it is beneficial to consider the geometry of the Pt(111) and Cu(111) surfaces. A plan view of a segment of a (111) surface of an fcc metal is shown in Figure 7(a). On both Pt(111) and Cu(111) the distance between hollow sites *a* and *b*, in Figure 7(a), is approximately 1.6 Å and the distance between hollow sites *a* and *c* is approximately 2.7 Å. Since the distance between two reactants in this type of transition state is about 1.8-2.0 Å it is clear that if one reactant is at hollow site *a* (for example C and N are both reactive at hollow sites) then both hollow sites *b* and *c* are unsuitable locations for the other reactant. Hollow site *b* is too close at 1.6 Å and hollow site *c* is too far away at 2.7 Å. Naturally the bridge site or near to the bridge site between these two hollow sites is a suitable location for the second reactant. With one reactant close to a hollow site and the other to a bridge site then the transition state for the C+O reaction can be readily understood. Carbon which is reactive at hollow sites is located at the hollow site. Oxygen which is reactive at bridge sites is located at the bridge site. In the C+N reaction, however, since both C and N are reactive at hollow sites then there are two possible transition state structures: (i) C is at the hollow site and N is at the bridge site;

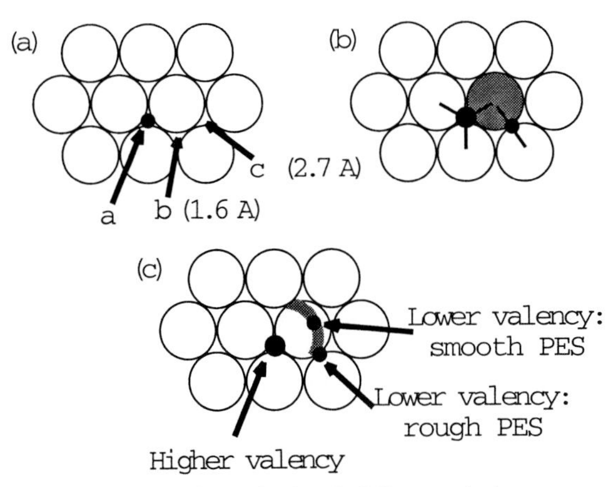

Figure 7. Illustrations of: (a) typical distances in Angstroms between adsorbates when chemisorbed at neighbouring three-fold hollow sites on Pt(111) and Cu(111); (b) bonding competition in a transition state. Both reactants compete for bonding with the shaded surface atom; (c) the structure of two common types of transition state on a (111) metal surface. In (a) to (c) the large white circles represent surface atoms, and the small black circles reactants.

and (ii) C is at the bridge site and N is at the hollow site. We find that the former is favoured (Figure 5). We have attempted but could not locate a transition state with N near a hollow site and C near a bridge site. In the C+H reaction, C is again at a three-fold hollow site and H is at a bridge site. This leads to an important suggestion: In these three atomic recombination reactions it is the reactant with the higher valency that is located close to the high coordination site while the reactant with the lower valency is close to the low coordination site. We believe that this may be a general finding since it holds, not only for the current reactions, but also for the dissociation reactions from the literature mentioned above.

It must be emphasised that in the transition state of the C+N reaction the observation that N resides at a bridge site was initially surprising because the calculated energy difference between a N atom chemisorbed at a three-fold hollow and bridge site of Pt(111) is 0.87 eV, whilst in the case of C this difference is only 0.78 eV. In other words, the potential energy surface for diffusion of N is more corrugated than that of C. This was confirmed by calculations on a four layer Pt slab in which the top layer of Pt atoms was allowed to relax. Thus, one may expect that C rather than N should be activated from the hollow site to a bridge site in order to achieve a transition state. This is clearly not the case and it shows, as with the hydrogenation reactions in section 3.2, that an important factor in determining the structure of a transition state is the valency of the reactants involved.

An obvious question is therefore: *why, in a transition state of this type, is the lower valency adsorbate activated to the bridge site?* The answer relates to the fact that this arrangement minimises repulsion between the adsorbates. In a transition state of this type, the adsorbates can be arranged in two ways: (i) the lower valency adsorbate is close to the bridge site with the other close to the hollow site; or (ii) the lower valency adsorbate is close to the hollow site with the other close to the bridge site. In either case there is one surface atom that is bonded to both adsorbates (the shaded atom in Figure 7(b)). The 'shared' surface atom is one of three atoms of the hollow site, but one of two of the bridge site (Figure 7(b)). The bonding of each adsorbate to this particular surface atom is weakened considerably. Consequently, a greater proportion of the adsorbate-surface bond is weakened for the bridge site adsorbate than the hollow site adsorbate: It can be approximately considered that one of two chemical bonds for the bridge site adsorbate are weakened while the bond strength of one of three is reduced for the hollow site adsorbate. We know that absolute chemisorption energy decreases with valency, i.e., low valency adsorbates bind more weakly to a surface than high valency adsorbates, as shown in Tables 4 and 5 for a variety of small molecules on Pt(111). It is, therefore, preferable to have the lower valency adsorbate, the one with the weaker chemisorption energy, on the bridge site. Clearly, it is better to lose a greater proportion of the chemisorption energy of the weaker adsorbate rather than the stronger adsorbate.

Further insight into the structure of transition states can be gained by comparing the transition state of the C+H reaction on Cu(111) with those of the C+H, CH+H and N+H reactions on Pt(111). As we have seen in the transition state of the C+H reaction on Cu(111), C is at a hollow site and H is close to a bridge site (Figure 5). In the transition states of the three reactions on Pt(111) C, CH and N are close to a hollow site whilst H in each case is close to a *top* site (Figure 2(a), (b) and (e)). This difference in the location of H can be traced back to the individual potential energy surfaces for diffusion of H. On Pt(111) both experimental and theoretical studies, including ours (Table 4), reveal that H has a very smooth potential energy surface.[17,18] In agreement with this, we find that H binds at each of the four high symmetry sites of Pt(111) to within a 0.1 eV energy difference. When H reacts with another adsorbate on Pt(111) it, therefore, approaches in the most reactive direction (the p-d bonding direction) about the chemisorbed reactant, which is over the top site of a metal atom. The potential energy surface for H diffusion on Cu(111) is, however, much more corrugated. H exhibits quite a strong preference for adsorption at three-fold hollow sites. We find, in fact, in agreement with the recent DFT calculations of Stromquist *et al.*[23] that there is approximately a 0.5 eV difference between H adsorption at the fcc and top sites. It is this dislike for H adsorption at top sites that prevents H from making transition states at these sites on Cu(111). Similarly, in the C+O and C+N reactions on Pt(111) the strong dislike for O and N adsorption at top sites prevents these adsorbates from accessing transition states at top sites. In general, therefore, as illustrated in Figure 7(c), if the potential energy surface of the activated adsorbate is quite smooth then it will be located close to a top site. If, on the other hand, the potential energy surface of the activated adsorbate is corrugated and it has a strong dislike for the top site then it will be located at a bridge site.

4. Discussion

Perhaps the most important result in this study is the identification of the relationship between valency and the location of transition states. For the nine hydrogenation reactions on Pt(111) it was found that as the valency of the reactants decreases so too does the tendency of the adsorbates to access transition states at sites of higher coordination number.

It is interesting at this stage to investigate how far this relationship between transition state and valency may extend. By comparing the present results on Pt with those from the literature we find that it holds for a variety of reactants on a variety of transition metal surfaces. The best example is CH_3 hydrogenation; or the reverse of this reaction, CH_4 dissociation. DFT calculations have indicated that this reaction proceeds through a transition state which has CH_3 located close to a top site on Ru(0001),[24] Rh(111),[25] Ni(111),[26] and, we report here, Cu(111). Indeed in the first application of DFT to a reaction involving a subsurface species we demonstrated that the hydrogenation of CH_3 by a subsurface H on Ni(111) also proceeds through such a transition state.[26] The C+H reaction has also been examined with DFT on several transition metal surfaces. Reaction pathways have been determined for this reaction on Ru(0001), Rh(111), and Pd(111) and in every case C, as expected, is located close to a three-fold hollow site in the transition state.[27] An interesting and illuminating example is the CH_2 hydrogenation reaction on Ni(111).[28] This reaction proceeds through a transition state which has CH_2 located close to a bridge site. This is the same location for the transition state as on Pt(111) despite the fact that CH_2 initially chemisorbs at a three-fold hollow site on Ni(111) whilst on Pt(111) it favours a bridge site. This example highlights the important observation that the transition state-valency relationship is independent of the preferred chemisorption site of the reactants. Having demonstrated that this relationship holds on metal surfaces other than Pt for CH_X hydrogenation reactions, we see that it may also be generally applicable to certain NH_X hydrogenation reactions. In the transition states of the N+H and NH_2+H reactions on Ru(0001), for example, N and NH_2 are located, as on Pt(111), at three-fold hollow and top sites, respectively.[29]

It must be emphasised that although this relationship has so far been followed rigorously on Pt(111) and holds on other metal surfaces it is by no means rigorously followed on all metal surfaces. The transition state for the CH_2+H reaction on Cu(111)[11] is, for example, located at a three-fold hollow site, not a bridge site as on Pt(111) (and Ni(111)). In addition DFT calculations have indicated that the NH+H reaction on Ru(0001) proceeds preferentially through a hollow site transition state.[29] It appears that this framework applies best to reactions involving adsorbates of low valency (monovalent) and high valency (tri/tetravalent). Explaining this more subtle nature of the relationship is key and requires detailed investigations of the electronic structures, in particular the adsorbate p orbitals, of each system.

It is of further interest to consider the adsorbates and types of reactions to which this relationship might apply. Obviously it works well for hydrogenation reactions involving CH_X, NH_X and OH_X species, especially those on Pt(111). Recently we have found that this relationship can also be applied to sulphur hydrogenation reactions on Pt(111). The transition states for the S+H and SH+H reactions on Pt(111) are, as expected, located at bridge and top sites, respectively.[30] As discussed this relationship applies also to the C+O and C+N reactions presented above. Of these two classes of reaction, oxidation reactions have received the greatest attention. CO oxidation in particular has been extensively examined within our group and in transition states of the CO+O reaction on Pt(111),[4] Ru(0001),[5] Rh(111),[5] and Pd[31] the O atom is located close to a bridge site. In fact, on Pd it has been found that the relationship applies to CO oxidation transition states on both the (111) and (100) surfaces of Pd with O located close to a bridge site in the transition site. This example gives an indication that the relationship may extend to surfaces with symmetries other than hexagonal.

5. Conclusion

In conclusion, DFT has been used to determine reaction pathways for catalytic reactions on Pt(111) and Cu(111). On Pt(111) all the elementary steps in the hydrogenation of carbon to methane, nitrogen to ammonia, and oxygen to water were investigated. Also on Pt(111) the reactions of C with O and N were studied, and on Cu(111) the reaction of C with H was examined. From a consideration of these reactions and others from the literature we find that as the valency of a reactant decreases, so too does the coordination number of the site at which the transition state is located. Monovalent adsorbates tend to make transition states close to top sites; divalent adsorbates near bridge sites; and tri- and tetra-valent adsorbates near three-fold hollow sites. This relationship has been shown to apply to a variety of reactions on a variety of substrates. However, it is not followed rigorously on all substrates. On this basis we generalise our findings and extract the following simple rules for catalytic reaction pathways on transition metal surfaces:

(i) Isoelectronic adsorbates tend to go to similar sites to form transition states.
(ii) The higher the valency of the adsorbate, the greater its tendency to access a transition state close to a high coordination site.

It remains to be seen how far this framework can be extended. In particular it will be interesting to investigate how far this framework extends to other types of reactions, reactions on different surfaces, reactions on different crystal facets etc. Of paramount importance is to understand the influence adsorbate p orbitals have on reaction pathways. Given the solid foundation of this framework we remain, however, optimistic that the two rules will be followed for many catalytic reactions and may make it possible to predict transition state structures and reaction pathways for a wide variety of reactions in heterogeneous catalysis.

6. References

1. Payne, M.C.; Teter, M.P.; Allan, D.C.; Arias, T.A; Joannopoulos, J.D. *Rev. Mod. Phys.* **1992,** 64, 1045.
2. Somorjai, G.A. *Introduction to Surface Chemistry and Catalysis,* John Wiley and Sons: New York 1994.
3. Mavrikakis, M.; Hammer, B.; Nørskov, J.K. *Phys. Rev. Lett.* **1998,** 81, 2819.
4. Alavi, A.; Hu, P.; Deutsch, T.; Silvestrelli; P.L.; Hutter, J. *Phys. Rev. Lett.* **1998,** 80, 3650.
5. Zhang, C.J.; Hu, P.; Alavi, A. *J. Am. Chem. Soc.* **1999,** 121, 7931.
6. Vanderbilt, D. *Phys. Rev. B* **1990,** 41, R7892.
7. Perdew, J.P.; Chevary, J.A.; Vosko, S.H.; Jackson, K.A.; Pederson, M.R.; Singh, D.J.; Fiolhais, C. *Phys. Rev. B* **1992,** 46, 6671.
8. Bleakley, K.; Hu, P. *J. Am. Chem. Soc.* **1999,** 121, 7644.
9. Monkhorst, H.J.; Pack, J.D. *Phys. Rev. B* **1976,** 13, 5188.
10. Michaelides, A.; Hu, P. *J. Chem. Phys.* **2000,** 114, 513.
11. Michaelides, A.; Hu, P. *J. Chem. Phys.* **2001,** 114, 2523.
12. Michaelides, A.; Hu, P.*J. Am. Chem. Soc.* **2000,** 122, 9866.
13. Michaelides, A.; Hu, P. *J. Chem. Phys.* **2001,** 114, 5792.
14. Michaelides, A.; Hu, P. *Surf. Sci.* **1999,** 437, 362.
15. Michaelides, A.; Hu, P. *J. Chem. Phys.* **2000,** 112, 6006.
16. Zhang, C.J.; Hu, P. *J. Am. Chem. Soc.* **2000,** 122, 2134.
17. Richter, L.J.; Ho, W. *Phys. Rev. B* **1987,** 36, 9797.
18. Papoian, G.; Nørskov, J.K.; Hoffmann, R. *J. Am. Chem. Soc.* **2000,** 122, 4129.
19. Mortensen, J.J.; Morikawa, Y.; Hammer, B.; Nørskov, J.K. *J. Catal.* **1997,** 169, 85.
20. Morikawa, Y.; Mortensen, J.J.; Hammer, B.; Nørskov, J.K. *Surf. Sci.* **1997,** 386, 67.
21. Hammer, B. *Faraday Disc.* **1998,** 110, 323.
22. Hammer, B *Phys. Rev. Lett.* **1999,** 83, 3681.
23. Stromquist, J.; Bengtsson, L.; Persson, M.; Hammer, B. *Surf. Sci* **1998,** 397, 382.
24. Ciobîcá, I.M.; Frechard, F.; van Santen, R.A.; Kleyn, A.W.; Hafner *J. Phys. Chem. B* **2000,** 104, 3364.
25. Michaelides, A.; Hu, P. In preparation.
26. Michaelides, A.; Hu, P.; Alavi, A. *J. Chem. Phys.***1999,** 111, 1343.
27. Liu, Z.-P.; Hu, P. *J. Chem. Phys.* Submitted.
28. Michaelides, A.; Hu, P. *J. Chem. Phys.* **2000,** 112, 8120.
29. Zhang, C.J.; Liu, Z.-P.; Hu, P. Submitted.
30. Michaelides, A.; Hu, P. In preparation.
31. Zhang, C.J.; Hu, P. *J. Am. Chem. Soc.* **2001,** 123, 1168.
32. Perdew, J.P.; Burke, K.; Ernzerhof, M. *Phys. Rev. Lett.* **1996,** 77, 3865.
33. Zhang, Y.K.; Yang, W.T. *Phys. Rev. Lett.* **1998,** 80, 890.

THE ADSORPTION OF ACETYLENE AND ETHYLENE ON TRANSITION METAL SURFACES

C.G.P.M. BERNARDO, J.A.N.F. GOMES *
Cequp / Departamento de Química, Faculdade de Ciências, Universidade do Porto, Rua do Campo Alegre, 687, 4150 Porto, Portugal

Abstract

The density functional theory and the cluster model approach enable the quantitative computational analysis of the adsorption of small chemical species on metal surfaces. Two studies are presented, one concerning the adsorption of acetylene on copper (100) surfaces, the other concerning the adsorption of ethylene on the (100) surfaces of nickel, palladium and platinum. These studies support the usefulness of the cluster model approach in studies of heterogeneous catalysis involving transition metal catalysts.

1. Introduction

The adsorption of small unsaturated hydrocarbons, such as ethylene and acetylene, on transition metal surfaces is of considerable scientific interest due to their involvement in several elementary catalytic reactions. One of the prerequisites to a

M.A. Chaer Nascimento (ed.), Theoretical Aspects of Heterogeneous Catalysis, 217–240.

more detailed understanding of these catalytic reactions of ethylene or acetylene on metal surfaces is the unequivocal determination of the adsorption sites, orientation and geometric structure of the adsorbed substrate molecules. A large variety of experimental techniques are available nowadays to gain some knowledge on the adsorption sites, orientation and geometric structure of adsorbed molecules on metal surfaces and a significant experimental effort has been made in recent years to investigate the adsorption of ethylene and acetylene on transition metal surfaces. However, in general, experimental data alone are not sufficient to obtain a consistent and complete picture of the structure of these adsorption complexes. Therefore, numerous theoretical investigations have also been carried out in recent years.

2. The adsorption of C_2H_2 on transition metal surfaces

The experimental study of the low temperature adsorption of acetylene on transition metal surfaces has lead to the identification of two main types of structures (A and B) shown in Fig. 1.

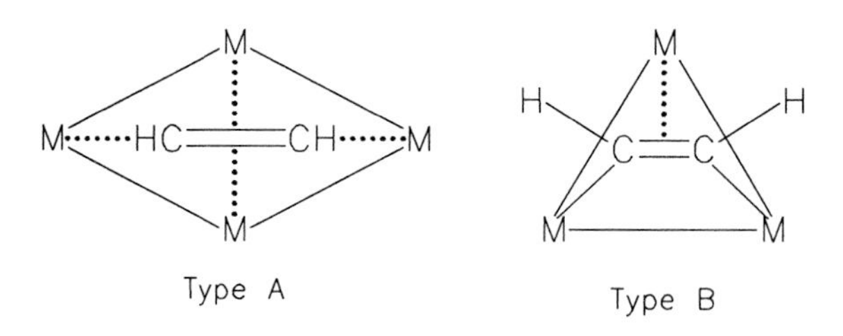

FIGURE 1. Structures resulting from adsorption of acetylene on metal surfaces

Although there is a certain degree of uncertainty as to the nature of each structure, it is believed that structure A involves a bonding with four metal atoms. Structure B is believed to originate from a σ bonding of C=C to two metal atoms and a π-bonding to the third one. However, in practice this is an oversimplification and some other adsorption modes have already been suggested and investigated, and are shown in Fig. 2.

The adsorption of acetylene on metal surfaces is commonly described using a frontier orbital approach, developed by Dewar [1] and Chatt and Duncanson [2]. In this model the bonding to the metal surface is described as a donation of acetylene π-electrons into the metal and back-donation from the metal into the acetylene antibonding π^* orbital.

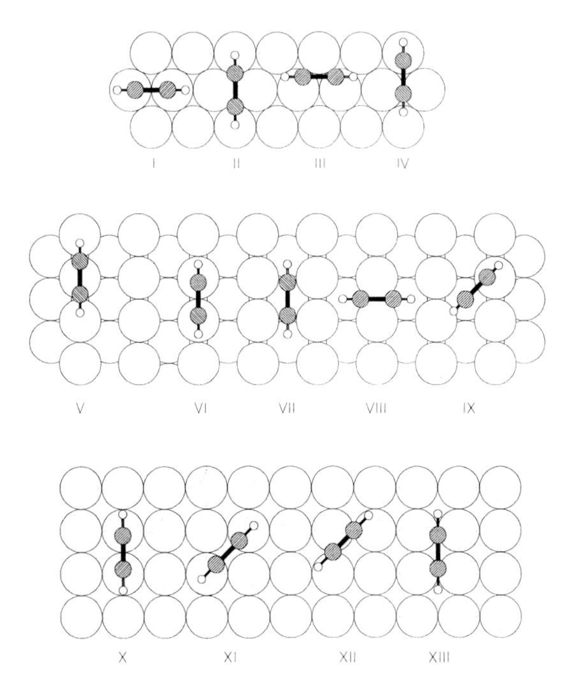

FIGURE 2. Adsorption modes of acetylene on single-crystal metal surfaces.

Table 1 summarizes the main types of structures that are formed upon adsorption of C_2H_2 on some of the most extensively studied transition metal surfaces. As Table 1 shows the adsorption mode of acetylene depends on the electronic structure of the surface. For example on (111) surfaces, the μ-bridging adsorption mode (II on Fig. 2.) was found to be the prefered on copper [6-8] while on palladium the threefold-hollow adsorption mode (III) is the most favorable [10-12]. The same happens on (110) surfaces, for which the trough adsorption (VII) is the most favored on nickel [5] while for example on copper [9] the adsorbed acetylene molecule adopts a low symmetry (most likely C_1) adsorption geometry (IX on Fig.2).

As shown in Table 1, the adsorption mode of acetylene has already been unequivocally determined on several metal surfaces both by experimental and by theoretical techniques. However, on some metal surfaces a general consensus about the adsorption mode of C_2H_2 still does not exist. One such surface is the Cu(100) surface for which the conclusions obtained by several experimental studies have been in general contradictory.

The contradictory available experimental results regarding the "C_2H_2 / Cu(100)" adsorption system served as a motivation for the first of the theoretical studies presented in this paper. The aim of this study was to compile theoretical results for the adsorption of acetylene on four different adsorption sites on the (100) surface of copper in order to determine the prefered adsorption site and to compare our results with the available experimental results.

Metal	Research Type[a]	Prefered Adsorption Mode [b]	Adsorption Energy (kJ.mol-1)	Distance C_2H_2-surface (Å)	Distance CC (Å)	Angle CCH (degree)
Ni(111)	E	II [c]	----	----	----	----
	T	II [d]	Ni4: 155 [d] Ni14: 209 [d]	Ni4: 1.27 [d]	Ni4: 1.51[d]	Ni4: 119.5 [d]
Ni(110)	E	VII [e]	----	----	----	----
	T	VII [e]	----	----	1.42 [e]	55 [e]
Cu(111)	E	II [f, g]	----	1.41 [g]	1.48 [g]	----
	T	II [h]	26 [h]	----	1.36 [h]	60 [h]
Cu(110)	E	IX [i]	----	----	----	----
	T	IX [i]	----	----	----	----
Pd(111)	E	III [j, k]	----	----	1.3 ± 0.05 [j]	117 [j]
	T	III [l, k]	309 [l]	1.47 [l]	1.33 [l]	129

TABLE 1. Structures that are formed upon adsorption of C_2H_2 on some transition metal surfaces. [a] E stands for Experimental and T stands for Theoretical, [b] See Fig. 2, [c] Ref.[4], [d] Ref.[3], [e] Ref.[5], [f] Ref.[6], [g] Ref.[7], [h] Ref.[8], [i] Ref.[9], [j] Ref.[11], [k] Ref.[12], [l] Ref.[10].

2.1. THE ADSORPTION OF C_2H_2 ON A Cu(100) SURFACE

In recent years, many papers have been published dealing with the interaction of acetylene with the low index single crystal copper (111)[6-8, 13-15], (110)[9,16,17] and (100)[17-29] surfaces.

The adsorption mode of acetylene on the (111) surfaces of copper has already been unequivocally determined both by experimental [6,7,13] and theoretical techniques [8,14,15]. These studies have shown that acetylene adsorbs with the CC axis almost parallel to the surface over a bridge site with the two carbon atoms pointing towards adjacent 3-fold hollow sites (adsorption mode II on Fig. 2).

The adsorption mode of acetylene on a Cu(110) surface is more complex and the available experimental information concerning this system, is fairly scarce. In a recent paper [9], an experimental and a theoretical study of this adsorption system was presented. Using HREELS and ARUPS the authors concluded that acetylene adopts a low symmetry (most likely C_1) adsorption geometry on Cu(110) (IX on Fig. 2). Then, using ab initio Hartree-Fock cluster calculations the authors tested and confirmed this result. These calculations also showed that the C_2H_2 molecule was essentially sp^2 hybridized and the internal structure of the molecule was relatively insensitive to the adsorption site.

The adsorption of acetylene on the Cu(100) surface has also been the subject of several investigations during the last decades. However, the conclusions obtained have been in general very contradictory. Avery [17] based on his HREELS results, concluded that, when adsorbed on the Cu(100) surface, the acetylene molecule was highly distorted having a bond order of ~ 1.5. He suggested that the most probable adsorption site was a twofold-bridge site. Arvanitis et al [18 – 20] utilizing SEXAFS and NEXAFS found that at 60 K, the acetylene molecule was coordinated to three Cu surface atoms in a threefold hollow site, with the C-C bond lying parallel to the surface at 1.5 Å above the surface. They determined a CC bond length of 1.42 Å, indicating a significant stretching of the gas-phase CC bond length (1.21 Å). The threefold hollow site was also favored by Hartree-Fock cluster calculations [21] and by molecular orbital calculations based on the tight-binding approach within the extended Hückel formalism [22]. Marinova et al [23] using HREELS also observed that at low temperatures (140 K) the acetylene adsorbs molecularly on a Cu(100) surface, but is strongly distorted with CC and CH stretching frequencies indicating a bond order of 1.5 and a rehybridization close to sp^3. On this study [23], Marinova et al suggested the threefold hollow and the diagonal fourfold hollow adsorption modes as being the most favorable ones. In a very recent STM study, Stipe et al [24, 25] found the acetylene to be centered above the fourfold hollow site with the molecular plane aligned across the diagonal of the square formed by four Cu atoms. The molecular plane is parallel to either the copper [100] or [010] axis and is perpendicular to the surface. In this configuration, the acetylene is expected to be coordinated to four Cu atoms. As far as we known, no supporting theoretical study has been reported to date.

2.1.1 Theoretical details

The interaction of the acetylene molecule with the (100) surface of copper was studied using the cluster model approach. All the calculations have been performed using the Density Functional Theory. The BLYP method included in the Gaussian 98 [30] package was used. This method combines the gradient corrected exchange functional of Becke [31] with the gradient corrected correlation functional of Lee *et al* [32].

To model the copper (100) surface a two-layer cluster of C_{4V} symmetry, with 5 copper atoms in one layer and 4 copper atoms in the other layer, has been used. In this cluster, all the 9 metal atoms were described by the LANL2DZ basis set. The LANL2DZ basis set treats the $3s^2$ $3p^6$ $3d^{10}$ $4s^1$ Cu valence shell with a double zeta basis set and treats all the remainder inner shell electrons with the effective core potential of Hay and Wadt [33]. The non-metallic atoms (C and H) were described by the 6-31G** basis set of double zeta quality with p polarization functions in hydrogen atoms and d polarization functions in carbon atoms. The use of copper clusters of similar size has already proved to be very useful in the study of the adsorption of other species, such as $CH_3O^\bullet$ [34,35], H_2CO [36], CO_2H [37] and C_2H_4 [14,15] on copper surfaces.

In order to obtain a closed shell system and to make the calculations computationally less expensive, in all these calculations an overall charge of (-1) has been considered for the C_2H_2-Cu_9 system.

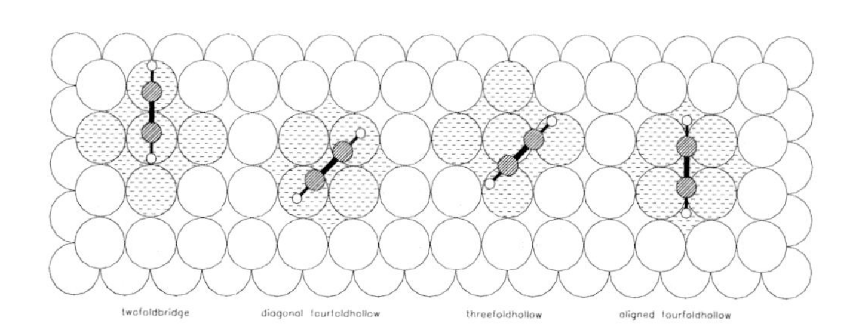

FIGURE 3. Schematic top view of the several C_2H_2-Cu_9 cluster systems considered in the present study: twofold-bridge-$C_2H_2/Cu_9(5,4)$, diagonal-fourfold-hollow-$C_2H_2/Cu_9(4,5)$, threefold-hollow-$C_2H_2/Cu_9(5,4)$ and aligned-fourfold-hollow-$C_2H_2/Cu_9(4,5)$. Only the shaded metal atoms were included in the calculations. The unshaded metal atoms were included in the picture only to better illustrate the several studied adsorption modes of acetylene on a Cu(100) surface.

In all the calculations, no surface reconstruction nor surface relaxation have been considered, that is, the metal cluster geometry was kept frozen. The nearest-neighbour distance was taken from the bulk and is 2.55612 Å. This cluster forms a compact section of the corresponding ideal Cu(100) surface.

In this study four different adsorption sites have been considered, namely the twofold-bridge, the threefold hollow, the diagonal fourfold hollow and the aligned fourfold hollow sites, see Fig. 3.

To study the adsorption on the twofold-bridge and threefold hollow sites we used a $Cu_9(5,4)$ cluster as shown in Fig. 4(a), where the numbers inside brackets indicate the number of metal atoms in the first and second layers respectively. To study the adsorption on the aligned-fourfold-hollow and diagonal-fourfold-hollow sites we used the same cluster but on an inverted position, that is, we used a $Cu_9(4,5)$ cluster as shown in Fig. 4(b).

In order to reduce the size of the calculations, in the study of all the adsorption modes the CC bond of the acetylene was forced to be parallel to the surface. This approach is fully supported by the available experimental information. Also in all the calculations the acetylene molecule was placed in a plane perpendicular to the surface.

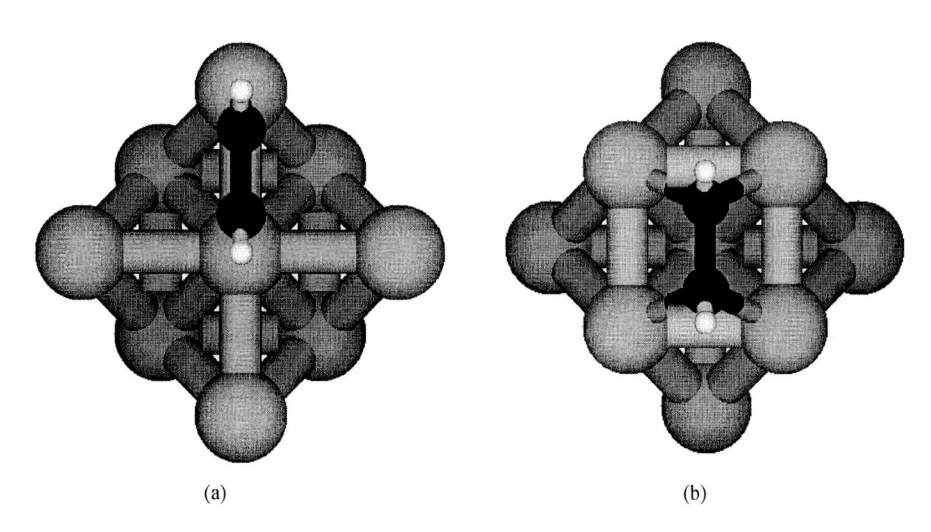

FIGURE 4. Top view of two of the C_2H_2/Cu_9 adsorption systems considered in the present study: (a) the aligned fourfold hollow-$C_2H_2/Cu_9(4,5)$ system and (b) the twofold bridge-$C_2H_2/Cu_9(5,4)$ system. All the metal atoms are treated with the LANL2DZ basis set.

In the study of the adsorption on the aligned-fourfold-hollow and diagonal-fourfold-hollow sites, the symmetry of the overall system Cu_9-C_2H_2 was C_{2V}. In the study of the adsorption on the twofold-bridge and threefold-hollow sites the symmetry of the overall system Cu_9-C_2H_2 was C_S. In the study of the adsorption on the threefold hollow site the plane of the molecule is perpendicular to the surface and contains the point of interception of the bisecting lines of the angles of the triangle formed by the 3 copper atoms directly involved in the bonding to the C_2H_2 molecule. In this adsorption mode, the CC bond of acetylene is parallel to the line joining the nucleus of the two non-nearest-neighbour copper atoms of the triangle, see Fig. 3.

It is well known that the determination of the adsorption energies is one of the most critical aspects of the cluster modeling of surfaces [38]. For small clusters, usually the adsorption energies oscillate with the cluster size even if the bonding nature remains the same [39]. Another related problem already reported on the literature is that for some systems the optimised geometric parameters of the system show that there is a bond between the adsorbate and the metal cluster, but energetic considerations show that the overall system "cluster + adsorbate" is unbound with respect to the dissociation limit but with local minima on the energy surface. This phenomenon has already been reported on several papers [9, 10] and has been investigated in detail by Clotet et al [10] for the Cu(111) surface and is known to involve a crossing or avoided crossing of electronic states. It has been suggested that the use of naked metal clusters in calculating good chemisorption energies of adsorbates requires the cluster to be in a prepared bonding state. The argument used

is that the excitations to higher states are involved in the preparation. For the infinite surface, this value is close to zero, but for a small cluster, the required excitation energy should be added to the calculated chemisorption energy of the ground state of the cluster. This idea has been recently used by Triguero et al [40] to calculate the surface chemisorption energy of acetylene and ethylene to cluster models of the copper (100), (110) and (111) surfaces. According to these authors, in these systems there are two triplet states: a triplet surface state and a triplet adsorbate state, which are coupled to produce a net singlet state for the two metal-carbon bonds. This is the "spin-uncoupling mechanism", which was inspired by the similarity of the geometric structure of the gas-phase π - π^* triplet excited acetylene and the adsorbed species. It must however be noted that, for the C_2H_2-Cu(100) adsorption system Triguero et al [40] have only considered the adsorption on the twofold-bridge site which, according to the theoretical evidence to be presented in the present paper and also according with the experimental information available in the literature, is not the most favored adsorption site.
A similar situation has been found in the present work for some of the C_2H_2 – Cu_9 adsorption systems considered. This presents a major difficulty in interpreting the adsorbate binding energies obtained for these systems. Given these difficulties, in this study no adsorbate binding energies will be calculated. Instead, when using energetic considerations we will refer to the total energy (in Hartrees) of the C_2H_2-Cu_9 system, because this energy is independent of the reference system used.

2.1.2. Results and discussion

2.1.2.1.Free C_2H_2 molecule. The free acetylene molecule was optimized at the BLYP/6-31G** level. Then, vibrational frequencies for the optimised free C_2H_2 molecule have been calculated. The computed geometry, C-C Mulliken overlap population and vibrational frequencies are listed in Table 2 along with the corresponding experimental values [41, 42]. Both of these results show that the acetylene molecule in its ground state is linear with a $D_{\infty h}$ symmetry.

2.1.2.2 The C_2H_2-Cu_9 adsorption systems. Full geometry optimizations (only subject to the symmetry constrains mentioned in section 2.1.1.) of the several C_2H_2 – Cu_9 adsorption systems have been performed. In these full geometry optimizations we began by using as starting geometries (guess geometries) the geometries obtained in previous theoretical cluster studies [8,9,13,14] of the adsortion of acetylene on Cu(111) and Cu(110) surfaces. Table 3 summarizes the computed optimized geometries, total energies and Mulliken overlap populations of the CC bond for the four studied adsorption modes. In this table the C_2H_2-surface distance is defined as being the perpendicular distance between the CC bond and a plane containing the nucleus of the metal atoms of the upper layer.

C_2H_2	Theoretical values	Experimental values
Distance (CC) (Å)	1.2153	1.204 [a]
Distance (CH) (Å)	1.0714	1.058 [a]
Angle(HCC)(degree)	180.000	180.0 [a]
C-C (M.O.P) (a.u.)	1.8944	-----
$\rho_{as}(CH)$	472	612 [b]
$\delta_{as}(CH)$	472	612 [b]
$\rho_s(CH)$	744	729 [b]
$\delta s(CH)$	744	729 [b]
$\nu(CC)$	2019	1974 [b]
$\nu_{as}(CH)$	3370	3287 [b]
$\nu_s(CH)$	3463	3374 [b]

TABLE 2. Optimized geometry, C-C Mulliken overlap population (M.O.P.) and vibrational frequencies (cm^{-1}) for the free C_2H_2 molecule. [a] Ref.[41], [b] Ref. [42]

The results in Table 3 show that the geometric and electronic structure of the acetylene molecule are highly perturbed upon adsorption. Compared to the calculated values of the free acetylene molecule we find that the CC bond is elongated by 0.1861 Å (diagonal fourfold-hollow) and 0.1909 Å (aligned fourfold-hollow) and the CH bonds are bent upwards by 63.3° (diagonal fourfold-hollow) and by 62.9° (aligned fourfold-hollow). The length of the CC bond in chemisorbed acetylene is between the values of 1.33 Å for gas phase ethylene and 1.54 Å for gas phase ethane. This shows that upon adsorption the hybridisation of the acetylene molecule changes from sp (in the gas phase) to between sp^2 and sp^3.
Considering as reference the energy of the lowest spin multiplicity bare cluster (-1765,239412 Hartrees) and the energy of the lowest spin multiplicity gas phase acetylene molecule (-77,292855 Hartrees), it is observed that on the twofold-bridge and threefold-hollow adsorption modes, after full geometry optimizations the energy of the overall system is higher than the sum of the energies of the separated fragments (-1842,532267 Hartrees), that is, the acetylene molecule is energetically unbound with respect to the separated fragments but with a local minimum in the potential energy surface. This phenomenum was mentioned in section 2.1.1 has already been extensively studied and documented in the literature [9,10,40].
The results on Table 3 show clearly that the adsorption of the acetylene molecule on the cavity formed by four copper atoms (namely the adsorption on the aligned and diagonal fourfold hollow sites) is much more favored than the adsorption on the two other sites considered. Based only on energetic arguments, and considering the large energetic difference obtained between the adsorption on the fourfold hollow sites (aligned and diagonal) and the adsorption on the other adsorption sites, the threefold-hollow and the twofold-bridge sites were excluded from the list of possible adsorption sites. It must be noted that the difference in stability between the aligned

fourfold-hollow site and the next more stable adsorption site (the twofold-bridge) is greater than 78 kJ.mol^{-1}.

	Diagonal fourfold-hollow	Aligned fourfold-hollow	Twofold-bridge	Threefold-hollow
Total Energy (Hartrees)	-1842.56149	-1842.55797	-1842.52800	-1842.52251
Distance C_2H_2–surface (Å)	1.6332	1.5902	1.8835	1.6879
Distance CC (Å)	1.4014	1.4062	1.3622	1.3352
Distance CH (Å)	1.1112	1.1128	1.1075	1.0966
Angle CCH (degrees)	116.6749	117.0573	122.2947	129.8815
CC M.O.P.	0.7441	0.5899	0.8913	0.9722

TABLE 3. Total energies of the system C_2H_2/Cu_9, optimized geometries and Mulliken overlap populations (M.O.P.) of the CC bond for C_2H_2 adsorbed on the diagonal fourfold hollow, aligned fourfold hollow, twofold bridge and threefold hollow sites. The distances are given in Å, the angles in degrees and the total energies in Hartrees.

The difference in stability between the diagonal fourfold-hollow site and the aligned fourfold-hollow site is only 9.2 kJ.mol^{-1}. This difference is too small to allow us to conclude unambiguously which is the most stable adsorption mode. This served as a motivation to the calculation of the vibrational frequencies for acetylene adsorbed on these two most stable adsorption sites. The results obtained, along with the available experimental results, are presented in Table 4.

The vibrational frequency results obtained show that for the adsorption in the diagonal fourfold-hollow site there is a better general agreement between the calculated results and the available experimental results. This is even more clear if we consider only two of the most relevant vibrational modes used in the study of the nature of the bonding C_2H_2-metal, namely the ν(CC) and the ν(C-Cu). For both the ν(CC) and the ν(C-Cu) normal modes, the theoretical results obtained for adsorption on the diagonal fourfold hollow site are much closer to the corresponding experimental values than the theoretical results obtained when considering the adsorption on the aligned fourfold-hollow mode.

Adsorption mode	Experimental [a]	Diagonal fourfold hollow	Aligned fourfold hollow
ν(C-Cu)	420	392	361
ρs(CH)	630	589	468
ρas(CH)	------	937	988
δs(CH)	950	1051	1035
δas(CH)	1140	1190	1179
ν(CC)	1320	1300	1277
νas(CH)	-------	2838	2821
νs(CH)	2880	2884	2869

TABLE 4. Calculated vibrational frequencies of acetylene adsorbed on the diagonal fourfold hollow and aligned fourfold hollow sites. [a] Ref. [23]

2.1.3. Conclusions

Recent Scanning Tunneling Microscopy results have shown that the adsorption of an acetylene molecule on the Cu(100) surface occurs on a fourfold hollow site, being the diagonal adsorption mode the most favored one. Our theoretical results are in excellent agreement with these Scanning Tunneling Microscopy results in that our results also show clearly and unambiguously that the acetylene molecule adsorbs on a fourfold hollow site, being the diagonal orientation the most favored one.

3. The adsorption of C_2H_4 on transition metal surfaces

The adsorption of ethylene has now been studied experimentally on many different metal surfaces using a large variety of spectroscopic techniques. Although the type of bonding differs depending on the electronic and geometric structure of the surface, all studies to date conclude that, for low adsorbate coverages, the CC bond in adsorbed ethylene lies parallel to the surface. In all cases, the main interaction with the metal surface is through the electrons in the CC bond, causing an elongation of the carbon-carbon bond, which is accompanied by a bending of the hydrogens away from the surface. This type of bonding is commonly described by the Dewar-Chatt-Duncanson model [1,2] originally proposed to explain ethylene coordination in organometallic complexes. According to this model, the filled ethylene π-orbital donates electron density into an empty metal orbital, and the empty antibonding (π*) molecular orbital of ethylene accepts electron density from the filled metal orbitals. These donations and back-donations of electron density, cause the ethylene molecule to rehybridize from sp^2 to somewhere between sp^2 and sp^3. Experimentally, the measurement of the vibrational frequency corresponding to the ν_2(CC) stretching mode, has been perhaps the most straightforward method for determining the state of hybridization of the adsorbed molecule, due to the high

sensitivity of the CC stretching frequency to the hybridized state of the two carbon atoms. For example, ν(CC) is 1623 cm^{-1} for gaseous C_2H_4 and is 993 cm^{-1} for gaseous C_2H_6, which represent sp^2 and sp^3 hybridized systems, respectively.
For most metals, two general types of low-temperature spectra have been reliably identified and have been assigned to two distinct nondissociative adsorption modes: the di-σ-adsorption mode(I) and the π-adsorption mode(II), see Fig. 5.

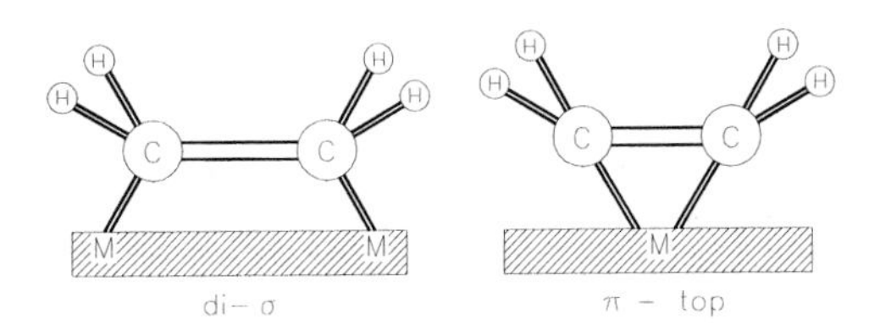

FIGURE 5. Structures resulting from the adsorption of ethylene on metal surfaces.

In the di-σ adsorption mode the ethylene molecule interacts directly with two metal atoms and in the π-adsorption mode the ethylene molecule interacts directly with a single metal atom. In some cases both types of species coexist, in others one type occurs on one crystal face of a given metal and the other on a different one. Both types of species have CC bond orders between 2 and 1, with the π-complexes having values nearer 2 and the di-σ complexes nearer 1.

3.1 THE ADSORPTION OF C_2H_4 ON THE (100) SURFACES OF PLATINUM, PALLADIUM AND NICKEL

The metals of group 10 (Ni, Pd and Pt) are very important catalysts in a large variety of chemical reactions such as hydrogenations/dehydrogenations, isomerizations and total oxidations. For this reason, numerous studies of the adsorption of ethylene on the surfaces of these transition metals have been performed during the last decades. Considering the order of these elements down the group, for a given adsorbate a gradual change of the adsorbate-surface interaction, might be expected to follow the order Ni → Pd → Pt. However, experimentally this is not observed. For example, for a given olefin the ease of hydrogenation over metal catalysts decreases in the order Pd > Pt > Ni and the ease of isomerisation tends to decrease in the order Pd > Ni > Pt. From these data it is clear that there are significant differences in the interaction of olefins with these metal surfaces. These differences served as a motivation for the study of the adsorption of ethylene on the Pt, Pd and Ni surfaces that we present below.
Whereas the adsorption of ethylene on the (111) and (110) surfaces of platinum, palladium and nickel has already been extensively studied both by experimental

[43–55] and by theoretical [51, 56-61] techniques, the adsorption on the corresponding (100) crystalographic surfaces has received considerably less attention [62-69]. Even so, some experimental studies have been performed on the ethylene adsorption on platinum (100) [62,63], palladium (100) [64,65] and nickel (100) [66-68] clean surfaces.

In what concerns to theoretical studies of the adsorption of ethylene on the (100) surfaces of these metals, only one study [69] of the system C_2H_4/Ni(100) was found in the literature. In this theoretical study, Xu et al used the Xα method to evaluate chemisorption properties.

The desire of obtaining a better understanding of the relative reactivities of the nickel, palladium and platinum surfaces towards ethylene reactions and the dearth of theoretical studies on the adsorption of C_2H_4 on the (100) surfaces of these metals, has served as motivation for our study of these adsorption systems.

It is known that in theoretical studies of the adsorption of chemical species on metal surfaces, the adsorption properties of the adsorbates, are influenced by the methodology, by the cluster size and shape and by the basis sets used in the corresponding study. So a reliable comparative theoretical study of the adsorption of ethylene on the Ni, Pd and Pt surfaces can only be obtained if the same conditions in terms of methodology, cluster size and shape and basis sets are used in the study of the adsorption on the three metal surfaces.The aim of this study was to compile theoretical results for the adsorption of ethylene on the di-σ and π-top adsorption sites on the (100) surfaces of platinum, palladium and nickel under the same conditions mentioned above, in order to compare and understand the trends in the adsorption of this species on these three metal surfaces.

3.1.1. Theoretical details

In the present work, the interaction of the ethylene molecule with the (100) surfaces of platinum, palladium and nickel is studied using the cluster model approach. All these metals have a face centered cubic crystal structure. The three metal surfaces are modelled by a two-layer $M_9(5,4)$ cluster of C_{4V} symmetry, as shown in Fig. 6, where the numbers inside brackets indicate the number of metal atoms in the first and second layer respectively. In the three metal clusters, all the metal atoms are described by the large LANL2DZ basis set. This basis set treats the outer 18 electrons of platinum, palladium and nickel atoms with a double zeta basis set and treats all the remainder electrons with the effective core potential of Hay and Wadt [33]. The non-metallic atoms (C and H) are described by the 6-31G** basis set of double zeta quality with p polarization functions in hydrogen atoms and d polarization functions in carbon atoms. In all the clusters, the nearest-neighbour distances were taken from the bulk and are 2.77483 Å for platinum, 2.75114 Å for palladium and 2.49184 Å for nickel. These clusters form compact sections of the corresponding ideal surfaces.

In this study the B3LYP hybrid method proposed by Becke [31] and included in the Gaussian 98 [30] package was used. This method includes a mixture of Hartree-Fock and DFT exchange terms associated with the gradient corrected correlation functional of Lee et al [32]. In all the calculations the metal cluster geometry was kept frozen.
In the study of the di-σ adsorption mode, the adsorbate was placed with its CC axis parallel to the surface and with its carbon atoms bridging nearest neighbour metal atoms at the surface (overall symmetry: C_S), see Fig. 6(a). In the study of the π-top adsorption mode, the overall system "adsorbate/metal substract" was forced to have C_{2V} symmetry, in order to reduce the size of the calculation, see Fig. 6(b).

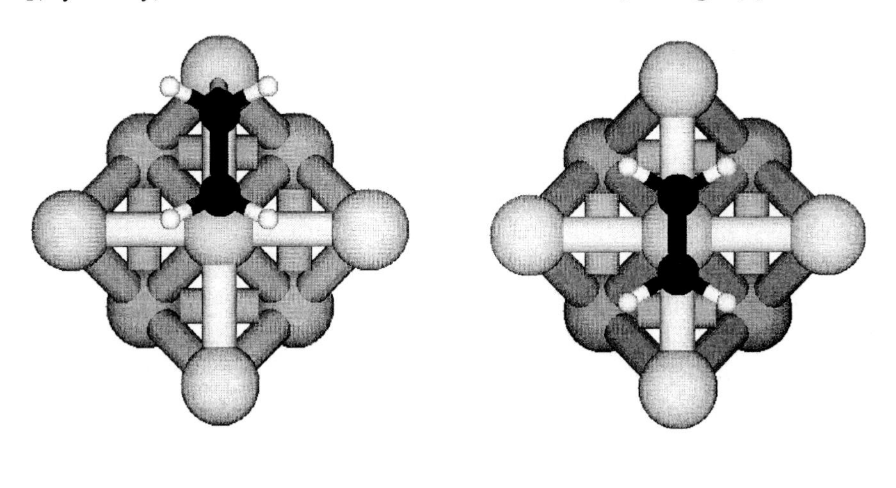

(a) (b)

FIGURE 6. Top views of the two adsorption modes studied on the (100) surfaces of platinum, palladium and nickel. The metal surfaces were modelled by a two-layer $M_9(5,4)$ cluster. (a) di-σ adsorption mode. (b) π adsorption mode.

In our calculation of the adsorption energies, the energies of the lowest spin multiplicity bare clusters and the energy of the lowest spin multiplicity gas phase ethylene molecule have been used as reference. We know that this method is not unambiguous. However, the alternative ways of calculating adsorption energies already suggested in the literature are not always straightforward and are also often ambiguous.

3.1.2 Results and discussion

3.1.2.1. Free C_2H_4 molecule. The free ethylene molecule was optimized at the B3LYP/6-31G** level. The computed geometry and C-C Mulliken overlap population are listed in Table 5 along with the corresponding experimental values [70].

In this study, we will always refer to the angle "(CH_2) – C" as being the angle between the HCH plane and the CC bond. This is exemplified in Fig. 7. Vibrational frequency calculations were also performed, and the results are also shown in Table 5 along with the corresponding experimental values [71]. There is a good agreement between the theoretical results in this table and the corresponding experimental results.

C_2H_4	Theoretical values	Experimental values
Distance (CC) (Å)	1.3306	1.337±0.003 [a]
Distance (CH) (Å)	1.0868	1.086±0.003 [a]
Angle (HCC) (degrees)	121.818	121.35±1 [a]
Angle ((CH_2) – C) (°)	180.000	180.00 [a]
C-C (M.O.P) (a.u.)	1.3682	-----
ν_{10}, $\rho(CH_2)$	831 (799)	826 (I.R.) [b]
ν_8, $\omega(CH_2)$	961 (924)	940 (R) [b]
ν_7, $\omega(CH_2)$	976 (938)	949 (I.R.) [b]
ν_4, $\tau(CH_2)$	1069 (1028)	Forbidden [b]
ν_6, $\rho(CH_2)$	1240 (1192)	1222 (R) [b]
ν_3, $\delta(CH_2)$	1387 (1333)	1342 (R) [b]
ν_{12}, $\delta(CH_2)$	1482 (1425)	1444 (I.R.) [b]
ν_2, $\nu(CC)$	1714 (1648)	1623 (R) [b]
ν_{11}, $\nu_s(CH_2)$	3147 (3025)	2989 (I.R.) [b]
ν_1, $\nu_s(CH_2)$	3162 (3040)	3026 (R) [b]
ν_5, $\nu_a(CH_2)$	3223 (3098)	3103 (R) [b]
ν_9, $\nu_a(CH_2)$	3248 (3122)	3105 (I.R.) [b]

TABLE 5. Optimized geometry, C-C Mulliken overlap population (M.O.P.) and vibrational frequencies (cm^{-1}) of the free C_2H_4 molecule. Theoretical vibrational frequency values inside brackets are scaled by a factor of 0.9614 [c]. [a] Ref.[70], [b] Ref.[71], [c] Ref.[72]

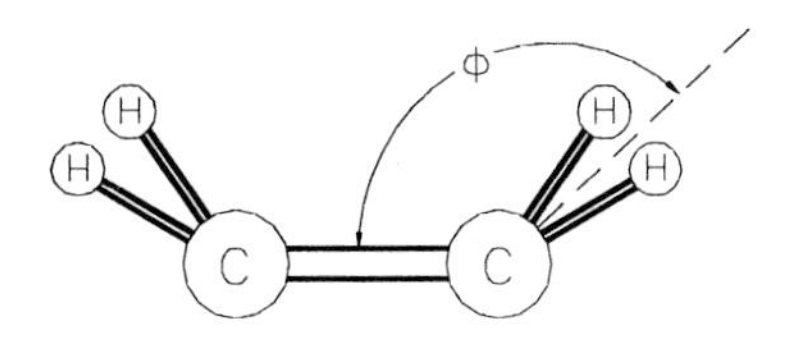

FIGURE 7. The angle (CH_2) – C

3.1.2.2. C_2H_4 / Pt(100). Among all the single crystal platinum surfaces, the adsorption on the Pt(111) surface has been by far the most extensively studied both by experimental and theoretical methods. Ge and King [73] have studied the adsorption of C_2H_4 on Pt(111) employing a plane wave basis set with norm-conserving pseudopotentials and the Perdew-Wang exchange and correlation energy. They obtained adsorption energies of 121.8 and 53.4 kJ.mol^{-1} for the di-σ and π modes.

	C_2H_4 / Pt(100) [a]		C_2H_4 / Pt(111)	
	di-σ adsorption	π-top adsorption	di-σ adsorption	π-top adsorption
Adsorption Energy (kJ.mol^{-1})	-223.8	-149.5	-150.7 [b] 107.3 / 121.8 [c] 108.7 [d]	35.5 [b] 53.4 [c] 59.9 [d]
Distance C-M [e] (Å)	2.0830 (1.9815) [e]	2.1992 (2.0813) [e]	2.06 [b] 2.125 / 2.097 [c] 2.144 [d]	2.70 [b] 2.219 [c] 2.275 [d]
Distance CC (Å)	1.4900	1.4212	1.51 [b] 1.479 / 1.480 [c] 1.483 [d]	1.35 [b] 1.397 [c] 1.399 [d]
Distance CH (Å)	1.0937	1.0851	1.093 / 1.093 [c] 1.095 [d]	1.085 [c] 1.089 [d]
Angle CCH (°)	114.344	119.375	106.9 / 111.9 [c] 121.6 [d]	121.5 [c] 120.2 [d]
Angle ((CH_2)–C) (°)	137.736	158.017	138.2 [d]	160.5 [d]
C – C M.O.P. (a.u.)	0.6651	0.7075	-------	------
C – Pt [f] M.O.P. (a.u.)	0.6217 / 0.5271	0.4034 / 0.4034	-------	------
$Q_{adsorbate}$ (Mulliken) (a.u.)	+ 0.1314	+ 0.3925	-------	------

TABLE 6. Adsorption energies (kJ.mol^{-1}), optimized geometries, C-C and C-Pt Mulliken overlap populations (M.O.P.) and total adsorbate charges, Q, for C_2H_4 adsorbed on the di-σ and top sites of the Pt(100) surface (present work) and Pt(111) surface (results from the literature). [a] Present work; [b] Ref.[74]; [c] Ref.[73] (the two values in di-σ refer to Pt cluster with top layer fixed and relaxed respectively); [d] Ref.[75], [e] Inside brackets is the perpendicular distance between the CC bond and a plane containing the nucleus of the metal atoms of the first layer; [f] The two values refer to the M.O.P. between each carbon atom and its nearest metal atom. Due to the shape of the cluster used to perform these calculations, in the study of the di-σ adsorption the C_2H_4/Pt_9 system has C_s symmetry and so the carbon atoms have sligthly different electronic populations. In the study of the π adsorption, the C_2H_4/Pt_9 system has C_{2V} symmetry and the two carbon atoms have identical electronic populations.

Kua et al [74] and Watson et al [75] have also studied the adsorption of ethylene on Pt(111) using DFT methods and also concluded that the di-σ adsorption is the most stable. No recent theoretical studies of the adsorption of C_2H_4 on a Pt(100) surface have been found in the literature.

Table 6 summarizes the computed geometries, adsorption energies and some electronic parameters obtained in the present study for the di-σ and π adsorption modes of the C_2H_4 molecule on a Pt(100) surface. In this Table are also presented the results of some recent theoretical studies on Pt(111).

The results show that the ethylene molecule binds very strongly on a Pt(100) surface, and its geometric and electronic structure are highly perturbed by the adsorption process. Compared to the calculated value of the CC bond in the free ethylene molecule (1.3306 Å) we find an elongation of the CC bond of 0.1594 Å (di-σ) and 0.0906 Å (π). The H-C-H plane is bent upward by 42.3° (di-σ) and by 22.0° (π). As shown in Table 6, these geometrical changes are very similar to those reported on previous theoretical studies of the adsorption of C_2H_4 on Pt(111) and indicate a significant rehybridization of the carbon atoms from sp^2 toward sp^3, in particular in the di-σ adsorption mode. The C-C Mulliken overlap population dropped from 1.3682 (gas phase) to 0.6651 (di-σ) and 0.7075 (π). The chemisorbed ethylene molecule is positively charged on both adsorption modes, +0.1314 (di-σ) and +0.3925 (π) and this indicates that there is a net transfer of electrons from the adsorbate molecule to the metal cluster and that in both adsorption modes the σ-donation is greater than the d-π* backdonation. The results in Table 6 also show that the adsorption energy is much higher (-223.8 vs –149.5 $kJ.mol^{-1}$) when ethylene adsorbs on the di-σ mode. This is in agreement with the experimental finding that when ethylene adsorbs on Pt(100) at low temperatures (T≅ 100K) di-σ complexes are formed [62,63].

C_2H_4 / Pt(100)	di-σ adsorption	π-top adsorption	Experimental [a]
ν_s (C-Pt)	437	340	450
ν_4, $\tau(CH_2)$	803	902	-----
ν_2, ν(CC)	1027	1202	984
ν_7, $\omega(CH_2)$	1062	979	-----
ν_3, $\delta(CH_2)$	1473	1517	-----
ν_1, $\nu_s(CH_2)$	3062	3159	2936

TABLE 7. Calculated vibrational frequencies (cm^{-1}) for C_2H_4 adsorption on the di-σ and π adsorption modes, on a Pt(100) surface. [a] Ref.[62]

Vibrational frequencies have been calculated, from the minima geometries, for the two adsorption modes, and the results are shown in Table 7. For reasons of simplicity, we will always refer to the vibrational modes of the adsorbed ethylene molecule by the mode numbers derived for the free molecule.

The ν(C-Pt) stretching frequency is higher for adsorption on the di-σ mode than on the π mode, indicating a stronger bond in the former case. The calculated vibrational frequencies for the di-σ adsorption mode agree fairly well with the available experimental data [62] but the calculated values for the π-top adsorption mode differ significantly from the experimental data [62]. This strengthens the idea that, on a Pt(100) surface, at low temperatures, the ethylene molecule adsorbs only as a di-σ complex.

3.1.2.3. C_2H_4 / Pd(100). Several experimental studies of the adsorption of ethylene on palladium surfaces have been found on the literature [47-49]. According with these studies on the (111) and (110) surfaces, the ethylene molecule adsorbs preferably in the π-mode.

In the only known experimental studies on the adsorption of C_2H_4 on a Pd(100) surface, at low temperatures, Stuve et al [64,65] using TPRS and EELS concluded that both di-σ and π-bonded forms are stable and coexist on this surface at 80 K. Table 8 lists the results obtained for C_2H_4 adsorbed on a Pd(100) surface on the di-σ and π adsorption modes. According to the results shown in Table 8, the adsorption energy of the C_2H_4 molecule is larger on the di-σ mode than on the π mode, by approximately 73 $kJmol^{-1}$.

C_2H_4 / Pd(100)	di-σ adsorption	π-top adsorption
Adsorption Energy ($kJ.mol^{-1}$)	-81.0	-32.2
Distance C_2H_4-surface (Å)	1.9651	2.0382
Distance CC (Å)	1.4556	1.4126
Distance CH (Å)	1.0942	1.0865
Angle CCH (°)	116.240	119.863
Angle ((CH_2) – C) (°)	144.157	160.715
C – C M.O.P. (a.u.)	0.7413	0.7571
C – Pd [a] M.O.P. (a.u.)	0.5508 / 0.4388	0.3976 / 0.3976
$Q_{adsorbate}$ (Mulliken) (a.u.)	+ 0.1954	+ 0.3385

TABLE 8. Adsorption energies ($kJ.mol^{-1}$), optimized geometries, C-C and C-Pd Mulliken overlap populations (M.O.P.) and total adsorbate charges, Q, for C_2H_4 adsorbed on the short-bridge and top sites of the Pd(100) surface. No theoretical results of the adsorption of C_2H_4 on palladium surfaces have been found in the literature. [a] The two values refer to the M.O.P. between each carbon atom and its nearest metal atom.

When adsorbed on the di-σ mode, the perpendicular C–Pd surface distance is 1.9651 Å, the CC distance is 1.4556 Å and the H-C-H plane is bent upward by 35.8°. On this adsorption mode, the CC bond elongation is accompanied by a drop of the CC Mulliken overlap population from 1.3682 (gas phase) to 0.7413. The π-metal forward donation outweighs the metal-to-π* back donation of electrons and so the

adsorbed ethylene is positively charged (on both adsorption modes). No experimental geometric results for adsorption of C_2H_4 on Pd(100) have been found in the literature.
The calculated vibrational frequencies are listed in Table 9.

C_2H_4 / Pd(100)	di-σ adsorption	π-top adsorption	Experimental [a]
ν_s (C – Pd)	391	329	390
ν_4, $\tau(CH_2)$	796	886	-----
ν_7, $\omega(CH_2)$	955	920	920
ν_2, $\nu(CC)$	1109	1210	1135
ν_3, $\delta(CH_2)$	1483	1525	1455
ν_1, $\nu_s(CH_2)$	3056	3143	2980

TABLE 9. Calculated vibrational frequencies (cm^{-1}) for C_2H_4 adsorption on the di-σ and π adsorption modes, on a Pd(100) surface. [a] Ref.[64,65]

The agreement between the calculated results for both adsorption modes and the experimental results [64,65], is good. However, the agreement between both kinds of results is better in the case of the di-σ adsorption mode. In this particular case, the calculated values of some of the most relevant vibrational modes (namely the ν_s(C-Pd) and the ν_2(CC)) are even in excellent agreement with the available experimental results. This strengthens the idea that on a Pd(100) surface the ethylene molecule adsorbs preferentially on the di-σ mode.
New experimental vibrational data, seems highly desirable in order to gain more insight into the details of the adsorption geometry of ethylene on Pd(100) and to determine if at 80K the π-bonded form of adsorbed ethylene really exists.

3.1.2.4. C_2H_4 / Ni(100). The optimized geometrical and electronic parameters obtained for the di-σ and π adsorption modes of C_2H_4 on a Ni(100) surface are presented in Table 10. The calculated vibrational frequencies are listed in Table 11 along with some available experimental frequencies.
The system C_2H_4/Ni(100) has already been studied theoretically by Xu et al [69]. On their work, these authors, using the Xα method and modelling the surface with a Ni_5 cluster, concluded that the electronic structure of the adsorbed ethylene molecule was similar on both adsorption modes.
Experimentally, some contradictory results have been obtained in the study of the adsorption of C_2H_4 on a Ni(100) surface, and general agreement on the bonding mode of ethylene on Ni(100), still does not exist. Zaera et al [66], studied the adsorption of ethylene on Ni(100) using TPD, LID and HREELS. Based on their TPD and LID results, the authors concluded that ethylene chemisorbs strongly at low temperatures and decomposes upon crystal heating. However, on their HREELS study, they interpreted a peak at 1575 cm^{-1} as corresponding to the CC stretching

mode and based on this they concluded that the geometry of the molecule was only slightly perturbed by the adsorption process. More recently, Sheppard [76] reanalysed the results of Zaera et al, and concluded that the bands at 1575cm^{-1} were experimental artifacts (possibly, arising from a small fraction of vinylidene species) and that the CC stretching frequency had a much lower value (ν(CC) $\cong$ 1100cm^{-1}). According to Sheppard [76], at low temperatures the ethylene molecule chemisorbs on a Ni(100) surface in the di-σ adsorption mode. Earlier results obtained by Lehwald et al [67] had also supported the idea that the ν(CC) stretching mode had a value of about 1130 cm^{-1}, and that the adsorbed molecule was highly distorted (relatively to its gas phase geometry).

	C_2H_4 / Ni(100) [a]		C_2H_4 / Ni(111) [b]	
	di-σ adsorption	π-top adsorption	di-σ adsorption	μ-bridging
Adsorption Energy (kJ.mol^{-1})	-170.2	-79.0	- 54.3	- 12.5
Distance C_2H_4–surface (Å)	1.8689	1.9075	1.89	1.81
Distance CC (Å)	1.4620	1.4340	1.49	1.58
Distance CH (Å)	1.0954	1.0876	-----	-----
Angle CCH (°)	116.404	118.497	114.5	113.0
Angle ((CH_2) – C) (°)	143.334	152.442	-----	-----
C – C M.O.P. (a.u.)	0.6775	0.7486	-----	-----
C – Ni [c] M.O.P.(a.u.)	0.6526 /0.4349	0.4361 / 0.4361	-----	-----
$Q_{adsorbate}$(Mulliken)(a.u)	+ 0.0820	+ 0.1864	-----	-----

TABLE 10. Adsorption energies (kJ.mol^{-1}), optimized geometries, C-C and C-Ni Mulliken overlap populations (M.O.P.) and total adsorbate Mulliken charges, Q, for C_2H_4 adsorbed on the di-σ and π adsorption modes, on a Ni(100) surface. [a] Present study, [b] Ref.[3], [c] The two values refer to the M.O.P. between each carbon atom and its nearest metal atom.

The theoretical results obtained in the present work, strongly support the conclusions of Sheppard [46] and of Lehwald et al [67]. The results in Table 10 show that the molecule binds strongly on a Ni(100) surface, the geometry and the electronic structure of the molecule are highly perturbed by the adsorption process, and the di-σ adsorption mode is more stable than the π mode, by approximately 91 kJmol^{-1}. The calculated CC bond length for the di-σ mode, 1.46 Å, is identical to the experimental value determined by NEXAFS [68]. An analysis of the electronic structure shows that on both adsorption complexes the degree of the σ-donation is superior to that of the d-π^* backdonation. This same conclusion was obtained by Xu et al [69] on their Xα study. The observed net flow of electrons from the adsorbate to the surface is responsible for the formation of the Ni-C bonds and for the lengthening and weakening of the CC bond. The lengthening of the CC bond is accompanied by a drop of the CC Mulliken overlap population from 1.3682 (gas phase) to 0.6775 (di-σ) and 0.7486 (π).

C_2H_4 / Ni(100)	di-σ adsorption	π adsorption	Experimental
ν_s(C – Ni)	388	284	350[b]
ν_{10}, $\rho(CH_2)$	885	847	840[b]
ν_7, $\omega(CH_2)$	890	833	-----
ν_2, ν(CC)	1090	1146	1130[a]
ν_3, $\delta(CH_2)$	1476	1501	1390[a] , 1385[b]
ν_1, $\nu_s(CH_2)$	3042	3124	3010[a] , 3000[b]

TABLE 11. Calculated vibrational frequencies (cm^{-1}) for C_2H_4 adsorption on the di-σ and π adsorption modes, on a Ni(100) surface. [a] Ref. [67], [b] Ref. [66]

The results in Table 11 show that the ν(CC) stretching mode has a value of about 1100cm^{-1}, which is in excellent agreement with the conclusions of Sheppard and of Lehwald et al. The agreement between the other available experimental frequencies and the theoretical results is also good.

3.1.3 Conclusions

According to the results obtained in this study, the adsorption energy of the ethylene molecule on a metallic surface depends strongly on the electronic structure of that surface and on the adsorption site. The results show that, upon adsorption the degree of distortion of the ethylene molecule (relatively to its gas phase geometry), as well as the binding strength to the metal surface increases in the order Pd $\rightarrow$ Ni $\rightarrow$ Pt. These results are in agreement with EELS results which showed that ethylene is more strongly bound and more distorted on Ni(111) and Pt(111) than on Pd(111) [47].

Our calculations also support the picture, already suggested by several experimental studies, of a significantly distorted adsorbate on the three metal surfaces: there is a lengthening of the CC bond and a rehybridization of the carbon atoms from sp^2 toward sp^3. On these three surfaces, the σ-donation is stronger than the d-π* backdonation, leading to positively charged species on the surface. The results obtained, also show that on platinum, palladium and nickel (100) surfaces the ethylene molecule adsorbs preferentially on the di-σ mode. This conclusion is based not only on the adsorption energies (whose calculation is known to be the major problem of the cluster model approach) but also on a comparision between the calculated vibrational frequencies and the available experimental results.

In what concerns to the calculation of the adsorption energies in the present study, we known that our method of doing so is not unambiguos. However, as we have already mentioned, there is a qualitative correspondence between the degree of distortion of the ethylene molecule (relatively to its gas phase geometry) and the corresponding calculated binding energy to the metal surface. So, it is our believe that any improvement of the reference energies for the various M_9 systems would not change our conclusions for the qualitative comparision of the binding energies.

4. General Conclusions

In the present paper we have presented some studies concerning the adsorption of acetylene on copper (100) surfaces, and the adsorption of ethylene on the (100) surfaces of nickel, palladium and platinum. In all these studies we have used a cluster with the same shape and size. Despite the limited size of the clusters used, some very interesting features of the systems "adsorbate – metal surfaces" were determined, namely adsorbate geometries, adsorption energies and vibrational frequencies.
The results presented support the usefulness of the cluster model approach in the study of chemical processes on solid surfaces.

Acknowledgments

Financial support from the Fundação para a Ciência e Tecnologia (Lisbon) and project PRAXIS/PCEX/C/QUI/61/96 is acknowledged. C.G.P.M.B. thanks PRAXIS for a scholarship (BM/14665/97).

References

[1] M. Dewar, *Bull. Soc. Chim. Fr.* **18** (1951) C79

[2] J. Chatt, L.A. Duncanson, *J. Chem. Soc.* (1953) 2939

[3] A. Fahmi and R.A. van Santen, *Surf. Sci.* **371** (1997) 53

[4] S. Bao, Ph. Hofmann, K.M. Schindler, V. Fritzsche, A.M. Bradshaw, D.P. Woodruff, C. Casado and M.C. Asensio, *Surf. Sci.* **307 – 309** (1994) 722 – 727

[5] M. Weinelt, W. Huber, P.Zebisch, H.P. Steinrück and P.Ulbricht, U.Birkenheuer, J.C. Boettger and N.Rösch, *J.Chem.Phys.* **102** (1995) 9709

[6] B.J. Bandy, M.A. Chesters, M.E. Pemble, G.S. McDougall and N. Sheppard, *Surf. Sci.* **139** (1984) 87

[7] S. Bao, K.M. Schindler, Ph. Hofmann, V. Fritzsche, A.M. Bradshaw and D.P. Woodruff, *Surf. Sci.* **291** (1993) 295

[8] K. Hermann, M. Witko, *Surf. Sci.* **337** (1995) 205

[9] J.R. Lomas, C.J. Baddeley, M.S. Tikhov and R.M. Lambert, *Langmuir* **13** (1997) 758

[10] A. Clotet and G. Pacchioni, *Surf. Sci.* **346** (1996) 91

[11] H. Hoffmann, F.Zaera, R.M. Ormerod, R.M. Lambert, J.M. Yao, D.K. Saldin, L.P.Wang, D.W.Bennett and W.T.Tysoe, *Surf. Sci.* **268** (1992) 1.

[12] J.C. Dunphy, M. Rose, S. Behler, D.F. Ogletree, M. Salmeron and P. Sautet, *Phys. Rev. B*, **57** (1998) 12705

[13] D. Fuhrmann, D. Wacker, K. Weiss, K. Hermann, M. Witko and Ch. Wöll, *J. Chem. Phys.* **108** (1998) 2651

[14] K. Hermann, M. Witko and A. Michalak, *Z. Phys. Chem.* **197** (1996) 219

[15] M. Witko and K. Hermann, *Appl. Catalysis A* **172** (1998) 85

[16] D.A. Outka, C.M. Friend, S. Jorgensen and R.J. Madix, *J. Am. Chem. Soc.* **105** (1983) 3468

[17] N. R. Avery, *J. Am. Chem. Soc.* **107** (1985) 6711

[18] D. Arvanitis, U. Döbler, L. Wenzel and K. Baberschke, *Surf. Sci.* **178** (1986) 686

[19] D. Arvanitis, K. Baberschke, L. Wenzel and U. Döbler, *Phys. Rev. Lett.* **57** (1986) 3175

[20] D. Arvanitis, L. Wenzel and K. Baberschke, *Phys. Rev. Lett.* **59** (1987) 2435

[21] P. Geurts and Ad Van Der Avoird, *Surf. Sci.* **103** (1981) 416

[22] Yat-Ting Wong and R. Hoffmann, *J. Chem. Soc. Faraday Trans.* **86** (1990) 553

[23] Ts.S. Marinova and P.K. Stefanov, *Surf. Sci.* **191** (1987) 66

[24] B.C. Stipe, M.A. Rezaei and W. Ho, *Phys. Rev. Lett.* **81** (1998) 1263

[25] B.C. Stipe, M.A. Rezaei and W. Ho, *Science* **280** (1998) 1732

[26] B.C. Stipe, M.A. Rezaei and W. Ho, *Phys. Rev. Lett.* **82** (1999) 1724

[27] L.J. Lauhon and W. Ho, *J. Chem. Phys.* **111** (1999) 5633

[28] H. Rabus, D. Arvanitis, M. Domke and K. Baberschke, *J. Chem. Phys.* **96** (1992) 1560

[29] J. Dvorak and Jan Hrbek, *J. Chem. Phys.B* **102** (1998) 9443

[30] M.J. Frisch, G.W. Trucks, H.B. Schlegel, G.E. Scuseria, M.A. Robb, J.R. Cheeseman, V.G. Zakrzewski, J.A. Montgomery, R.E. Stratmann, J.C. Burant, S. Dapprich, J.M. Millam, A.D. Daniels, K.N. Kudin, M.C. Strain, O. Farkas, J. Tomasi, V. Barone, M. Cossi, R. Cammi, B. Mennucci, C. Pomelli, C. Adamo, S. Clifford, J. Ochterski, G. A. Petersson, P.Y. Ayala, Q. Cui, K. Morokuma, D.K. Malick, A.D. Rabuck, K. Raghavachari, J.B. Foresman, J. Cioslowski, J.V. Ortiz, B.B. Stefanov, G. Liu, A. Liashenko, P. Piskorz, I. Komaromi, R. Gomperts, R.L. Martin, D.J. Fox, T. Keith, M.A. Al-Laham, C.Y. Peng, A. Nanayakkara, C. Gonzalez, M. Challacombe, P.M.W. Gill, B.G. Johnson, W. Chen, M.W. Wong, J.L. Andres, M. Head-Gordon, E.S. Replogle and J.A. Pople, Gaussian 98, Gaussian, Inc., Pittsburgh PA, 1998. Revision A.1.

[31] A.D. Becke, *J. Chem. Phys.* **98** (1993) 5648.

[32] C. Lee, W. Yang, R. G. Parr, *Phys. Rev. B.* **37** (1980) 785

[33] P. J. Hay and W. R. Wadt, *J. Chem. Phys.* **82** (1985) 270

[34] J.R.B. Gomes and J.A.N.F. Gomes, *J. Mol. Struct. (THEOCHEM)*, **463** (1999) 163

[35] J.R.B. Gomes and J.A.N.F. Gomes, *Surf. Sci.*, **443** (1999) 165

[36] J.R.B. Gomes and J.A.N.F. Gomes, *Surf. Sci.*, **432** (1999) 279

[37] J.R.B. Gomes and J.A.N.F. Gomes, *Electrochimica Acta*, **45** (1999) 653

[38] G. Pacchioni, P.S. Bagus and F. Parmigiani, Eds., Cluster Models for Surface and Bulk Phenomena, NATO ASI Series B, Vol. 283 (Plenum Press, New York, 1992).

[39] K. Hermann, P.S. Bagus and C.J. Nelin, *Phys. Rev. B* **35** (1987) 9467

[40] L. Triguero, L.G.M. Pettersson, B. Minaev and H. Agren, *J. Chem. Phys.* **108** (1998) 1193

[41] Tables of Interatomic Distances and Configuration in Molecules and Ions, The Chemical Society, Burlington House, London (1965)

[42] Herzberg G., in Molecular Spectra and Molecular Structure, Vol II: Infrared and Raman Spectra of Polyatomic Molecules; Van Nostrand Reinhold Company Inc.: New York, 1945

[43] A.M. Baro, S. Lehwald and H.Ibach, *J. Electron. Spectrosc. Related Phenomena*, (1980) 215

[44] H. Steininger, H. Ibach and S. Lehwald, *Surf. Sci.*, **117** (1982) 685

[45] R.J. Koestner, J. Stöhr, J.L. Gland and J.A. Horsley, *Chem. Phys. Lett.*, **105** (1984) 332

[46] E. Yagasaki and R.I. Masel, *Surf. Sci.*, **222** (1989) 430

[47] J.A. Gates and L.L. Kesmodel, *Surf. Sci.*, **120** (1982) L461

[48] L.P. Wang, W.T. Tysoe, R.M. Ormerod, R.M. Lambert, H. Hoffmann and F. Zaera, *J. Phys. Chem.*, **94** (1990) 4236

[49] M. Nishijima, J. Yoshinobu, T. Sekitani and M. Onchi, *J. Chem. Phys.*, **90** (1989) 5114

[50] W.T. Tysoe, G.L. Nyberg and R.M. Lambert, *J. Phys. Chem.*, **88** (1984) 1960

[51] U. Gutdeutsch, U. Birkenheuer, E. Bertel, J. Cramer, J.C. Boettger and N. Rösch, *Surf. Sci.*, **345** (1996) 331

[52] J.A. Stroscio, S.R. Bare and W. Ho, *Surf. Sci.*, **148** (1984) 499

[53] T.K. Sham and R.G. Carr, *J. Chem. Phys.*, **84** (1986) 4091

[54] S. Bao, Ph. Hofmann, K-M Schindler, V. Fritzsche, A.M. Bradshaw, D.P. Woodruff, C. Casado and M.C. Asensio, *J. Phys. Condens. Matter*, **6** (1994) L93

[55] E. Cooper and R. Raval, *Surf. Sci.*, **331-333** (1995) 94

[56] V. Maurice and C. Minot, *Langmuir*, **5** (1989) 734

[57] Q. Ge and D.A. King, *J. Chem. Phys.*, **110** (1999) 4699

[58] R.M. Watwe, B.E. Spiewak, R.D. Cortright and J.A. Dumesic, *J. Catal.*, **180** (1998) 184

[59] J.F. Paul and P. Sautet, *J. Phys. Chem.*, **98** (1994) 10906

[60] A. Fahmi and R.A. van Santen, *Surf. Sci.*, **371** (1997) 53

[61] Qi Minglong, C. Barbier and R. Subra, *J. Mol. Catal.*, **57** (1989) 165

[62] G.H. Hatzikos and R.I. Masel, *Surf. Sci.*, **185** (1987) 479

[63] G.H. Hatzikos and R.I. Masel, in J.W. Ward (Eds.), Catalysis 1987, Elsevier, Amsterdam, 1988, 883.

[64] E.M. Stuve and R.J. Madix, *J. Phys. Chem.*, **89** (1985) 105

[65] E.M. Stuve, R.J. Madix and C.R. Brundle, *Surf. Sci.*, **152-153** (1985) 532

[66] F. Zaera and R. B. Hall, *Surf. Sci.*, **180** (1987) 1

[67] S. Lehwald, H. Ibach and H. Steininger, *Surf. Sci.*, **117** (1982) 342

[68] F. Zaera, D.A. Fischer, R.G. Carr and J.L. Gland, *J. Chem. Phys.*, **89** (1988) 5335

[69] X. Xu, N.Q. Wang and Q.E. Zhang, *J. Mol. Struct.*, **247** (1991) 93

[70] H. C. Allen and E. K. Plyler, *J. Am. Chem. Soc.*, **80** (1958) 2673

[71] D. Van Lerberghe, I. J. Wright and J. L. Duncan, *J. Mol. Spectrosc.*, **42** (1972) 251

[72] A. P. Scott, L. Radom, *J. Phys. Chem.*, **100** (1996) 16502

[73] Ge Q. and King D.A., *J. Chem. Phys.* **110** (1999) 4699

[74] Kua J., Goddard III W.A., *J. Phys. Chem. B* **102** (1998) 9492

[75] Watson W.G., Wells R.P.K., Willock D.J. and Hutchings G.J., *J. Phys. Chem. B* **104** (2000) 6439

[76] N. Sheppard, *Ann. Rev. Phys. Chem.*, **39** (1988) 589

THEORETICAL APPROACHES OF THE REACTIVITY AT MgO(100) AND TiO_2(110) SURFACES.

Christian MINOT
Laboratoire de Chimie Théorique, UMR 7616 CNRS, Tour 23-22, Case courrier 137, Université P. et M. Curie, 75252 Paris, Cédex05, France

1. Introduction

In this paper, we will review the chemical behaviour of transition metal oxides which is of crucial importance for heterogeneous catalysis, adhesion and many technological applications. Among them, MgO(100) is the simplest surface, with a square unit-cell containing two ions with opposite charges; titanium oxides represent another important class of systems used for their catalytic properties either directly as catalyst or indirectly as support for other catalysts (metals such as Ni, Rh for the Fischer-Tropsch reaction or V_2O_5 for the reduction of NO_x) or as promotors[1]. The most stable surface for rutile is the (110) face.

We will present results based on *ab-initio* Periodic Hartree-Fock calculations[2] that we have done[3] and on other theoretical results that have been published from embedded clusters and density functional calculations, insisting more on trends than on numerical values that may vary with the methods of calculations. We will focus on the reactivity at the clean perfect surfaces.

Compared with pure metals whose surface sites are all made of the same atoms, differing only by structural environments, in most of the stable naked surfaces of metal oxides, the two ions, metallic cations and oxygen ions, coexist and are exposed to the adsorption. This is the case of MgO(100) a surface of type 1 according the classification by Tasker[4] and of TiO_2(110) a surface of type 2. Metallic cations are acidic sites available to adsorb a Lewis base whereas oxygen anions may fix a Lewis acid. This controls the adsorption of the different species. When they do not decompose, basic adsorbates fix on the surface metal sites (on top of a surface metal atom or bridging two of them) and the acidic ones on the oxygen ions (on top of one, bridging or capping several of them).

This simple and useful guideline implies however several remarks :

1) It has to be noted that the presence of charges at the surface sites is not enough to infer that the reactivity is indeed controlled by the acid-base properties. For example, the surface sites of the Si(111) surface that are equivalent for the unreconstructed surface, become different after reconstruction. The atoms at the surface are electron-rich or electron-poor : the former are deeper in the surface (restatoms and corner atoms) while the later are protuberant surface atoms, the adatoms (more precisely, those among them whose orbitals have the largest 3p character). Despite this distinction in charges, the sites of adsorption do not match the charge distribution[5]; the H and OH fragments arising from the water dissociation do not result from the adsorption of H^+ on electron

M.A. Chaer Nascimento (ed.), Theoretical Aspects of Heterogeneous Catalysis, 241–249.

rich sites (restatoms) and of OH^- on electron-poor sites (adatoms). The orientation is the opposite and is explained considering the adsorption of radicalar species.

In this example the gap is small, the silicon is a semiconductor. The promotion of electrons from the occupied states to the unoccupied states is more difficult for a metal oxide and this is the reason why the charge distribution plays a dominant role for the adsorption on metallic oxides. When the water molecule dissociates on the oxide surface, H^+ goes on electron rich sites (oxygen anions) and of OH^- on electron-poor sites (metallic cations). The orientation is that predicted by an heterolytic cleavage of an O-H bond.

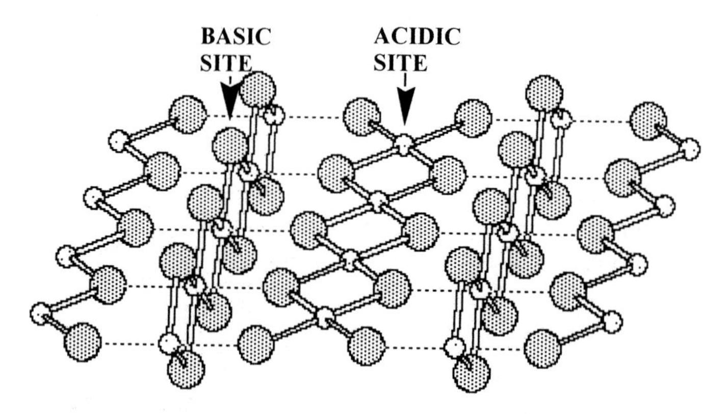

Figure 1. The (110) face of rutile results from the coupling of TiO_2 polymers and contains the basic and acidic sites of adsorption. Small white circles are titanium atoms and large grey circles are oxygen atoms. The in-plane oxygen atoms (from the basal plane) donate electrons to the out-of-plane polymers (doted lines) enriching the O atoms belonging to protruded ridges while the unsaturated titanium atoms become more acidic.

2) The reactivity is not directly related to the magnitude of the charge. The surface atoms that are the more reactive are not those that bear the largest charge. The Madelung field stabilises the charges and increases the stability of the ions. The largest charges are thus localised on the ions that have the largest coordination. These ions are poorly reactive. On the contrary, the ions that are the most reactive are poorly coordinated. The trend is the same for the effect of the coordination by the first neighbours (a very large effect) and that of the coordination by remote neighbours (a weaker effect). It follows that small aggregates are more reactive than large ones; that corrugated surfaces are more reactive than perfect surfaces and that defect sites are more reactive than regular sites. For a modelisation, the increase in size of a model to describe an adsorption site (larger clusters; embedding in an array of point charges, description of a slab by a larger number of layers) logically induces a decrease of the calculated heats of adsorption.

At the (100) surface of MgO, the ions are pentacoordinated; this high coordination decreases the reactivity. In this surface, the oxygen dianions are less reactive than those at the protruded ridges of the $TiO_2(110)$ surface that are dicoordinated. MgO is less basic than TiO_2.

In conclusion, charges are not good indices for the reactivity and considering coordination is a better tool. Frontier orbitals (the position of the projected DOS on the orbitals of the surface ions relative to the Fermi level) is a better index since their shift reflect the coordination. The "ion excess charges"[6] whose definition involves the coordination are also appropriate to predict the reactivity.

3) Acidity and basicity are relative properties. Many compounds are amphoteric and behave as acids or as bases according to a partner. Metal oxides are classified as acidic, amphoteric or basic. Experimentally, this classification corresponds to the adsorption of probe molecules[7, 8]. NH_3 is a base probe molecule that reacts with the electron deficient metal atoms (Lewis acid) or the protons adsorbed on the hydrated surface. CO_2 is usually considered as acidic and thus it is expected to adsorb more strongly on basic sites. According to this classification, TiO_2 belongs to an amphoteric species and MgO to a basic species. A general difficulty for such classifications is that the order can vary with the choice of the probe. The Hard and Soft Bases and Acids theory[9, 10] responds to the necessity to refine the model with a second scale : it is better to couple acids and bases of the same hardness than to mix them. In terms of HSBA theory, Ti^{4+}, more strongly charged, is an hard acid while Mg^{2+} is a soft acid. According to Klopman, the hardness is measured by the E* energy value, the energy of an orbital under the influence of its partner; those for Ti^{4+} and Mg^{2+} are 4.35 eV and 2.42 eV respectively. TiO_2 is therefore expected to adsorb bases more strongly than MgO and to preferentially adsorb hard bases. The adsorption of NH_3 (hard base) is particularly strong. On MgO that of CO_2 (soft base) is relatively favoured.

Since there is no unique scale, the scale of the acidities is not the inverse of the scale of basicities; one cannot infer that a surface which is basic has to be poorly acidic. The same oxide can be both more acidic (a property of the surface cations) and more basic (a property of the surface anions) than another one.

2. Adsorption without dissociation

2.1 MOLECULES

The adsorption of a molecule implies a reorganisation of its electrons to maximise the interaction. This decreases the internal cohesion of the molecule. Dissociation occurs when the interaction is large and the molecular adsorption when it is weak. Ti^{4+} is a stronger and harder acid than Mg^{2+} and dissociation is more frequent on TiO_2 than on MgO. The (110) face of TiO_2 is stable but it is also reactive due to the presence of dative bonds that polarise the surface. Indeed, this face results from the coupling of in-plane and out-of-plane polymers both oriented along the 001 direction. This coupling is made by a donation from the oxygen atoms in the plane to the titanium atoms of the out-of-plane polymer decreasing the electronic density of the in-plane polymers and increasing their acidic properties. At the same time, the O from the out-of-plane polymers (the protruded ridge) becomes more basic.

All the molecules that we have adsorbed on the metal oxide surfaces are adsorbed[3, 11-15] on the metal cations and the heat of adsorption is parallel to their proton affinities. The surface clearly appear to be predominantly acidic.

NH_3 is the usual probe for the acidic site[7, 8, 16]; it is a hard base that strongly adsorbs on Ti^{4+} with the lone pair oriented toward the metal[17-19]. A similar adsorption also takes place on top of Mg^{2+}, but the heat of adsorption is weaker[3, 13] by a factor 7.

H_2O is also strongly adsorbed on $TiO_2(110)$ and weakly on MgO(100)[20, 21]. For the former, dissociation is more favourable even still debated[14, 22, 23]; this is not the case for MgO. The adsorption modes differ. For $TiO_2(110)$, there is a donation of a σ electron pair to Ti^{4+} and the molecule is perpendicular to the surface plane. The interaction is under a charge control as expected for hard reactants; at the same time this geometry minimises the repulsion between Ti^{4+} and the protons from the water. For MgO, there is a donation of a p electron pair to Mg^{2+} and the molecule is parallel to the surface plane. The interaction is under an orbital frontier control as expected for soft reactants. At low coverage (θ=1/2 or less), there is a secondary interaction between the protons from the water molecule and the oxygen ions from the lattice. This might be considered as revealing a basic character of these oxygen atoms; however at full coverage, this interaction disappears and the water molecules rotate to build H-bonds within them. The oxide is then covered by a full monolayer of ice[3, 24]. The adsorption of molecules in hydrated media, if it does not lead to remove this cover is essentially an adsorption on ice. H_2S is a weaker base and the adsorption on Ti(110) looks like that of water on MgO (except that dissociation is easier). It lies flat, parallel to the surface.

CO_2 and SO_2 are frequently considered as acidic species. The adsorption of the former on MgO has been used to classify MgO as basic surface. The adsorption on O leads to $CO_3^=$ and $SO_3^=$ species that are nicely evoked to prepare the adsorbed species toward reactivity. For CO_2, this leads to monodentate or bidentate carbonates and to hydrogen carbonates (bicarbonates). Monodentate carbonates have been postulated together with vertical CO_2[1]. On MgO[25] and TiO_2 surfaces[26] the reaction with basic surface hydroxyl groups leads to the formation of bicarbonate ions HCO_3^-. Tsyganenko[27] Lamotte[28] and Hoggan[29] have proposed a donation from a lone pair of a surface hydroxyl to CO_2. The mechanism of CO_2 insertion in surface hydroxyl groups[30-32] has also been explained through the basic properties of CO_2. Knözinger[32]proposed that the formation of hydrogen carbonates involves the reaction of a CO_2 already adsorbed on strong acidic sites, the acidity of the carbon being increased by the adsorption. Similarly, the anhydrase-catalyzed hydration of CO_2 as also been explained either by acidic or basic properties of CO_2 : by a nucleophilic attack from an Zn^{2+}-bound OH^- to C [33, 34] (acidity of CO_2)or by a first step where the CO_2 is at first bound to the metal cation[35-37](basicity of CO_2).

Calculations for the CO_2 adsorption show that, on the naked surfaces, CO_2 has to be considered as a basic species that binds to the exposed metallic cation[38, 39]. On TiO_2 (110) CO_2 is perpendicular to the surface building a OCO...Ti bond while on MgO(100) CO_2 is flat and parallel to the surface bridging two adjacent Mg sites[24, 40]. The adsorption energies are nearly equal (7 kcal/mol. *vs.* 6.3 kcal/mol.) on MgO and TiO_2. On MgO (100), it is an interaction between soft reactants and two bonds are built.

For SO_2 which is a poor base, we have calculated with the VASP program[41-44] using the GGA approxiamtion its adsorption on a three-layers slab of MgO(100) (at $\theta=1/2$) and found a weak adsorption (4 kcal/mol.) when SO_2 bridges to Mg atoms (formation of 2 Mg-O bonds with a length of 2.77 Å) in agreement with ab initio cluster calculations[45]. Forcing the sulphur to remain on top of a surface oxygen does not lead to an adsorption (an optimum is found corresponding to 0 kcal/mol. and a long O-S distance, 3.5 Å). A very different reactivity is observed at step sites where the oxygen atoms loose coordination and become more reactive[45].

Hydration may modify the reactivity if the metallic sites are not available; the acidity of the surface hydroxyl is less than that of the exposed metallic cations. H_2O molecular adsorption on hydrated TiO_2(110) surface can result from these weak interactions (H-bonding with the surface hydroxyl groups)[22]Lindan, 1998 #223]. In this case, there is no direct interaction and the vibration frequencies should be recalculated on another model. Let us note that the basicity of the surface also varies; in the case of TiO_2(110) the lone pairs of a surface hydroxyl are less reactive than those of a naked surface oxygen; for MgO it is the reverse; the hydrated surface becomes more basic.

2.2 TRANSITION METALS

Independendently of their oxidation state, the metals adsorb on the surface oxygen dianions[46]. This is natural for a metal cation. It also happens when the metal remains neutral. There is no significant transfer of electron from the metal to the oxide. Alkali metals are also adsorbed on the oxygen dianions from reducible oxides; this is accompanied by an electron transfer and will be discussed in section 3.2. The nickel atom is on top of an O for MgO(100) and bridging two O for TiO_2(110). The heat of adsorption is weak and aggregation is favourable : adsorbed dimers or tetramers are more favourable than adsorbed atomic species. The dimer is bridging 2 surface oxygen atoms for MgO(100) and over a triangle of three oxygen atoms for TiO_2(110); this is obtained from CRYSTAL and VASP calculations[47] and from EXAFS experiments[48]. Even though the charge transfer is very weak, the binding mode implies a dative bond from the oxygen dianions to the metal. This bond involves the 4s(Ni) atomic orbital which, as major component of the vacant σ^* level, is empty. Since the global charge does not change, the decrease of the σ population is compensated by an increase of the 3d population. This suggests that, when the interaction is large enough, the optimal correspond to a singlet state ($d^{10}s^0$) with a short Ni-O distance. The alternative is a weaker bond (at larger distance) where the oxygen atoms remains in a triplet state. The calculations with an embedded cluster[49, 50] privilege the singlet state (and conclude that the adsorption kills the spin) or lead to equal adsorption energies for the two states. The CRYSTAL calculations for the atomic adsorption conclude that the spin is preserved. In the VASP calculations, an interaction takes place between the Ni atoms at $\theta=1/2$ and the spin is reduced. This is obtained with all the methods when the dimer is made : indeed the dimer in the gas phase is a triplet and the interaction with the surface is too weak to cancel the spin. The tetramer is a quintet[51].

3. Adsorption with dissociation

3.1 HETEROLYTIC CLEAVAGE

When an adsorbate dissociates on the surface, one has to consider the independent adsorption of the fragments and the cost of the fragmentation. Usually the dissociation is the consequence of a strong adsorption. It may also be the consequence of a weak binding : H_2S dissociates more easily than water[14, 52-54] and is adsorbed dissociatively even if the adsorption of the fragment is not so important. Brönsted acids give rise to protons that are easily adsorbed on the surface oxygen while the basic fragment goes on the surface cation. H^+ is a strongest Lewis acid. The adsorption can be analysed as resulting from an heterolytic cleavage even if the mechanism is not completely proofed.

The adsorption has to be large to overcome the cost of the dissociation. $TiO_2(110)$ is more acidic and more basic than MgO(100); the dissociation will therefore happen more easily. Defect sites are also more reactive and should participate more easily to the dissociation. For water, the competition between molecular and dissociative adsorption can be analysed as a competition for the site of adsorption on an hydroxylated surface. In this picture, the proton has the choice between to sites : the proton of an hydroxyl group and that from the lattice. The projected densities of states for the 2p(O) orbitals show that the OH group is more reactive (contribution at highest levels for the projected DOS) than the surface O in MgO (the O from the lattice are stabilised by the five neighbouring cations) but less reactive than the surface O in TiO_2 (the O pair engaged in a strong and covalent O-H bond is more stabilised than that forming a weaker Ti-O bond). Once again, at variance with the usual classification, the oxygen atoms of MgO appear as poorly basic due to the strong stabilisation by the Madelung field. For H_2S, whose cleavage is easier, MgO allows the dissociation but "exhibit the lowest reactivity" on a series of oxides, NiMgO(100) and ZnO(0001); according to Rodriguez[53-55], this is related to the size of the band gap. The band gap for MgO is rather large due to the strong interaction between each ion and its neighbours that shifts all the atomic levels away from the Fermi level.

Alcohols R-O-H could lead to two fragmentations : a basic cleavage, $R^+ + OH^-$, or an acidic cleavage, $RO^- + H^+$. This determines the site for the adsorption : On $TiO_2(110)$, in case of basic cleavage, the R group would go on a bridging oxygen atoms while the hydroxyl group would adsorb on the titanium site; in the acidic case the orientation would be reversed. The energies of the gas phase cleavages monitor the cleavage and determine the orientation. The basic gas phase cleavage is easier (less endothermic) in the case of ROH and the best orientation for the adsorption is $Ti_{lattice}$-OH and $O_{lattice}$-R. The corresponding orientation is obtained from CRYSTAL [56] and VASP [57] calculations. In the case of the sulphur compound, RSH, the acidic cleavage is the least expensive and the orientation is the opposite[56], $Ti_{lattice}$-SR and $O_{lattice}$-H.

Even if, except at defect sites[58, 59], the hydrogen does not easily dissociate on the surface oxides[1, 60], distributions of hydrogen atoms are involved in surface reactions (for instance the decomposition of the formic acid) and have motivated both experimental and theoretical works. The decomposition is made at high temperature on activated MgO powders[61, 62]. On a non reducible oxides, the heterolytic decomposition of H_2 generates an H^+ (proton) that binds to an oxygen from the lattice and an H^- (hydride) that should bind to the metal[63, 64]. This orientation avoids the promotion of electrons to the conduction band. Morokuma has suggested an electrostatically controlled mechanism where the charges on Mg^{2+} and O^{2-} cooperate to facilitate the heterolytic cleavage[65]. The M-H bond is very weak compared to the O-H bond and mechanism have been proposed to avoid the adsorption of H^- on the metal cation[66-69] in the decomposition of the formic acid, HCOOH $\rightarrow CO_2 + H_2$.

3.2 REDUCTIVE ADSORPTION

Since the O-H bond is much stronger than the M-H bond, hydrogen atoms are more likely to be adsorbed as protons on the oxygen atoms. When the oxide is not reducible, in the H_2 dissociation, protons are equilibrate by an equivalent amount of hydrides. In the case of reducible oxides, there is a better possibility : the excess of electrons should be transferred on the oxide, reducing the Ti^{+4} ions. For calculations, this leads to high spin states (1 unpaired electron per adsorbed atomic oxygen). The energy for this state is much better than that resulting for an heterolytic cleavage for $TiO_2(110)$ by CRYSTAL and VASP calculations[47]. Similar orientations are obtained with V_2O_5[70-72].

The adsorption of alkali metals also occurs with a large charge transfer inducing a reduction of a metal atom. For Na/$TiO_2(110)$, the adsorption site is above the protruded ridges [73] or capping three oxygen atoms (two from the ridge and one basal oxygen)[74, 75]; the electron is transferred to a hexacoordinated titanium atom from the sublayer that are below the pentacoordinated ones at the surface[76, 77]. For K/$TiO_2(110)$ these two sites are found at different coverages [78, 79]. Transition metals adsorb similarly. The reduction is clearly seen from the spin localisation. However, in this case, it is not accompanied by a large charge transfer.

4. Limits

We have reviewed theoretical approaches concerning perfect surface. Defects modify the adsorption. Hydration can strongly perturb the conclusions. If water saturates the adsorption sites, the admolecule should either displace an adsorbed molecule (and has to be more strongly adsorbed than water itself) or adsorb by hydrogen bonding on an ice layer. Such layers are formed on MgO(100) surfaces[3, 24, 80].

We have not commented the surface specificity, the coverage effects and the coadsorption effects and refer to previous papers. The adsorbate-adsorbate interaction can contribute to a stabilisation of the surface layer, suggesting sometimes that, at low coverage, island of adsorbates can be favourable. H_2O-H_2O[3, 20, 81] or NH_3-H_2O[17] lateral interactions can influence the adsorption mode. NH_3 leads to NH_4^+ and OH^- adsorbed when water is present. For CO/MgO(100) the lateral interactions at high coverage impose the bending of two thirds of the CO molecules[82-85].

The stable apolar surfaces of the metal oxides do not significantly reconstruct[86]. The largest reconstruction occurs for polar surfaces where layers of different ions alternate leading to unstability[87]. However, relaxations should be better investigated.

5. Acknowledgements

The author thanks A. Fahmi, J. Ahdjoudj, J. Leconte, A. Markovits and K. Skalli for the many fruitful discussions about metal oxides.

6. References

1. Raupp, G. B. and Dumesic, J. A. (1985) *J. Phys. Chem.*, **89**, 5240-5246.
2. Pisani, C., Dovesi, R. and Roetti, C. *Hartree-Fock Ab Initio treatment of Crystalline Systems 48*; Lecture Notes in Chemistry, Springer, Heildeberg, (1988).
3. Ahdjoudj, J., Makovits, A. and Minot, C. (1999) *Catalysis Today*, **50**, 541-551.
4. Tasker, P. W. (1979) *J. Phys. C*, **12**, 4977-4984.
5. Ezzehar, H., Stauffer, L., Leconte, J. and Minot, C. (1997) *Surf. Sci.*, **388**, 220-230.
6. vanSanten, R. A. *Theoretical Heterogeneous Catalysis 5*; World Scientific Lecture and Course in chemistry, World Scientific, Singapore (1991).
7. Auroux, A. and Gervasini, A. (1990) *J. Phys. Chem.*, **94**, 6371-6379.
8. Gervasini, A. and Auroux, A. (1991) *J. Therm. Analysis*, **37**, 1737-1744.
9. Pearson, R. G. (1968) *J. Chem. Ed.*, **45**, 581.
10. Klopman, G. (1968) *J. Amer. Chem. Soc.*, **90**, 223.
11. Minot, C., Fahmi, A. and Ahdjoudj, J.; (1995)*Periodic HF calculations of the adsorption of small molecules on TiO2*, Drymen, Scotland, Kluwer Academic Publishers.
12. Markovits, A., Ahdjoudj, J. and Minot, C. (1997) *Molecular Engineering*, **7**, 245-261.
13. Markovits, A., Ahdjoudj, J. and Minot, C. (1997) *Il nuovo cimento*, **19D**, 1719-1726.
14. Casarin, M., Maccato, C. and Vittadini, A. (1998) *J. Chem. Phys. B*, **102**, 10745-10752.
15. Ito, T., Kobayashi, H. and Tashiro, T. (1997) *Il Nuovo Cimento*, **19 D**, 1695.
16. Peri, J. B. (1965) *J. Phys. Chem.*, **69**, 231.
17. Markovits, A., Ahdjoudj, J. and Minot, C. (1996) *Surf. Sci.*, **365**, 649-661.
18. Allouche, A., Cora, F. and Girardet, C. (1995) *Chem. Phys.*, **201**, 59-71.
19. Lakhlifi, A. and Girardet, C. (1991) *Surf. Sci.*, **241**, 400.
20. Fahmi, A. and Minot, C. (1994) *Surf. Science*, **304**, 343-359.
21. Bredow, T. and Jug, K. (1995) *Surf. Sci.*, **327**, 398-408.
22. Brinkley, D., Dietrich, M., Engel, T., Farrall, P., Gantner, G., Schafer, A. and Szuchmacher, A. (1998) *Surf. Sci.*, **395**, 292-306.
23. Goniakowki, J. and Gillan, M. J. (1996) *Surf. Sci.*, **350**, 145-158.
24. Picaud, S. and Girardet, C. (1993) *Chem. Phys. Lett.*, **209**, 340.
25. Evans, J. V. and Whateley, T. L. (1967) *Trans. Farad. Soc.*, **63**, 2769.
26. Boehm, H. P. (1971) *Discuss. Faraday Soc.*, **52**, 264.
27. Tsyganenko, A. A. and Trusov, E. A. (1985) *J. Phys. Chem.*, **59**, 1554.
28. Lamotte, J., Saur, 0., Lavalley, J. C., Busca, G., Rossi, P. F. and Lorenzolli, V. (1986) *J. Chem. Soc. Faraday Trans. 1*, **82**, 3019.
29. Hoggan, P. E., Bensitel, M. and Lavalley, J. C. (1994) *J. Mol. Struct.(Theochem)*, **320**, 49-56.
30. Bensitel, M., Moravek, V., Lamotte, J., Saur, O. and Lavalley, J. C. (1987) *Spectrochimica acta,Part A*, **43**, 1487.
31. Bensitel, M.; *Caractérisation et etude de la sulfatation de Zr02, TiO2 et MgO par spectrometrie Infra-rouge*, Caen, France,(1988 (8 juillet))
32. Knözinger, H. (1976) *Adv. Catal.*, **25**, 234.
33. Liang, J. and Lipscomb, W. N. (1986) *J. Amer. Chem. Soc.*, **108**, 5051.
34. Liang, J.-Y. and Lipscomb, W. N. (1989) *Int. Journal of Quantum Chemistry*, **XXXVI**, 299-312.
35. Lindskog, S. *Zinc Enzymes* (T. G. Spiro), Wiley, N.Y. (1983).
36. Jacob, O., Cardenas, R. and Tapia, O. (1990) *J. Amer. Chem. Soc.*, **112**, 8692-8705.
37. Jacob, O. and Tapia, O. (1992) *Int. J. Quant. Chem.*, **42**, 1271-1289.
38. Pacchioni, G. (1993) *Surf. Sci.*, **281**, 207-219.
39. Markovits, A., Fahmi, A. and Minot, C. (1996) *J. Mol. Struct., Theochem.*,
40. Panella, V., Suzanne, J., Hoang, P. N. M. and Girardet, C. (1994) *J. phys. I France 4*, 905.
41. Kresse, G. and Hafner, J. (1993) *Phys. Rev. B*, **47**, 558.
42. Kresse, G. and Hafner, J. (1993) *Phys. Rev. B*, **48**, 13115.
43. Kresse, G. and Hafner, J. (1994) *Phys. Rev. B*, **49**, 14251.
44. Kresse, G. and Hafner, J. (1994) *J. Phys. Condens. Matter*, **6**, 8245.
45. Pacchioni, G., Ricart, J. M. and Illas, F. (1994) *Surf. Sci.*, **315**, 337-350.

46. Neyman, K. M., Vent, S., Pacchioni, G. and Rösch, N. (1997) *Il Nuovo Cimento*, **19 D**, 1743.
47. Skalli, M. K., Leconte, J., Markovits, A., Minot, C. and Belmadjoub, A. (2001) *to be submitted*,
48. Bourgeois, S., LeSeigneur, P., Perdereau, M., Chandesris, D., LeFèvre, P. and Magan, H. (1997) *Thin Solid Films*, **304**, 267-272.
49. Yudanov, I., Pacchioni, G., Neyman, K. and Rösch, N. (1997) *J. Phys. Chem. B*, **101**, 2786.
50. Lopez, N. and Illas, F. (1998) *J. Phys. Chem. B*, **102**, 1430-1436.
51. Giordano, L., Pacchioni, G., Ferrari, A. M., Illas, F. and Rösch, N. (2001) *to be published*,
52. Fahmi, A., Ahdjoudj, J. and Minot, C. (1996) *Surf. Science*, **352-354**, 529-533.
53. Rodriguez, J. A., Jirsak, T. and Chaturvedi, S. (1999) *J. Chem. Phys.*, **111**, 8077.
54. Rodriguez, J. A. and Maiti, A. (2000) *J. Chem. Phys. B*, **104**, 36309-3638.
55. Rodriguez, J. A., Chaturvedi, S., Kuhn, M. and Hrbeck, J. (1998) *J. Chem. Phys. B*, **102**, 5511.
56. Ahdjoudj, J., Fahmi, A. and Minot, C. (1996) *Surf. Sci.*, **352-354**, 529-533.
57. Bates, S. P., Gillan, M. J. and Kresse, G. (1998) *J. Phys. Chem. B*, **102**, 2017-2026.
58. Heinrich, V. and Kurtz, R. L. (1981) *Phys. Rev. B*, **22**, 6280-6287.
59. Göpel, W., Rocker, G. and Feierabend, R. (1983) *Phys. Rev. B*, **28**, 3427.
60. Nyberg, M., Nygren, M. A., Petterson, L. G. M., Gray, D. H. and Rohl, A. L. (1996) *J. Chem. Phys.*, 9054-9063.
61. Coluccia, S., Boccuzzi, F., Ghiotti, G. and Morterra, C. (1982) *J. Chem. Soc. Farad. Trans. I*, **78**, 2111-2119.
62. Ito, T., Watanabe, T., Tashiro, T. and Toi, K. (1989) *J. Chem. Soc., Farad. Trans. I*, **85**, 2381-2395.
63. Pisani, C. and D'Ercole, A. (2001) *Progress in Theoretical Chemistry and Physics vol 4, Kluwer, Dordrecht*, **to be published**,
64. Anderson, A. B. and Nichols, J. A. (1986) *J. Am. Chem. Soc.*, **108**, 4742-4746.
65. Anchell, J. L., Morokuma, K. and Hess, A. C. (1993) *J. Chem. Phys.*, **99**, 6004.
66. Dilara, P. A. and Vohs, J. M. (1993) *J. Chem. Phys.*, **97**, 12919-12923.
67. Nakatsuji, H., Yoshimoto, M., Hada, M., Domen, K. and Hirose, C. (1995) *Surf. Sci.*, **336**, 232-244.
68. Nakatsuji, H., Yoshimoto, M., Umemura, Y., Takagi, S. and Hada, M. (1996) *J. Chem. Phys.*, **100**, 694-700.
69. Ahdjoudj, J. and Minot, C. (1997) *Catalysis letters*, **46**, 83-91.
70. Hermann, K., Chakrabarti, A., Druzinic, R. and Witko, M. (1999) *Phys. Stat. Sol.(a).*, **173**, 195-208.
71. Witko, M., Tokartz, R. and Hermann, K. (1998) *Polish J. Chem.*, **72**, 1565-1583.
72. Witko, M., Hermann, K. and Tokartz, R. (1999) *Catalysis Today*, **50**, 553-565.
73. Onishi, H., Aruga, T., Ewawa, C. and Iwasawa, Y. (1988) *Surf. Sci.*, **199**, 597.
74. Nerlov, J., Christensen, S. V., Wichel, S., Pedersen, E. H. and Møller, P. J. (1997) *Surf. Sci.*, **371**, 321.
75. Sanz, J. F. and Zicovich-Wilson, C. M. (1999) *Chem. Phys. Lett.*, **303**, 111-116.
76. Albaret, T.; *Etude théorique des oxydes de titane*, UPMC-Paris VI,(2000)
77. Albaret, T., Finocchi, F., Noguera, C. and deVita, A. (2001) *to be submitted*,
78. Bredow, T., Apra', E., Catti, M. and Pacchioni, G. (1998) *Surf. Science*, **418**, 150-165.
79. Sanz, M. A., Calzado, C. J. and Sanz, J. F. (1999) *Int. J. Quant. Chem.*,
80. Heidberg, J., Redlich, B. and Wetter, D. (1995) *Ber. Bunsenges. Phys. Chem.*, **99**, 1333-1337.
81. Lindan, P. J. D., Harrison, N. M. and Gillan, M. J. (1998) *Phys. Rev. Lett.*, **80**, 762.
82. Gerlach, R., Glebov, A., Lange, G., Toennies, J. P. and Weiss, H. (1995) *Surf. Sci.*, **331-333**, 1490-1495.
83. Heidberg, J., Kampshoff, E., Kandel, M., Kühnemuth, R., Meine, D., Schönekäs, O., Suhren, M. and Weiss, H. (1993) *React. Kinet. Catal. Lett.*, **50**, 123.
84. Minot, C., VanHove, M. A. and Biberian, J. P. (1996) *Surf. Sci.*, **346**, 283-293.
85. Picaud, S., Briquez, S., Lakhlifi, A. and Girardet, C. (1995) *J. Chem. Phys.*, **102**, 7229.
86. Gustafsson, T., Garfunkel, E., Gusev, E. P., Häberle, P., Lu, H. C. and Zhou, J. B. (1996) *Surf. Rev. and Lett.*, **3**, 1561-1565.
87. Ahdjoudj, J., Martinsky, C., Minot, C., VanHove, M. A. and Somorjai, G. (1999) *Surf. Sci.*, **443**, 133-153.

SUBJECT INDEX

Progress in Theoretical Chemistry and Physics

1. S. Durand-Vidal, J.-P. Simonin and P. Turq: *Electrolytes at Interfaces.* 2000
   ISBN 0-7923-5922-4
2. A. Hernandez-Laguna, J. Maruani, R. McWeeny and S. Wilson (eds.): *Quantum Systems in Chemistry and Physics.* Volume 1: Basic Problems and Model Systems, Granada, Spain, 1997. 2000 ISBN 0-7923-5969-0; Set 0-7923-5971-2
3. A. Hernandez-Laguna, J. Maruani, R. McWeeny and S. Wilson (eds.): *Quantum Systems in Chemistry and Physics.* Volume 2: Advanced Problems and Complex Systems, Granada, Spain, 1998. 2000 ISBN 0-7923-5970-4; Set 0-7923-5971-2
4. J.S. Avery: *Hyperspherical Harmonics and Generalized Sturmians.* 1999
   ISBN 0-7923-6087-7
5. S.D. Schwartz (ed.): *Theoretical Methods in Condensed Phase Chemistry.* 2000
   ISBN 0-7923-6687-5
6. J. Maruani, C. Minot, R. McWeeny, Y.G. Smeyers and S. Wilson (eds.): *New Trends in Quantum Systems in Chemistry and Physics.* Volume 1: Basic Problems and Model Systems. 2001 ISBN 0-7923-6708-1; Set: 0-7923-6710-3
7. J. Maruani, C. Minot, R. McWeeny, Y.G. Smeyers and S. Wilson (eds.): *New Trends in Quantum Systems in Chemistry and Physics.* Volume 2: Advanced Problems and Complex Systems. 2001 ISBN 0-7923-6709-X; Set: 0-7923-6710-3
8. M.A. Chaer Nascimento: *Theoretical Aspects of Heterogeneous Catalysis.* 2001
   ISBN 1-4020-0127-4

KLUWER ACADEMIC PUBLISHERS – DORDRECHT / LONDON / BOSTON